MICROBIOLOGY RESEARCH ADVANCES

BIOFILMS IN BIOENGINEERING

MICROBIOLOGY RESEARCH ADVANCES

Additional books in this series can be found on Nova's website under the Series tab.

Additional e-books in this series can be found on Nova's website under the e-book tab.

MICROBIOLOGY RESEARCH ADVANCES

BIOFILMS IN BIOENGINEERING

MANUEL SIMÕES, M.D.

AND

FILIPE MERGULHÃO, M.D.

EDITORS

NOVA BIOMEDICAL

New York

For permission to use material from this book please contact us:
Telephone 631-231-7269; Fax 631-231-8175
Web Site: http://www.novapublishers.com

Library of Congress Cataloging-in-Publication Data

Biofilms in bioengineering / [edited by] Mmanuel Simues and Filipe Mergulhco (LEPAE-Department of Chemical Engineering, Faculty of Engineering, University of Porto, Porto, Portugal).
pages cm
Includes bibliographical references and index.
ISBN 978-1-62948-161-6 (hardcover)
1. Biofilms. 2. Biomedical materials. I. Simues, Manuel, 1978- editor of compilation. II. Mergulhco, Filipe, 1974- editor of compilation.
QR100.8.B55B5647 2013
610.28--dc23

2013034683

Published by Nova Science Publishers, Inc. † New York

Contents

Foreword

L. F. Melo
LEPAE, Department of Chemical Engineering, Faculty of Engineering, University of Porto, Rua Dr. Roberto Frias, Porto, Portugal

Biofilm science is a relatively new technical discipline focused on the understanding and modulation of the combination of biological and chemical processes as well as on the transport and interfacial processes that potentially affect the microbial accumulation and activity on both biotic and abiotic surfaces. Research on biofilms has progressed rapidly in the last decades. Due to the fact that biofilm studies required the development of new analytical tools in the experimental and the computer science fields, many recent advances have resulted from collaboration between microbiologists, microbial ecologists, medical doctors, pharmacologists, engineers and mathematicians. The scientific community has come to understand much more about the particular aspects of microbial biofilms through the use of a variety of microscopy and other physical methods, as well as new chemical, and molecular biology approaches [7].

Biofilms have been first documented in 1943 by Zobell, who reported the attachment of layers of microbial cells to bottle walls and the increase in the biological activity of batch suspended cultures when glass rods were added [22]. Further investigations revealed that this effect was even more pronounced under oligotrophic conditions when compared to the results obtained under high nutrient conditions [11, 22]. These conclusions emphasized the perception that the adhesion is a strategy of microorganisms to access nutrients from the surface and biofilm. The study of Characklis [5], about microbial slimes in industrial water systems, revealed their high cohesiveness as well as their strong resistance to disinfectants, but it was Costerton et al., [6] in 1978 who postulated the general theory of biofilm predominance. Only recently, however, have attempts been made to define genetic, physiological and ecological basis of such phenomena [2, 8, 14].

It is supposed that biofilm is the first form of communitarian life recorded on the planet, being estimated that most microorganisms on Earth are organized in biofilms and they even occur in extreme environments such as hydrothermal vents, nuclear power plants and disinfection pipelines [6].

Biofilms are formed ubiquitously in any interface (liquid or solid) in contact with water. The ubiquity of biofilms can cause significant problems of public health, medicine and industry concern [6, 9, 10, 15]. Accordingly, there has been a great deal of research to better understand biofilm development and to identify improved control strategies. In the health context, some diseases and adverse medical conditions are now recognized to be the result of a biofilm infection.

Biofilms are as versatile as they are ubiquitous. Intentional and unintentional biofilms concern a broad range of areas, comprising special attention in the industrial, environmental and biomedical areas [3, 4]. Biofilms can be beneficial or detrimental depending on where they are found. Biofilms used in pharmaceutical and food fermentation industries, wastewater treatment plants and natural biofilms present in lakes or rivers (which contribute to pollutant degradation) are examples of beneficial biofilms. On the other hand, biofilms that accumulate in cooling water towers and heat exchangers, membrane systems, filters, drinking water distribution systems, swimming pools, food processing equipment, paper manufacture industries, ship hulls, catheters, medical implants, tissues, teeth, lungs and contact lenses are harmful. One of the main problems of these biofilms is their potential impact in human health. Moreover, they may create problems for hygiene and cleaning, as well as being responsible for energy losses in heat transfer equipment, blockages in flow systems and microbial induced corrosion.

In engineered systems, detrimental biofilm formation is usually named as “biofouling” or “biological fouling”. Its costs have not been fully assessed, but dispersed data from different case studies suggest a quite significant financial burden brought upon by biofouling. According to Koch et al., [12] and Wang [21], microbial influenced corrosion accounts for about 20-30% of all corrosion costs in the USA, the latter representing around 3% of the county’s Gross National Product (GDP). Macrobiological fouling layers (barnacles, oysters, algae, etc) attach to biofilms on ship’s hulls, substantially increasing the drag forces against motion of the vessel. These layers may cause up to 30%-40% higher fuel consumption. Additionally, in order to mitigate biofouling unwanted effects, the sea water cooled industrial equipment operators (including thermal power plants) all over the world are estimated to spend around 10-15 billion dollars/year [17]. Azis et al., [1] indicates that the costs of biofouling in desalination plants may amount to 15 billion US$/year worldwide.

The general fouling costs (all types of fouling phenomena) have been estimated in several countries to represent 0.25% of their respective GDP-Gross National Product [13]. In view of the values reported above for biofouling costs in a diversified number of cases, a conservative educated guess would point out to biofouling costs being at least 20% of the overall fouling costs. Since the world GDP is around 71×10^{12} US$ [16], the total biofouling costs may reach 3.5×10^{10} US$/year worldwide.

Despite the huge economical losses entailed with biofilm formation in industrial or engineered systems, medical biofilms usually have a worse reputation due to the simple fact that biofilms can kill. It has been estimated that more than 60% of all microbial infections in humans are caused by biofilms [18] and that a significant part of these incidences corresponds to hospital-acquired infections that have been estimated to be greater than 2 million per year in American hospitals alone [17]. In fact, hospital-acquired infections have been considered the fourth leading cause of death in the US affecting about 10% of the hospital patients [17]. About 60-70% of the hospital-acquired infections are related to implanted medical devices that have become indispensable in modern-day medicine [17]. More than 5 million central venous catheters are inserted annually in the US and it expected that 2-12% will results in sepsis [19]. The cost per infection is estimated to be between $34,000 and $56,000 and the annual cost of caring for patients with catheter-associated bloodstream infections can reach $2.3 billion [19]. Another common biofilm-related infection results in ventilator associated pneumonia. It has been estimated that this condition alone is responsible for an expenditure of $1.5 billion per year corresponding to 20-34% of all hospital costs [20]. The problem becomes even more serious if one starts to consider the effects of antibiotic resistance which is more likely to occur in biofilms than in planktonic cells. In fact, it is calculated that infection treatment of resistant organisms costs twice as much as treating those that are susceptible to conventional treatment. It has been estimated that, for an average 500-bed hospital, the additional cost for treatment of biofilm infections that have developed some sort of antibiotic resistance would be up to $21 million (considering only sepsis, pneumonia, and cervical site infections) and that approximately 120 patients would die from these infections [20].

Having considered the economical and societal impacts of biofilms it is clear that biofilm control is a question of paramount importance. However, biofilm control strategies that either potentiate their development in beneficial applications or inhibit/eliminate them in detrimental cases have to be designed taking into consideration many factors, among which the definition of acceptable biofilm threshold levels. As it will be detailed in the following chapters, biofilms affect many aspects of our life. In some of these aspects, the presence of a biofilm impacts on the performance of a given system and therefore biofilm development to a certain level is tolerated (for instance in cooling water systems) or even promoted (in biofilm reactors). There are other scenarios, particularly in the health and food sectors, where biofilms have to be eliminated because they are simply too much of a risk. In these latter cases we are at war with these biofilms. In this respect, the information presented in the following chapters is intended to provide clues towards a better understanding of the biofilm development process that may eventually make us win some of these battles.

References

[1] Azis, P. K. A., Al-Tisan, I. & Sasikumar, N. (2001). Biofouling potential and environmental factors of seawater at a desalination plant intake. *Desalination, 135,* 69-82.

[2] Battin, T. J., Sloan, W. T., Kjelleberg, S., Daims, H., Head, I. M., Curtis, T.P. & Eberl, L. (2007). Microbial landscapes: new paths to biofilm research. *Nature Reviews Microbiology, 5,* 76-81.

[3] Bryers, J. D. (2000). Biofilm formation and persistence. *In*: Biofilms II: Process analysis and applications. Bryers, J. D. (ed). John Wiley & Sons, New York, pp. 45-88.

[4] Bryers, J. D. (2008). Medical Biofilms. *Biotechnology and Bioengineering*, *100*, 1-18.

[5] Characklis, W. G. (1973). Attached microbial growth-II. Frictional resistance due to microbial slimes. *Water Research, 7,* 1249-1258.

[6] Costerton, J. W., Geesey, G. G. & Cheng, K.-J. (1978). How bacteria stick. *Scientific American, 238,* 86-95.

[7] Denkhaus, E., Meisen, S., Telgheder, U. & Wingender, J. (2007). Chemical and physical methods for characterisation of biofilms. *Microchimica Acta, 158,* 1-27.

[8] Davey, M. E. & O`Toole, G. A. (2000). Microbial biofilms: from ecology to molecular genetics. *Microbiology and Molecular Biology Reviews, 64,* 847-867.

[9] Donlan, R. M. & Costerton, J. W. (2002). Biofilms: survival mechanisms of clinically relevant microorganisms. *Clinical Microbiology Reviews, 15,* 167-193.

[10] Hall-Stoodley, L., Costerton, J. W. & Stoodley, P. (2004). Bacterial biofilms: from the natural environment to infectious diseases. *Nature Reviews in Microbiology, 2,* 95-108.

[11] Heukelekian, H. & Heller, A. (1940). Relation between food concentration and surface for bacterial growth. *Journal of Bacteriology, 40,* 547-558.

[12] Koch, G. H., Brongers, M. P. H. Thompson, N. G. Virmani, V. P. & Payer, J. H. (2001). Corrosion costs and preventive strategies in the United States, Report FHWA-RD-01-156 (by CC Technologies Laboratories, Inc.) to the Federal Highway Administration.

[13] Mueller-Steinhagen, H., Malayeri, M. R. & Watkinson, A. P. (2005). Fouling of heat exchanger-new approaches to solve old problem. *Heat Transfer Engineering, 26,* 1-4.

[14] O'Toole, G. A., Pratt, L. A., Watnick, P. I., Newman, D. K., Weaver, V. & Kolter, R. (1999). Genetic approaches to study of biofilms. *Methods in Enzymology, 310,* 91-109.

[15] Raad, I., Hanna, H. & Maki, D. (2007). Intravascular catheter-related infections: advances in diagnosis, prevention, and management. *Lancet Infectious Diseases, 7,* 645-657.

[16] Report for Selected Countries and Subjects. (2013). *World Economic Outlook Database, April 2013*. International Monetary Fund. http://en.wikipedia.org/wiki/List_of_countries_by_GDP_(nominal). Retrieved April 16, 2013.

[17] Satpathy, K. K. & Nair, K. V. K. (1990) Impact of power plant discharge on the physicochemical characteristics of Kalpakkam coastal waters. *Mahasagar – Bulletin of National Institute of Oceanography*, 23, 117-125.

[18] Shunmugaperumal, T. (2010). Introduction and overview of biofilm. In: *Biofilm Eradication and Prevention* (John Wiley & Sons, Inc., Hoboken, New Jersey), pp. 3-35.

[19] Shunmugaperumal, T. (2010). Rationale for biofilm eradication from modern medical devices. In: *Biofilm Eradication and Prevention*. Shunmugaperumal, T. (Ed.), John Wiley & Sons. pp. 36-72.

[20] Thomas J. G., Litton, I. & Rinde H. (2006). Economic impact of biofilms on treatment costs. In: *Biofilms, Infection, and Antimicrobial Therapy*. Pace, J. L., Rupp, M. E. & Finch, R. G. (Eds). CRC Press, Boca Raton, FL. pp. 21-39.

[21] Wang, G. (2011). Biofouling and biocorrosion. Progress Report sponsored by US Department of Energy-Wind and Water Program.

[22] Zobell, C. E. (1943). The effect of solid surfaces upon bacterial activity. *Journal of Bacteriology, 46,* 39-56.

In: Biofilms in Bioengineering
Editors: Manuel Simões and Filipe Mergulhão

ISBN: 978-1-62948-161-6

Chapter I

Biofilm Formation, Development and Relevance

Patrizia Messi*
Department of Life Sciences,
University of Modena and Reggio Emilia, Italy

Abstract

Bacteria are able to grow adhered to almost every surface, forming architecturally complex communities termed biofilms. Biofilms are multicellular communities held together by a self-produced extracellular matrix. Biofilm development is a dynamic process that occurs in different sequential phases and involves a series of steps. Biofilms impact humans in many ways as they can form in natural, medical, and industrial settings. For instance, formation of biofilms on medical devices, such as catheters or implants often results in difficult-to-treat chronic infections. Not all biofilms cause problems both in the clinical and environmental context. Within humans, bacterial biofilms may also play a protective role and promote a probiotic effect when consumed through various fermented foods. Also within the industrial context there are several successful examples of the positive use of beneficial biofilms. Recent data concerning biofilm formation, evolution, structure and functions are reported in this chapter. Particular emphasis is given to discussing the different adverse effects and beneficial aspects of biofilms.

Keywords: Biofilm structure, development and propagative mechanisms; Adverse and beneficial aspects

* E-mail: patrizia.messi@unimore.it.

Introduction

Recent advances in microscopy and molecular technologies have made possible the direct observation of a wide variety of natural habitats, demonstrating that in natural, industrial and medical setting, the majority of microbes are associated with surfaces within complex bacterial communities (biofilms) and not as free-floating organisms [1]. These microbial structures occur in nearly every moist environment where sufficient nutrient flow is available and surface attachment can be achieved. Biofilm is usually defined as organized communities of microorganisms attached to a surface and embedded in a matrix of self-produced extracellular polymeric substances (EPS) [2, 3]. The new technologies have made possible to examine in detail such communities, and have led researchers to conclude that biofilms represent biological systems with a high level of organization where bacteria form structured, coordinated, functional communities. Consequently, the definition of biofilm has evolved over time, taking into consideration not only the morphological characteristics, but also other physiological attributes of these biofilm-forming organisms. For example the gene expression profiles of bacteria in these microbial consortia are quite different compared with the expression profiles of the same strains grown planktonically, and such genes include those responsible for regulation and/or expression of surface adhesion proteins, appendages such as fimbriae, pili or flagella, etc. Therefore, the useful definition of a biofilm is "microbially derived sessile community characterized by cells that are irreversibly attached to a substratum or interface or to each other, are embedded in a matrix of extracellular polymeric substances that they have produced, and exhibit an altered phenotype with respect to growth rate and gene transcription".

Advantages of Forming a Biofilm Versus Living As Individual Cells

Biofilm is an ancient prokaryotic adaptation [5] and represents a mode of growth that allows bacteria to survive in hostile environments and to colonize new niches. Because of their dynamic character, biofilm communities can continuously change in time and space, providing better survival and growth of the associated microorganisms. For this reason, in most natural environments biofilms are the prevailing microbial lifestyle.

Environmental advantages. Protection from the environment, nutrient availability and metabolic cooperativity and acquisition of new genetic traits are some reasons why the biofilm strategy has been adopted by so many microbes [6]. Within this structure the microorganisms are protected from chemical and natural antibacterial substances, environmental bacteriophages, and phagocytic amoebae. For example, after treatment with antibiotics or chemicals, biofilm cells survive and are often responsible for recurring symptoms and medical treatment failure [8]. Some phenomena are postulated to contribute to the biofilm defence, including incomplete antimicrobial penetration, slow or no growth of some of the biofilm cells, and the expression of biofilm-specific phenotypes [9, 10].

Phenotypic variation. The properties of biofilm cells have been shown to be distinct from their planktonic counterparts. There is evidence that gene expression may be regulated once a cell comes into contact with a surface. This up- and downregulation of genes has been demonstrated in *Pseudomonas aeruginosa* with genes associated with alginate synthesis (algC up-regulation) and in *Staphylococcus aureus* for genes encoding enzymes involved in glycolysis [11]. In some organisms, up to 22% of genes were up-regulated in the biofilm state and 16% were down-regulated [12]. Horizontal gene transfer in biofilm organisms by conjugation, transduction and transformation is also important for the evolution and genetic diversity of natural microbial communities [13,14,15]. With the cells embedded in a polysaccharide matrix, biofilms are an ideal place for exchanging genetic material at a greater rate than cells in the planktonic phase [16,17]. A biofilm community provides an ideal niche for transfer of mobile genetic elements (e.g., plasmids) that might encode beneficial traits, such as resistance to antibiotics and/or heavy metals. Conjugative plasmids might also enhance biofilm formation because plasmids repressed for horizontal gene transfer retained the ability to form biofilms even if the biofilm population included plasmid-free recipient cells. Plasmid recipient strains have been shown to demonstrate increased biofilm growth [18]. Conjugative transposons are probably the most promiscuous of all mobile elements. These elements integrate into the host cell genome and do not need to provide their own replication machinery or rely upon interactions with the host cell for stable replication and maintenance. Roberts et al., [17] investigating the horizontal gene transfer of a conjugative transposon Tn*5397* (tetracycline resistant) in a mixed species oral biofilm confirmed the transfer of genes in biofilm communities among related or unrelated strains. Virus-mediated transduction is another mode of gene transfer. Both chromosomal and plasmid DNA can be successfully transduced in natural water environments ranging from sewage plants to rivers and lakes and has been shown to confer increased survival mechanisms (e.g. antibiotic resistance) on the biofilm cells [19,20]. It was recently demonstrated that high transduction frequencies can also occur in marine environments [21], and therefore the impact of transduction on gene exchange in biofilms may be significant. Bacteriophage-mediated transfer of tetracycline resistance has been demonstrated in *S. aureus* biofilms [22], and other examples include the transfer of penicillin binding proteins (PBP) and IgA proteases [23,24,25]. Lastly, natural competence is a pervasive phenomenon among bacteria. Hence, it could be expected that transformation of DNA would occur efficiently in dense bacterial populations such as biofilms that are regulated by quorum sensing. *Streptococcus mutans* cells grown in a biofilm were transformed at a frequency up to 4×10^{-3} per cell and, in most cases, at rates of ten to 600-fold higher than planktonic *S. mutans* cells [26].

Biofilm Structure and Functions

The application of confocal scanning laser microscopes (CSLM) to biofilm research has revealed the elaborate three-dimensional structure. It is now recognised that biofilms are heterogeneous, structurally organised microbial communities containing microcolonies encased in an EPS matrix and loosely connected to nearby microcolonies by water channels and voids [27,28], these last responsible for transport of nutrients, oxygen, genes and even antimicrobial agents [4].

Microcolonies. The microcolonies that constitute the biofilm can be composed of single-species populations or multimember communities of bacteria, depending on the environmental parameters under which they are formed [29]. Having docked at a surface, cells begin to grow relatively quickly and form a microcolony. At this stage, as the microcolonies dispersed over the surface begin to grow, first of all in the horizontal mode across the surface, and then out into the liquid phase, the three-dimensional, multispecies consortia biofilm begins to develop. In addition, secondary colonisers, i.e. those that lack the ability to initially attach to a surface, may preferentially attach to the primary colonisers by co-adhesion and contribute to community development resulting in a biofilm with a thickness in the range of 30–50 µm.

EPS. Biofilms are mainly composed of bacterial cells, organized in microcolonies, and EPS, microbial products located on or outside the cell surface. If biofilms can be metaphorically called the "city of microbes" [30], the EPS represent the "house of the biofilm cells" [31]. The production of extracellular matrix appears to overlap with all stages that occur after initial surface adhesion, but it is considerably more than simply the glue for biofilms. The EPS matrix is a dynamic system, constructed by the organisms and responding to environmental changes. It enables the cells to function in a manner similar to multicellular organisms [32]. In fact, the EPS determine the immediate conditions of life of biofilm cells living in this microenvironment by affecting porosity, density, water content, charge, sorption properties, hydrophobicity, and mechanical stability [33]. Rather, it appears to be a highly sophisticated system, which endows the biofilm mode of life with particular, successful features [31]. Given the great diversity of microbial communities that form biofilms, as well as the different environments that they inhabit, it is difficult to make generalizations about biofilm EPS structure and physiological function. Detailed composition analysis of an EPS is difficult, as EPS is often a complex mixture of polysaccharides, proteins, glycoproteins, and glycolipids, DNA, and humic acid substances [34]. Furthermore, even though carbohydrates have been identified as one of the major components of EPS, the biochemical properties of these compounds remain elusive due to their complex structures and unique monomer linkages [35,36,37,38]. The nature of carbohydrates within EPS is a dynamic function of the microbial community composition; this is reflected by the complexity of carbohydrates containing diverse sugar residues extracted from biofilm matrices [36,38,39]. The biofilm matrix not only facilitates structural organization and protects the microbial community [40], but also influence predator-prey interactions, as demonstrated in a system consisting of a predatory ciliate and yeast cells, in which grazing led to an increase in biofilm mass and viability, with EPS as the preferred food source [41]. Several studies indicate that EPS are not necessarily required for the initial attachment of microbial cells to surfaces [42], but their production is essential for the development of the biofilm matrix, providing the framework into which microbial cells are inserted [43]. EPS can account for 50–90% of the total organic carbon of the biofilm and is highly hydrated because it can incorporate large amounts of water into its structure by hydrogen bonding [44]. This hydrated EPS will protect the biofilm from desiccation. Briefly, EPS can be considered the primary matrix material of the biofilm that serves as a storage facility for nutrients and can entrap other microbes and non-cellular materials, such as minerals, crystals and corrosion products.

Interstitial water channels. It has been widely reported that biofilms consist of clumps of cells separated by interstitial voids and channels [45]. The interstitial voids or channels are the lifeline of the system, since they provide a means of circulating nutrients as well as exchanging metabolic products with the bulk fluid layer [46], and it was also discovered that water could flow through the channels [47]. Probably the channels are a vital part of the biofilm structure and function, and therefore there are likely to be mechanisms for the formation as well as the maintenance of these structures. Proximity of cells within the microcolony (or between microcolonies) provides an ideal environment for creation of nutrient gradients, exchange of genes, and quorum sensing. Bacteria in biofilms, in fact, coordinate behavior by cell–cell communication using secreted chemical signals. Cell signalling allows bacteria to sense and phenotypically respond to their environment, for example, by assessing cell density (also termed quorum sensing or QS) or environmental cues.

Quorum sensing. Bacterial gene regulation in response to cell density, known as "quorum sensing," is a regulatory response thought to occur in bacterial communities through the sensing of extracellular signal molecules called autoinducers that are produced by members of the community [48]. These cell-to-cell signals are also involved in the development of a bacterial biofilm [49], and have recently been demonstrated to play a role in cell attachment and detachment from biofilms [50]. At sufficiently high population densities these signal molecules, homoserine lactones in Gram-negative bacteria and peptides in Gram-positive bacteria, reach concentrations required for activation of genes involved in biofilm differentiation. Signals can be produced and secreted by the bacterial community itself, in which case the molecules are termed autoinducers. Autoinducers accumulate extracellularly and the concentration of autoinducer can be correlated with population density. At high concentrations, autoinducers trigger signal transduction cascades that lead to multicellular responses in the bacterial population. This mechanism of cell–cell communication in bacteria controls a large number of developmental processes included those related to biofilm formation [51], but in addition to quorum-sensing molecules, a diversity of other signals trigger biofilm formation. These include secondary metabolites such as antibiotics, pigments, and siderophores [52]. At subinhibitory concentrations many antibiotics function not to kill cells, but rather as signals that trigger changes in gene expression [53].

Biofilm Formation and Evolution: Role of Cellular Factors and Environmental Conditions

Biofilms can form on a vast array of abiotic and biotic surfaces, including living tissues, indwelling medical devices, industrial or potable water system piping, or natural aquatic systems. The solid-liquid interface between a surface and an aqueous medium (e.g., water, blood) provides an ideal environment for the attachment and growth of microorganisms. Biofilms can be formed by a single bacterial species, although in nature, biofilms exist primarily as complex multi-species communities of bacteria in which each species fills an ecological niche within the biofilm depending on its metabolism and morphology [54]. Biofilm structure is affected by both cellular factors (microbial composition and biology) and environmental conditions (environmental parameters).

Cellular factors may affect biofilm formation. Cell surface hydrophobicity, presence of fimbriae and flagella, and production of EPS all influence the rate and extent of attachment of microbial cells. The hydrophobicity of the cell surface is important in adhesion because hydrophobic interactions tend to increase with an increasing nonpolar nature of one or both surfaces involved (i.e., the microbial cell surface and the substratum surface). A hydrophobic cell will be more able to overcome the initial electrostatic repulsion with the solid surface and adhere more readily. Most bacteria are negatively charged but still contain hydrophobic surface components [55]. Hydrophobic non-flagellar appendages such as fimbrae play a role in cell attachment by overcoming the electrostatic repulsion barrier that exists between the cell and the substratum. Lipopolysaccharides (LPS) and exo-polysaccharides (EPS) also play a role in the attachment process. Briefly, cell surface structures such as fimbriae, other proteins, LPS, EPS, and flagella all clearly play an important role in the attachment process. In particular, cell surface polymers with nonpolar sites such as fimbriae, other proteins, and components of certain gram-positive bacteria (mycolic acids) appear to dominate attachment to hydrophobic substrata, while EPS and lipopolysaccharides are more important in attachment to hydrophilic materials. Flagella are important in attachment also, although their role may be to overcome repulsive forces rather than to act as adsorbents or adhesives.

Environmental conditions, such as nutrient availability, hydrodynamics, surface and interface properties can influence biofilm structure. The amount of nutrients present in the medium can affect the rate of biofilm formation. Biofilms tend to form more readily in the presence of ample nutrients [56]. The concentration of cations, glycoproteins, proteins, and organic molecules at the surface can provide a relatively nutritious zone for bacteria compared to the bulk aqueous environment [57]. In addition, fluid flow in the boundary region near the surface can be considered negligible which allows bacteria to approach the surface. Once near the surface, it will either approach the surface by Brownian motion or move by chemotaxis towards the surface in response to the chemical concentration gradient [57]. One function of the biofilm is to anchor cells in a friendly, nutrient rich environment. Phosphorus is a particularly important nutrient. Cells saturated with phosphate have a higher tendency to flocculate and adhere due to their increased hydrophobicity, while those cells depleted in phosphate are more hydrophilic and less likely to adhere [58]. Temperature can have an effect on biofilm formation. Temperatures at the high end of a culture's growth range can enhance biofilm formation. Depending upon the species involved, high temperature increases the rate of cell growth, EPS production, and surface adhesion, all of which enhance biofilm formation [57]. The surface onto which cells will attach has an important impact on biofilm formation. The attachment will occur most readily on surfaces that are rougher [59]. Shear forces are lower near a rough surface, and there is a larger surface area to which cells can adhere. Porous materials also work well for biofilm formation. Shear forces are very low inside pores even under conditions where bulk fluid velocity is high. Pores provide a protected environment for cells to attach and grow. Biofilm formation also tends to increase with the hydrophobicity of the surface material [3]. Biofilms form much more rapidly on Teflon and other plastics than glass or metal. Possibly this is due to differences in hydrophobicity of the surfaces and ionic charges [60].

Hydrodynamics. In theory, the flow velocity immediately adjacent to the substratum–liquid interface is negligible [60]. This zone of insignificant flow is termed the hydrodynamic boundary layer whose dimensions are inversely related to the water velocity. Once a microbial cell has reached the hydrodynamic boundary layer, then the processes that are involved in attachment can begin. Cells behave as particles in a liquid, and the rate of settling and association with a submerged surface will largely depend on the velocity characteristics of the liquid. Studies carried out to determine the effect of flow velocity on biofilm structure showed that under various hydrodynamic conditions, such as laminar and turbulent flows, biofilm structures are altered in response to flow conditions. Biofilms grown under laminar flow were found to be patchy and consisted of roughly circular cell clusters separated by interstitial voids. Biofilms grown in the turbulent flow cells were also patchy, but these biofilms consisted of patches of ripples and elongated 'streamers' which oscillated in the flow [61].

Phases of Biofilm Development

Biofilm formation is a developmental process in which bacteria undergo a regulated lifestyle switch from a nomadic unicellular state to a sedentary multicellular state where subsequent growth results in structured communities and cellular differentiation. It is generally accepted that biofilm development is a dynamic process of growth and detachment (or shedding) of bacterial cells and aggregates that occurs in different sequential phases and involves a series of steps. Figure 1 shows a diagrammatic representation of cyclic steps involved in the formation of an active biofilm.

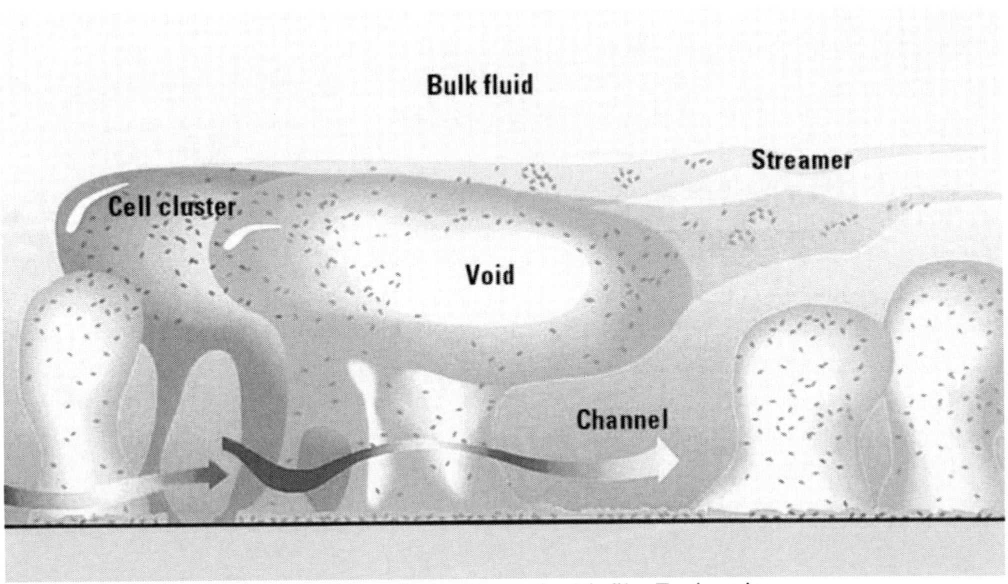

Reproduced with permission from: Peg Dirckx, Center for Biofilm Engineering.

Figure 1. The structure of slime. At the microscopic level, biofilms are heterogeneous in structure and include characteristics such as cell clusters, void spaces, water channels, and slime streamers that are affected by flow of the fluid surrounding the biofilm.

Surface conditioning. The life of a biofilm starts with the planktonic or free floating cell, and the formation of a biofilm virtually occurs on any submerged surface in any environment where the bacteria are present and can interact. Firstly, surfaces of attachment are conditioned by adsorption of organic and inorganic nutrients that influence subsequent bacterial attachment [2,62]. A material surface exposed in an aqueous medium will inevitably and almost immediately become conditioned or coated by polymers from that medium, and the resulting accumulation of molecules at the solid-liquid interface on surfaces (commonly referred to as *conditioning film*) leads to a higher concentration of nutrients compared to the fluid phase that will affect the rate and extent of microbial attachment. This increase in nutrient level favours biofilm formation and is also dependent on the type of the competitive culture associated with the biofilm [63]. The conditioning also alters the physiochemical properties of the surface viz., surface free energy, changes in hydrophobicity and electrostatic charges [64], which may also affect the sequence of microbial events. It is also established that adsorption of certain proteins to surfaces plays an important role in microbial adhesion, inhibiting [65, 66] or favouring [67] the process of attachment.

Surface attachment - Adhesion of cells. The second step in the formation of biofilms is the attachment of microorganisms to the conditioned surface. Attachment is a complex process regulated by diverse characteristics, such as the effects of the substratum, conditioning films forming on the substratum, hydrodynamics of the aqueous medium, characteristics of the medium, and various properties of the cell surface. Initial contact between the bacterial cells and the substratum is represented by weak interactions where microbial cells, still showing Brownian motion, can easily be removed by the fluid shear forces e.g. merely by rinsing (*reversible adhesion*). This transient adherence can result in a return to planktonic existence, as various repulsive forces (van der Waals attraction forces, electrostatic forces and hydrophobic interactions) can prevent the bacterial cells in making a direct contact with the surface. However, the contact still occurs due to the production of surface appendages by the bacteria such as flagella, fimbriae, pili and the EPS fibrils [68,69], leading to a stable surface association (*irreversible adhesion*), and a subsequent switch to biofilm development. In the irreversible attachment bacterial cells attached reversibly to surfaces produce EPS due to stimulation of membrane-bound sensory proteins of the bacterial cell [70], which allows for the development of cell-to-cell bridges that, in turn, cement the cells to the surface [2, 71]. Various short-range forces involved include dipole-dipole interactions, hydrogen, ionic and covalent bonding and hydrophobic interactions. The polymeric fibrils form a bridge between the bacterial cell and the substratum and this enables the irreversible association with the surface. The above mentioned physiochemical properties of the bacterial cell surface are also important in determining the adhesion of cells, and may possibly provide a competitive advantage for one organism where a mixed community is involved. Many studies have proposed a hypothetical developmental model for biofilm formation, in particular for the differences found in the attachment process, that can be generalized for many different bacterial species. This model has been adjusted by Lemon et al., [72] to fit either of two general modes of unicellular lifestyle: nonmotile and motile, and on the whole, in both cases, the initial encounters with a surface usually lead to transient adherence, as above mentioned. In non-motile species (Figure 2), when conditions are propitious for biofilm formation, individual bacteria appear to increase the expression of adhesins on their outer surface, i.e., they increase their "stickiness". This increased stickiness promotes both cell-cell adherence and cell-surface adherence when these bacteria encounter a surface [73]. In particular in nonmotile bacteria, not only the production of EPS plays a critical role in the initiation of biofilm formation but also the changes in cell surface, due to the secretion of large extracellular proteins of variable length within a biofilm population [74]. The first member of this group of proteins is the cell wall associated protein of 2,276 amino acids named Bap, for biofilm-associated protein, which shows global organizational similarities to surface proteins of gram-negative (*Pseudomonas aeruginosa* and

Salmonella enterica serovar *Typhi*) and gram-positive (*Enteroccocus faecalis*) microorganisms [75]. In the case of motile species (Figure 3), when conditions favour biofilm formation, individual bacteria localize to a surface and initiate a dramatic lifestyle switch. Motility is lost and bacteria begin to produce an extracellular matrix that holds the cells together. The presence of flagella is extremely important in the initiation of biofilm formation as, for a number of motile organisms, motility is the driving force that overcomes repulsive forces between the bacteria and the surface.

Colonization. The final stage in biofilm establishment is surface colonization [62]. The irreversibly attached bacterial cells grow and divide by using the nutrients present in the conditioning film and the surrounding fluid environment, forming microcolonies that are considered to be the basic organizational units of a biofilm [76]. Successively, microcolonies enlarge and coalesce to form a layer of cells covering the surface. Entrapment of other planktonic cells in the EPS also occurs, resulting in the formation of a biofilm. The colonization of a surface by one bacterium, i.e. primary colonizers, is also often found to influence the attachment of others to the same surface, i.e. secondary colonizers. During this period, the attached cells also produce additional polymer (EPS), which helps in the anchorage of the cells to the surface and to stabilize the colony from the fluctuations of the environment [77]. The surface microenvironment gets altered as the primary colonizers get attached, grow and divide and produce the EPS [78]. The completed biofilm has a complex architecture, consisting of biofilm bacteria in EPS-enclosed microcolonies interspersed with less dense regions of the matrix that include highly permeable water channels carrying nutrients and waste products [76, 79].

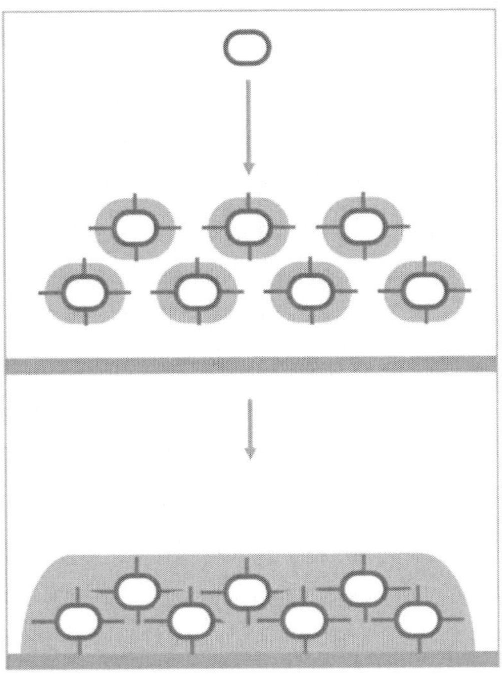

Reproduced with permission from reference 72.

Figure 2. General model for biofilm formation: attachment process by nonmotile bacteria.

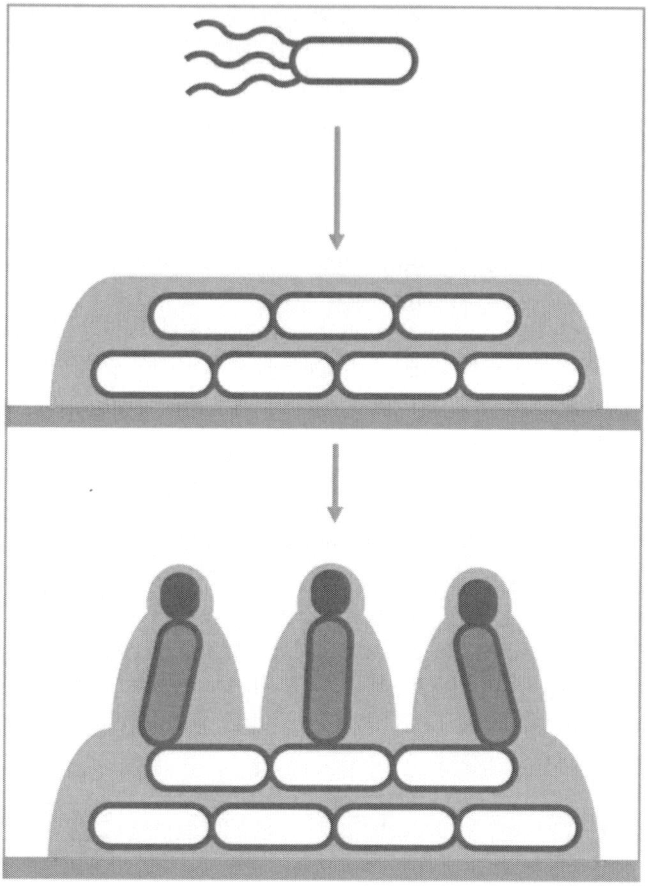

Reproduced with permission from reference 72.

Figure 3. General model for biofilm formation: attachment process by motile bacteria.

Detachment. The detachment of bacterial cells from biofilms is of fundamental importance to the dissemination of infection and to contamination in both clinical [80] and public health settings [81, 82, 62]. The spontaneous detachment of cells from bacterial biofilms has been divided into two processes, erosion and sloughing, based on the magnitude and frequency of the detachment event [83, 84]. Erosion is the continual detachment of single cells and "small portions of the biofilm," whereas sloughing is the "rapid, massive loss of biofilm" [84]. It was initially suggested that turbulent shear forces may be responsible for detachment of clumps of biofilm cells and subsequent transfer to other surfaces for attachment. This type of detachment mechanism only seems to be accurate for biofilms that have grown under laminar shear forces and are more likely to detach when shear forces become more turbulent [85]. However, recent studies have suggested that detachment is an active process that is highly regulated by the attached cell populations [86], a strategy by which bacteria proactively colonize new niches before space and nutrients become limited.

Regarding how biofilm bacteria disseminate into other areas for further surface colonization, several strategies have been suggested in recent studies [87 and references therein]. An active process that is regulated by the attached cell populations, and is often termed 'dispersion' or 'dissolution' [88] suggests that cells located at the periphery of the biofilm are released into the surrounding environment, return to the planktonic state and find new surfaces for biofilm development, a strategy adopted for example by *P. aeruginosa* [71]. A further mechanism of dispersion has been shown in bacterial cells that display 'swarmer' motility, as *Proteus mirabilis*, the causative agent in many urinary and catheter-associated infections [89,90]. It has also been proposed that certain bacteria may alter various surface components, such as glycolipids, peptidolipids, lipopolysaccharides and lipoteichoic acids, which may, in turn, alter cell surface hydrophobicity, facilitating release of a surface-bound cell. Reportedly, *Escherichia coli* and *P. aeruginosa* cells de-adhere from biofilms by increasing their cell-surface hydrophobicity [91]. Lastly, quorum sensing (cell-to-cell signalling) has also been suggested to mediate the exodus of cells from crowded biofilms of *Serratia* spp [92].

Adverse Effects and Beneficial Aspects of Biofilms

Human interference with nature and changes in living conditions have created niches allowing colonization and growth of microorganisms, resulting in an intensification of human exposure to potentially or effectively pathogenic bacteria. Biofilms are often considered problematic from the human perspective due to the damage associated with their presence (rusting, dental plaque formation, and the disease cystic fibrosis) [93].

Bacterial biofilms pose serious problems in industry, medicine, and dentistry, and attempts to control biofilms with biocides often prove ineffective. It is widely recognized that bacteria colonizing a surface as a biofilm can be much more resistant to antimicrobial agents than are their planktonic counterparts. Within biofilms bacteria tolerate higher levels of antibiotics than comparable planktonic cells do, and can be up to 1,000-fold more resistant to antibiotic treatment than the same organism grown planktonically [94,95,96,97]. Several studies have examined the basis for biofilm antibiotic resistance in more detail, and given that some properties of biofilms depend on the component microorganisms and environmental conditions, the level of antibiotic resistance may vary and the factors that give rise to the increased resistance may also differ. Many factors, such as physiological changes due to reduced growth rates, reduced diffusion of antibiotics through the biofilm, the production of enzymes degrading antimicrobial substances, and the activation of stress responses in bacteria within the biofilm, probably contribute to biofilm-associated antibiotic resistance. The exopolysaccharide matrix may also act to various degrees as a diffusion barrier, molecular sieve and adsorbent [98], and therefore, the production of excess amounts of exopolysaccharide by the bacteria during biofilm formation and growth may protect the innermost cells by binding with antimicrobial substances and quenching their effect as they diffuse through it [99,100,101].

Clinically, biofilms are important as the source of persistent infections. Types of surfaces within clinical settings that may be prone to biofilm development are numerous and include wounds, teeth, gastrointestinal tract, indwelling medical devices, water systems, dialysis equipment and endoscopes [2,3,102,103,104,105]. Biofilms are increasingly being recognized by the public health community as an important source of bacterial pathogens for all classes of patients, especially immunocompromised individuals and those with indwelling medical devices [2]. Indwelling medical devices are often primarily colonized by a single bacterial species, but as the biofilm develops, multi-species consortia quickly develop [3]. The role of biofilms in the contamination of medical implants has been well established. Devices that are prone to biofilm formation include artificial voice prostheses, replacement joints, prosthetic heart valves, cardiac pacemakers, cerebrospinal fluid shunts, endotracheal tubes, urinary catheters, peritoneal dialysis catheters, central venous catheters, contact lenses and implanted prosthetic devices for erectile dysfunction [3,71, 104.105,106, 107,108]. Millions of catheters (e.g., central line, intravenous, and urinary catheters) are inserted into patients every year, and these implants serve as a potential surface for biofilm formation. Biofilm formation can also occur on contact lenses, and these biofilms are thought to contribute to keratitis [109,110,111]. Overall, it is thought that upwards of 60% of all nosocomial infections are due to biofilms. These biofilm-based infections can increase hospital stays by 2 to 3 days and cost upwards of $1 billion per year in added costs [112]. Although the role of biofilms in implant infections has been demonstrated in numerous systems, the role of biofilms in nonimplant diseases is less well established. One example of a disease in which biofilms are thought to play a prominent role is the occurrence of lung infections by *P. aeruginosa* in patients with cystic fibrosis (CF). Another example of a likely biofilm-mediated infection is chronic ear infection (otitis media). All these infections are often caused by bacteria able to form biofilms [113]. In addition, it has been a quandary for some time why patients with chronic otitis media do not appear to have infections when tested by routine culture methods. This is likely due to the fact that biofilm bacteria can be difficult to culture by routine methods [1]. Periodontitis is another example of a biofilm-mediated disease that results in chronic inflammation of the tissue supporting the gums and can eventually lead to tooth loss. Biofilms are also responsible for dental caries as well as a variety of other infections and diseases [114]. Dental unit water systems have been proposed to expose patients to opportunistic pathogens from biofilms lining the walls of tubing in dental unit water lines [3,115]. These biofilms are reported to contain counts exceeding 10^5 colony-forming units/mL of mostly harmless saprophites, such as *Moraxella* spp., *Flavobacterium* spp., *Micrococcus* spp. and *Actinomyces* spp. [115]. However, opportunistic pathogens, such as *P. aeruginosa*, as well as other potential pathogens, such as *Legionella* spp. and *Mycobacteria* spp., have also been isolated from these biofilms. [3,115,116]. Although rare, infections in patients related to dental unit water lines have been recorded. For example, cancer patients reportedly contracted *P. aeruginosa* infections while undergoing dental work [117]. A further concern of colonized dental unit water lines is the potential for cross-contamination of patients with infectious microorganisms [115,118], and the presence of biofilms in dental unit water lines makes disinfection between patients a necessity [119]. Even though bacteria are the best-studied microorganisms with regard to biofilm formation, other microorganisms have also been shown to be capable of this phenomenon. Fungal pathogens are emerging as novel opportunistic agents responsible for many nosocomial infections. Many strains of *Candida* spp. including *C. albicans*, *C. parapsilosis*, *C. dubliniensis*, *C. krusei*, *C. glabrata*, *C. tropicalis* and *C. lusitaniae,* have been identified as possible agents in yeast-associated infections [120,121,122]. Furthermore, the development of biofilms by *Candida* spp. strains on medical-related devices, such as indwelling intravascular catheters, has been closely related to disease development [120,123,124]. As bacteria, *Candida* spp. biofilm forms in similar phases and are highly resistant to treatment with antifungal agents [120].

Industrially, biofilms are beneficial in many cases and detrimental in many others. However, biofilms also pose important economic problems. For example, mixed-species biofilms containing sulphur-reducing and -oxidizing bacteria on industrial metal surfaces may result in corrosion due to cathodic depolarization and acid production [88]. Biofilm formation represents a serious risk factor in food processing environments also [125,126]. The bacteria attached on food or contact surfaces, may be released and contaminate food and raw materials for food production [127,128,129,130], and lead to serious hygienic problems and economic losses due to pathogenic/opportunistic and food spoilage bacteria. The role of biofilm in persistence of pathogens in food products is documented in a number of reports and many new organisms like *Listeria monocytogenes*, *Yersinia enterocolitica*, *Campylobacter jejuni* and *Escherichia coli* O157:H7 have been added to the list [131,132,133,134]. The composition of the substrate also influences biofilm formation. For example, Buna-N and Teflon seals have been implicated as important sites for biofilm formation [135,136]. Herald and Zottola [137] observed the attachment of *Listeria monocytogenes* to stainless steel through the involvement of fibrils. The pathogen also attached to glass, polypropylene and rubber [135]. The number of bacteria recovered from these surfaces was high and dependent on the length of exposure time. These bacteria may also act as a source for post-pasteurization contamination [138]. The formation of biofilms in drinking water distribution systems leads to decrease in water velocity and carrying capacity, clogging of pipes, increase in energy utilization and decrease in efficiency of operations [139,140]. This development is mainly due to the traces of nutrients in the water supply and even the high levels of residual chlorine may also not prevent the biofilm formation [141,142]. In recent years special attention has been paid to *Legionella pneumophila*, an opportunistic waterborne germ responsible for the Legionnaires' disease, a common life-threatening atypical pneumonia in humans [143]. The pathology is transmitted via aerosols generated from water systems and the main sources of infection are represented by hot distribution systems and cooling towers [144,145]. Cooling towers provide good environments for the multiplication of pathogenic microorganisms, and adhesion, retention and growth on solid surfaces play important roles in the resident bacterial community [146]. Biofilms readily grow on the extensive and diverse surfaces present in cooling tower reservoirs and fill material, and both harbour and protect pathogens [147,46,148,5]. In this context microbial ecological interactions in biofilms are particularly important for all the pathogens that infect amoebae, including *Legionella* species [149,150,151]. Most of the microbial interactions, which are crucial for *Legionella* survival and growth in aquatic environments, occur within the biofilms which, besides protozoa, represent one of the most important key factors for *Legionella* ecology [150,148]. Besides favouring the persistence of *Legionella* in water systems by providing a shelter and a possible source of nutrients, it is also possible that biofilms may enhance the emergence of new genetic or phenotypic variants of the pathogen [152]. Biofilm protection would confer a remarkable ecological advantage to these variants, determining their selection and colonization, and this phenomenon may be enhanced by biocidal treatments. The hypothesis of the emergence of different phenotypes within the biofilms is supported by the observation that gene expression in other germs may be regulated once a cell comes into contact with a surface [153]. Biofilm formation in cooling towers also poses serious problems by causing equipment damage through corrosion and by decreasing energy efficiency by clogging hydraulic systems and increasing heat transfer resistance across fouled surfaces [154,155].

Lastly, not all biofilms cause problems both in the clinical and environmental context. Within humans, bacterial biofilms may also play a protective role. For example, in the human gut, biofilms consisting of mixed consortia of commensal bacteria attached to gut epithelial cells form a formidable barrier against foodborne pathogens [156]. Lactic acid bacteria and *Bifidobacterium* spp. constitute as a major part of the natural microflora [157,158] and, when present in sufficient number, create a healthy equilibrium between the beneficial and potentially harmful microflora in the gut [159,160,161]. These organisms also promote a probiotic effect when consumed through various fermented foods [162,163]. Within the industrial context there are several successful examples of the positive use of biofilms that are called "beneficial biofilms". Biofilm formation by *B. subtilis* and related species permits the control of infection caused by plant pathogens, the reduction of mild steel corrosion, and the exploration of novel compounds [164]. Among the first successful biocontrol agents used against insects and pathogens were members of the genus *Bacillus* [165,166]. Commercial strains of *B. subtilis* have been marketed as biocontrol agents for fungal diseases of crops. The commercial biofungicide, Serenade (Bayer CropScience, Germany), which contains a *B. subtilis* strain, is reported to be effective against a variety of pathogenic bacteria, including *Erwina, Pseudomonas*, and *Xanthomonas* strains. Beneficial biofilms are used as inhibitor of mild steel corrosion. It is known that bacteria corrode metals. Sulfate-reducing bacteria (SRB) are responsible for the corrosion of cast iron, carbon steel, stainless steel, and some alloys [167]. Beneficial biofilms are capable of decreasing the above-described demages. A gramicidin-S-producing *Bacillus brevis* 18-3 biofilm, for example, has been shown to reduce this biological corrosion [168]. Mixed-species biofilms are useful in bioremediation processes of human and manufacturing wastes, as the treatment of wastewater for removal of organics [169,170] and heavy metals [171]. In many natural environments, the maintenance of water quality is brought about by the microbial metabolism in biofilms. The bacteria present in these biofilms biodegrade many of the toxic compounds and thereby minimize the build up of pollutants, thus acting as pollutant monitors [172]. Such systems, using mixed microbial consortia have found application in fluidized beds and trickling filters for sewage and waste water management and in water purification plants and also in waste gas treatment [173,174.175]. The organic nutrient-trapping capability of biofilms helps in reducing the organic content of the waste waters before they are released into the water streams or used for irrigation purposes. In recent years, microbial biofilms have also received considerable attention from the viewpoint of bioremediation of various industrial effluents [176,177] and in the nitrification process for the treatment of high strength nitrogen fertilizer waste water [178,179]. Biofilms represent a natural form of cell immobilization. The immobilized microorganisms have also been successfully employed in bioreactors to improve the productivity and stability of the fermentation processes [180,181]. The biofilms also find application in the manufacturing of industrial products like acetic acid, ethanol and polysaccharides and in addition for metal ore leaching [182,183]. Biofilm development by fungi, such as *Aspergillus* spp., has also been utilized within the industrial context for production of various compounds, such as citric acid and phosphates for fertilizers [184,185].

Conclusion

The role of biofilms in implant infections and nonimplant diseases is widely recognized, and in industrial context also these microbial structures pose important economic problems. Within biofilms, microorganisms are much more resistant to antimicrobial compounds than are their planktonic counterparts, and attempts to control biofilms with biocides often prove ineffective. New strategies aimed at biofilm-specific targets are therefore necessary. The most straightforward approaches generally focus on impregnating catheter material with classical antimicrobial agents. A series of nonbactericidal methodologies, employing biological substances are recently described [188]. For example the adhesion process can be impaired by compounds that generally perturb the physiochemical adhesive forces needed for biofilm attachment. A biological approach as the disruption of microbial quorum sensing by other organisms has begun to reveal new opportunities for therapeutic or industrial applications [189]. For example the use of synthetic mimic compounds was found to damage the architecture of *P. aeruginosa* biofilm [190] or counteract biofilm formation by *E. coli* [191]. Lytic bacteriophages have been proposed as new class of anti-biofilm agents [192]. Bacteriophage (phage) has been used to treat infectious diseases in humans and there is interest in phage to control biofilms, as some, propagated in their bacterial host, produce depolymerases that hydrolyze the biofilm extracellular polymers. Other strategies focused on the use of surfaces conditioned with bacterial compounds as surfactants [193] to prevent microbial colonization by pathogens of material in direct contact with food. Lastly, although it is obviously important to control harmful biofilm formation, future studies should be directed to the possible use of beneficial biofilms formed by industrial bacteria. Biofilms formed in a small-scale model by bacteriocin producer LAB (lactic acid bacteria) have been proposed as biopreservative for the capability to influence the survival and the multiplication of *L. monocytogenes* [194]. Further development in this area will certainly be required to fully explore these complex systems of the microbial world. A deeper understanding of the ecology of microbial biofilms will allow the identification of compounds with novel mechanisms of action to antagonize biofilm bacteria in a wide variety of settings.

References

[1] Costerton, J. W., Lewandowski, Z., Caldwell, D. E., Korber, D. R., Lappin-Scott, H. M. (1995). Microbial biofilms. *Ann. Rev. Microbiol*, *49*, 711–745.

[2] Donlan, R.M. (2002) Biofilms: microbial life on surfaces. *Emerg Infect Dis*, *8,* 881–890.

[3] Donlan, R.M., & Costerton, J.W. (*2002*) Biofilms: survival mechanisms of clinically relevant microorganisms. *Clin. Microbiol Rev, 15,* 167–193.

[4] Prakash, B., Veeregowda, M., Krishnappa, G. (2003). Biofilms: a survival strategy of bacteria. *Curr Sci, 85,* 1299–1305.

[5] Hall-Stoodley, L., Costerton, J.W., Stoodley, P. (2004) Bacterial biofilms: from the environment to infectious disease. *Nat. Rev. Microbiol*, *2,* 95–108.

[6] Davey, M.E. & O'toole, G.A. (2000). Microbial biofilms: from ecology to molecular genetics. *Microbiol. Mol. Biol. Rev*, *64*, 847-67.

[7] Li, Y.H., Lau, P.C., Lee, J.H., Ellen, R.P. and Cvitkovitch, D.G. (2001) Natural genetic transformation of *Streptococcus mutans* growing in biofilms. *J. Bacteriol, 183*, 897–908.

[8] Nickel, J.C., Ruseska, I., Wright, J.B., Costerton, J.W. (1985). Tobramycin resistance of *Pseudomonas aeruginosa* cells growing as a biofilm on urinary catheter material, *Antimicrob Agents Chemother*, *27*, 619–624.

[9] Stewart, P.S. & Costerton, J.W. (2001). Antibiotic resistance of bacteria in biofilms, *Lancet 358*, 135–138.

[10] Mah, T.F. & O'Toole, G.A. (2001). Mechanisms of biofilm resistance to antimicrobial agents. *Trends Microbiol*, *9*, 34–39.

[11] Becker, P., Hufnagle, W., Peters, G., Herrmann, M. (2001). Detection of differential gene expression in biofilm-forming versus planktonic populations of *Staphylococcus aureus* using micro-representational-difference analysis. *Appl Environ Microbiol*, *67*, 2958–2965.

[12] Prigent-Combaret, C., Vidal, O., Dorel, C., Lejeune, P. (1999). Abiotic surface sensing and biofilm-dependent regulation of gene expression in *Escherichia coli*. *J. Bacteriol, 181*, 5993–6002.

[13] Ghigo J.M. (2001). Natural conjugative plasmids induce bacterial biofilm development, *Nature*, *412*, 442–445.

[14] Roberts, A.P., Mullany, P., Wilson, M. (2001). Gene transfer in bacterial biofilms. *Methods Enzymol*, *336*, 60–65.

[15] Fux, C.A. (2005). Survival strategies of infectious biofilms, *Trends Microbiol*, *13*, 34–40.

[16] Hausner, M. & Wuertz, S. (1999). High rates of conjugation in bacterial biofilms as determined by quantitative in situ analysis. *Appl Environ Microbiol*, *65*, 3710–3713.

[17] Roberts, A.P., Pratten, J., Wilson, M., Mullany, P. (1999). Transfer of a conjugative transposon, Tn5397 in a model oral biofilm. *FEMS Microbiol. Lett*, *177*, 63–66.

[18] Beaudoin, D.L., Bryers, J.D., Cunningham, A.B., Peretti, S.W., (1998). Mobilization of broad host range plasmid from *Pseudomonas putida* to established biofilm of *Bacillus azotoformans*. *Biotechnol. Bioengin*, *57*, 272–279.

[19] Lebaron, P., Bauda, P., Lett, M.C., Duval-Iflah, Y., Simonet, P., Jacq, E., Frank, N., Roux, B., Baleux, B., Faurie, G., Hubert, J.C., Normand, P., Prieur, D., Schmitt, S., Block, J.C. (1997). Recombinant plasmid mobilization between *E. coli* strains in seven sterile microcosms. *Can J Microbiol*, *43*, 534–540.

[20] Miller, R.V. (2001). Environmental bacteriophage-host interactions: factors contribution to natural transduction. *Antonie Van Leeuwenhoek*, *79*,141–147.

[21] Chiura, H. X. (1997). Generalized gene transfer by virus-like particles from marine bacteria. *Aquat Microb. Ecol*, *13*, 75–83.

[22] Schwartz, T., Kohnen, W., Jansen, B., Obst, U. (*2003* Detection of antibiotic-resistant bacteria and their resistance genes in wastewater, surface water, and drinking water biofilms. *FEMS Microbiol Ecol*, *43*, 325-335.

[23] Poulsen, L.K., Ballard, G., Stahl, D.A. (1993). Use of rRNA fluorescence in situ hybridization for measuring the activity of single cells in young and established biofilms. *Appl Environ Microbiol*, *59*,1354–1360.

[24] Hakenbeck, R. (1998). Mosaic genes and their role in penicillin-resistant *Streptococcus pneumoniae*. *Electrophoresis*, *19*, 597–601.

[25] Hakenbeck, R., Coyette, J. (1998). Resistant penicillin-binding proteins. *Cell Mol. Life Sci*, *54*, 332–340.

[26] Molin S. & Tolker-Nielsen T. (2003) Gene transfer occurs with enhanced efficiency in biofilms and induces enhanced stabilization of the biofilms structure. *Curr. Opin. Biotechnol*, *14*, 255–261.

[27] Costerton, J.W., Lewandowski, Z., DeBeer, D., Caldwell, D., Korber, D., James, G., (1994). Biofilms, the customized microniche. *J. Bacteriol*, *176*, 2137–2142.

[28] Lewandowski, Z. Structure and function of biofilms. In: Evans LV, editor. *Biofilms: recent advances in their study and control*. Amsterdam: Harwood Academic Publishers; 2000; p. 1–17.

[29] Koo, H., Xiao, J., Klein, M. I., Jeon, J. G. (2010). Exopolysaccharides produced by *Streptococcus mutans* glucosyltransferases modulate the establishment of microcolonies within multispecies biofilms. *J. Bacteriol*, *192*, 3024-3032.

[30] Watnick, P. & Kolter, R. (2000). Biofilm, city of microbes. *J. Bacteriol*, *182*, 2675-2679.

[31] Flemming, H.C., Neu T.R., Wozniak D.J. (2007). The EPS Matrix: The "House of Biofilm Cells. *J Bacteriol*, *189*, 7945–7947.

[32] Flemming, H.C., & Wingender, J. (2001). Relevance of microbial extracellular polymeric substances (EPSs)--Part I: Structural and ecological aspects. *Water Sci Technol*, *43*, 1-8.

[33] Flemming, H.C., & Wingender, J. (2002). Extracellular polymeric substances: structure, ecological functions, technical relevance, p. 1223-1231. In Bitton G. editor. *Encyclopedia of environmental microbiology*. New York, NY; Wiley; vol. 3.

[34] Laue, H., Schenk, A., Hongqiao, L., Lambertsen, L., Neu, T. R., Molin, S., Ullrich, M. (2006). Contribution of alginate and levan to biofilm formation by *Pseudomonas syringae*. *Microbiology*, *152*, 2909-2918.

[35] Beech, I., Hanjagsit, L., Kalaji, M., Neal, A. L., Zinkevich, V. (1999). Chemical and structural characterization of exopolymers produced by *Pseudomonas* sp. NCIMB 2021 in continuous culture. *Microbiology*, *145*, 1491-1497.

[36] Manca, M. C., Lama, L., Improta, R., Esposito, E., Gambacorta, A., Nicolaus, B. (1996). Chemical composition of two exopolysaccharides from *Bacillus thermoantarcticus*. *Appl Environ Microbiol*, *62*, 3265-3269.

[37] Morikawa, M., Kagihiro, S., Haruki, M., Takano, K., Branda, S., Kolter, R., Kanaya, S. (2006). Biofilm formation by a *Bacillus subtilis* strain that produces gamma-polyglutamate. *Microbiology*, *152*, 2801-2807.

[38] Verhoef, R., de Waard, P., Schols, H.A., Ratto, M., Siika-aho, M., Voragen, A.G. (2002). Structural elucidation of the EPS of slime producing *Brevundimonas vesicularis sp*. isolated from a paper machine. *Carbohydr Res*, *337*, 1821-1831.

[39] Wrangstadh, M., Szewzyk, U., Ostling, J., Kjelleberg, S. (1990). Starvation-specific formation of a peripheral exopolysaccharide by a marine *Pseudomonas* sp., strain S9. *Appl Environ Microbiol*, *56*, 2065-2072.

[40] Laue, H., Schenk, A., Hongqiao, L., Lambertsen, L., Neu, T. R., Molin, S., Ullrich, M. (2006). Contribution of alginate and levan to biofilm formation by *Pseudomonas syringae*. *Microbiology*, *152*, 2909-2918.

[41] Joubert, L.M., Wolfaardt, G.M., Botha, A. (2006). Microbial exopolymers link predator and prey in a model yeast biofilm system. *Microb. Ecol*, *52*, 187-97.

[42] Allison, D.G., & Sutherland, I.W. (1987). Role of exopolysaccharides in adhesion of freshwater bacteria. *J. Gen. Microbiol*, *133*, 1319–1327.

[43] Danese, P.N. (2000). Exopolysaccharide production is required for development of *Escherichia coli* K12 biofilm architecture. *J. Bacteriol.*, *182*, 3593–3596.

[44] Schaule, G., Griebe, T., Flemming, H.C.. Steps in biofilm sampling and characterisation in biofouling cases. In: Flemming, H.C., Szewyk, U., Griebe, T., editor. *Biofilms—Investigative Methods and Applications.* Technomic, Basel, 2000; pp. 1–22.

[45] DeBeer, D., Stoodley, P. Roe, F., Lewandowski, Z. (1994). Effects of biofilm structures on oxygen distribution and mass transfer. *Biotechnol. Bioengin*, *43*, 1131 - 1138.

[46] Costerton, J. W. (1995). Overview of microbial biofilms. *J. Ind. Microbiol*, *15*, 137-140.

[47] Stoodley, P., D. DeBeer, Z. Lewandowski. (1994). Liquid flow in biofilm systems. *Appl. Environ. Microbiol. 60*, 2711-2716.

[48] Hammer, B.K., Bassler, B.L. (2003). Quorum sensing controls biofilm formation in *Vibrio cholerae*. *Mol Microbiol*, *50*, 101-104.

[49] Davies, D. G., Parsek, M. R., Pearson, J. P., Iglewski, B. H., Costerton, J. W., Greenberg, E. P. (1998). The involvement of cell-to-cell signals in the development of a bacterial biofilm. *Science*, *280*, 295-298.

[50] Zhu, J. & Mekalanos J.J. (2003). Quorum Sensing-Dependent Biofilms Enhance Colonization in *Vibrio cholerae*. *Develop Cell*, *5*, 647-656.

[51] Camilli, A. & Bassler, B.L. (2006). Bacterial small-molecule signaling pathways. *Science*, *311*, 1113–1116.

[52] Lopez, D, Vlamakis, H, Kolter, R. (2010). Biofilms. *Cold Spring Harb Perspect Biol*, 2010; 2:a000398.

[53] Yim, G., Wang, H.H., Davies, J.(2007). Antibiotics as signalling molecules. *Philos Trans R Soc Lond B Biol Sci*, *362*: 1195–1200.

[54] Stoodley, P., Sauer, K., Davies, D.G., Costerton, J.W. (2002). Biofilms as complex differentiated communities. *Annu. Rev. Microbiol*, *56*,187-209.

[55] Rosenberg, M., & Kjelleberg, S. (1986). Hydrophobic interactions: role in bacterial adhesion. *Microbial Ecol*, *9*, 353-393.

[56] Cowan, M.M., Warre, T.M., Fletcher, M. (1991). Mixed species colonization of solid surfaces in laboratory biofilms. *Biofouling*, *3*, 23-34.

[57] Annachhatre, A.P., & Bhamidimarri, S.M.R. (1992). Microbial attachment and growth in fixed-film reactors: process startup considerations. *Biotech. Adv*, *10*, 69-91.

[58] Bücks, J., Mozes, N., Wandrey, C., Rouxhet, P.G. (1988). Cell adsorption control by culture conditions. *Appl Microbiol Biotechnol*, *29*, 119-128.

[59] Characklis, W.G., McFeters, G.A., Marshall, K.C. Physiological ecology in biofilm systems. In Characklis WG, Marshall KC. Editors. *Biofilms. Volume 37.* New York: John Wiley and Sons; 1990: 67-72.

[60] Brading, M.G., Boyle, J., Lappin-Scott, H. (1995). Biofilm formation in laminar flow using *Pseudomonas fluorescens* EX101. *J. Ind. Microbiol*, *15*, 297–304.

[61] Stoodley, P., Boyle, J.D., Dodds, I., Lappin-Scott H.M. (1998). Influence of hydrodynamics and nutrients on biofilm structure. *J. Appl. Microbiol Symp Ser*, *85*, 19S–28S.

[62] Zottola, E.A., & Sasahara, K.C. (1994). Microbial biofilms in the food processing industry - should they be of concern? *Int. J. Food Microbiol*, *23*:125-148.

[63] Jeong, D.K. and Frank, J.F. (1994). Growth of *Listeria monocytogenes* at 21°C in biofilms with microorganisms isolated from meat and dairy environments. *Lebensm-Wiss Technol*, *27*, 415–424.

[64] Dickson, J.S. & Koohmaraie, M. (1989). Cell surface charge characteristics and their relationship to bacterial attachment to meat surfaces. *Appl. Environ Microbiol*, *55*, 832–836.

[65] Fletcher, M. (1976). The effect of proteins on bacterial attachment to polystyrene. *J. Gen Microbiol*, *94*, 400–404.

[66] Helke, D.M., Somers, E.B., Wong, A.C.L. (1993). Attachment of *Listeria monocytogenes* and *Salmonella typhimurium* to stainless steel and Buna-N in the presence of milk and milk components. *J. Food Prot*, *56*, 479–484.

[67] Al-Makhlafi, H., Nasir, A., McGuire, J., Daeschel, M.A. (1995). Adhesion of *Listeria+monocytogenes* to silica surfaces after sequential and competitive adsorption of bovine serum albumin and β-lactoglobulin. *Appl Environ Microbiol*, *61*, 2013–2015.

[68] Jones, G.W. and Isaacson, R.E. (1983). Proteinaceous bacterial adhesins and their receptors. *CRC Crit. Rev. Microbiol*, *10*, 229–260.

[69] Hancock, I.C. Microbial cell surface architecture. In: Mozes, N., Handley, P.S., Busscher, H.J., Rouxhet P.G. Editors. *Microbial Cell Surface Analysis*. Weinheim, Federal Republic of Germany: VCH Publishers; 1991; pp. 23–59.

[70] Boyd, A. & Chakrabarty, A.M. (1995). *Pseudomonas aeruginosa* biofilms: role of the alginate exopolysaccharide. *J. Indust Microbiol*, *15*, 162–168.

[71] Hall-Stoodley, L., Costerton, J.W. and Stoodley, P. (2004). Bacterial biofilms: from the natural environment to infectious diseases. *Nat. Rev. Microbiol*, *2*, 95–108.

[72] Lemon, K.P., Earl, A.M., Vlamakis, H.C., Aguilar, C., Kolter, R. (2008). Biofilm development with an emphasis on *Bacillus subtilis*. *Curr. Top Microbiol Immunol*, *322*, 1–16.

[73] Gotz, F. *Staphylococcus* and biofilms. (2002). *Mol. Microbiol*, *43*, 1367–1378.

[74] Latasa, C., Solano, C., Penades, J.R., Lasa, I. (2006). Biofilm-associated proteins. *C R Biol*, *329*, 849–857.

[75] Cucarella, C., Tormo, M.A., Knecht, E., Amorena, B., Lasa, I., Foster, T.J. Penadés, J.R. (2002). Expression of the biofilm-associated protein interferes with host protein receptors of *Staphylococcus aureus* and alters the infective process. *Infect Immun*, *70*, 3180-3186.

[76] Costerton, J.W., Lewandowski, Z., De Beer, D., Caldwell, D., Korber, D., James, G. (1994). Biofilms, the customised microniche. *J. Bacteriol*, *176*, 2137–2142.

[77] Characklis, W.G., Marshall, K.C. *Biofilms*. John Wiley, New York, USA. 1990.

[78] Lappin-Scott, H.M. and Costerton, J.W. (1989). Bacterial biofilms and surface fouling. *Biofouling*, *1*, 323–342.

[79] Hall-Stoodley, L. and Stoodley, P. (2002). Developmental regulation of microbial biofilms. *Curr. Opin. Biotechnol*, *13*, 228–233..

[80] Nickel, J.C., Costerton, J.W., McLean, R.J., Olson, M. (1994). Bacterial biofilms: influence on the pathogenesis, diagnosis and treatment of urinary tract infections. *J. Antimicrob Chemother*, *33*, 31-41.

[81] Stoodley, P., Wilson, S., Hall-Stoodley, L., Boyle, J.D., Lappin-Scott, H.M., Costerton, J.W. (2001). Growth and Detachment of Cell Clusters from Mature Mixed-Species Biofilms. *Appl. Environ. Microbiol*, *67*, 5608–5613.

[82] Walker, J.T., Mackerness, C.W., Mallon, D., Makin, T., Williets, T., Keevil, C.W. (1995). Control of *Legionella pneumophila* in a hospital water system by chlorine dioxide. *J. Ind. Microbiol*, *15*, 384-90.

[83] Bryers J D. Modeling biofilm accumulation. In: Bazin M J, Prosser J I, editors; Bazin M J, Prosser J I, editors. *Physiological models in microbiology*. Vol. 2. Boca Raton, Fla: CRC Press, Inc.; 1988. pp. 109–144.

[84] Characklis W G. Biofilm processes. In: Characklis W G, Marshall K C, editors; Characklis W G, Marshall K C, editors. *Biofilms*. New York, N.Y: John Wiley & Sons, Inc.; 1990. pp. 195–231.

[85] Hall-Stoodley L. & Stoodley, P. (2005). Biofilm formation and dispersal and the transmission of human pathogens. *Trends Microbiol*, *13*, 7–10.

[86] Parsek, M.R. & Fuqua, C. (2004). Biofilms 2003: emerging themes and challenges in studies of surface-associated microbial life. *J. Bacteriol*, *186*, 4427–4440.

[87] Lindsay, D. & von Holy, A. (2006). Bacterial biofilms within the clinical setting: what healthcare professionals should know. *J. Hosp Infect*, *64*, 313-25.

[88] Gjermansen, M., Ragas, P., Sternberg, C., Molin, S., Tolker-Nielsen, T. (2005). Characterization of starvation-induced dispersion in *Pseudomonas putida* biofilms *Environ Microbiol*, *7*, 894–904.

[89] Sabbuba, N., Hughes, G., Stickler, D.J. (2002). The migration of *Proteus mirabilis* and other urinary tract pathogens over Foley catheters. *BJU Int*, 89, 55-60.

[90] Rather, PN. (2005). Swarmer cell differentiation in *Proteus mirabilis*. *Environ Microbiol*, *7*, 1065-1073.

[91] Neu, T.R. (1996). Significance of bacterial surface-active compounds in interactions of bacteria with interfaces. *Microbiol Rev*, *60*, 151-166.

[92] Parsek, M.R. & Greenbur, E.P. (2005). Sociomicrobiology: the connections between quorum sensing and biofilms. *Trends Microbiol*, *13*, 27-33.

[93] Costerton, J.W., Stewar, P.S., Greenberg, E.P. (1999). Bacterial biofilms: a common cause of persistent infections. *Science*, *284*, 1318-1322.

[94] Petrocci, M.S. Surface-active agents: quarternary ammonium compounds. In: Block, S.S. Editor. *Disinfection, Sterilization and Preservation*, 3rd ed. Lea and Febiger, Philadelphia, PA, USA; 1983; pp. 309–329.

[95] Mustapha, A. & Liewen, M.B. (1989). Destruction of *Listeria monocytogenes* by sodium hypochlorite and quarternary ammonium sanitizers. *J Food Prot*, *52*, 306–311.

[96] Frank, J.F. & Koffi, R.A. (1990). Surface adherent growth of *Listeria monocytogenes* is associated with increased resistance to surfactant sanitizers and heat. *J. Food Prot, 53*, 550–554.

[97] Krysinski, E.P., Brown, L.J., Marchisello, T.J. (1992). Effect of cleaners and sanitizers on *Listeria monocytogenes* attached to product contact surfaces. *J. Food Prot*, *55*, 246–251.

[98] Hall-Stoodley L. & Stoodley, P. (2009). Evolving concepts in biofilm infections. *Cell Microbiol*, *11*, 1034–1043.

[99] Farber, B.F., Kaplan, M.H., Clogston, A.G. (1990). *Staphylococcus epidermidis* extracted slime inhibits the antimicrobial action of glycopeptide antibiotics. *J. Infect Dis*, *161*, 37–40.

[100] Hoyle, B.D., Jass, J., Costerton, J.W. (1990). The biofilm glycocalyx as a resistance factor. *J. Antimicrob Chemother*, *26*, 1–6.

[101] Stewart, P.S. (1996). Theoretical aspects of antibiotic diffusion into microbial biofilms. *Antimicrob Agents Chemother*, *40*, 2517–2522.

[102] Pajkos, A., Vickery, K., Cossart, Y. (2004). Is biofilm accumulation on endoscope tubing a contributor to the failure of cleaning and decontamination? *J. Hosp Infect*, *58*, 224-229.

[103] O'Donnell, M.J., Tuttlebee, C.M., Falkiner, F.R., Coleman, D.C. (2005). Bacterial contamination of dental chair units in a modern dental hospital caused by leakage from suction system hoses containing extensive biofilm. *J. Hosp Infect*, *59*, 348-360.

[104] Capoccia, D., Montanaro, L., Arciola, C.R. (2006). The significance of infection related to orthopedic devices and issues of antibiotic resistance. *Biomaterials*, *27*, 2331-2339.

[105] Tenke, P., Kovacs, B., Jäckel, M., Nagy, E. (2006). The role of biofilm infection in urology. *World J Urol*, *24*, 13-20.

[106] Morris, N.S. & Stickler, D.J. (1998). Encrustation of indwelling urethral catheters by *Proteus mirabilis* biofilms growing in human urine. *J. Hosp Infect*, *39*, 227–234.

[107] Silverstein, A. and Donatucci, C.F. (2003). Bacterial biofilms and implantable prosthetic devices. *Int. J. Impot Res*, *15*, S150–S154.

[108] Donlan, R.M. Biofilms: a source of infection? In: W. Rutala, Editor, *Disinfection, sterilisation and antisepsis: principles, practices and new research*, Association for Professionals in Infection Control, Washington. 2004; pp. 219–226.

[109] Elder, M.J., Stapleton, F., Evans, E., Dart, J.K.G. (1995). Biofilm-related infections in ophtalmology. *Eye*, *9*, 102-109.

[110] Gorlin, A.I., Gabriel, M.M., Wilson, L.A., Ahearn, D.G. (1996). Effect of adhered bacteria on the binding of *Acanthamoeba* to hydrogel lenses. *Arch. Ophthalmol*, *114*, 576-580.

[111] McLaughlin-Borlace, L., Stapleton, F., Matheson, M., Dart, J.K.G. (1998). Bacterial biofilm on contact lenses and lens storage cases in wearers with microbial keratitis. *J. Appl Microbiol*, *84*, 827-838.

[112] Archibald, L.K., & Gaynes, R.P. (1997). Hospital acquired infections in the United States: the importance of interhospital comparisons. *Nosocomial Infect*, *11*, 245-255.

[113] Dingman, J.R., Rayner, M.G., Mishra, S., Zhang, Y., Ehrlich, M.D., Post, J.C., Ehrlich, G.D. (1998). Correlation between presence of viable bacteria and presence of endotoxin in middle-ear effusions. *J. Clin. Microbiol*, *36*, 3417-3419.

[114] Bowden, G.H.W., & Li Y.H. (1997). Nutritional influences on biofilm development. *Adv. Dent. Res.*, *11*, 81-99.

[115] Rowland, B.M. (2003). Bacterial contamination of dental unit waterlines: what is your dentist spraying into your mouth? *Clin. Microbiol Newsletter*, *25*, 73–77.

[116] Bradshaw, D.J., Marsh, P.D., Allison, C., Schilling, K.M. (1996). Effect of oxygen, inoculum composition and flow rate on development of mixed-culture oral biofilms, *Microbiology*, *142*, 623–629.

[117] Shearer, B.G. (1996). Biofilm and the dental office, *J. Am. Dent. Assoc*, *127*, 181–189.

[118] Berlutti, F., Testarelli, L., Vaia, F., De Luca, M., Dolci, G. (2003). Efficacy of anti-retraction devices in preventing bacterial contamination of dental unit water lines. *J. Dent*, *31*, 105–110.

[119] Montebugnoli, L., Chersoni, S., Prati, C., Dolci, G. (2004). A between-patient disinfection method to control water line contamination and biofilm inside dental units, *J. Hosp. Infect*, *56*, 297–304.

[120] Jabra-Rizk, M.A., Falkler, W.A., Meiller, T.F. (2004) Fungal biofilms and drug resistance. *Emerg. Infect Dis*, *10*, 14-19.

[121] Patterson, T.F. (2005). Advances and challenges in management of invasive mycoses. *Lancet*, *366*, 1013-1025.

[122] Soni, N., & Wagstaff, A. (2005) Fungal infection. *Curr. Anaesth Crit Care*, *16*, 231-241.

[123] Kumamoto, C.A. (2005). A contact-activated kinase signals *Candida albicans* invasive growth and biofilm development. *Proc. Natl. Acad. Sci. USA*, *102*, 5576-5581.

[124] Laffey, S.F. & Butler, G. (2005). Phenotype switching affects biofilm formation by *Candida parapsilosis*. *Microbiology*, *151*, 1073-1081.

[125] Carpentier, B. & Cerf, O. (1993). Biofilms and their consequences, with particular reference to hygiene in the food industry. *J. Appl. Bacteriol*, *75*, 499–511.

[126] Shi, X. & Zhu, X. (2009). Biofilm formation and food safety in food industries. *Trends Food Sci & Technol*, *20*, 407-413.

[127] Campbell, S., Duckworth, S., Thomas, C., McMeekin, T. (1987). A note on adhesion of bacteria to chicken muscle connective tissue. *J. Appl. Bacteriol 63*, 67-71.

[128] Dickson, J. S. (1990). Survival and growth of *Listeria monocytogenes* on beef tissue surfaces as affected by simulated processing conditions. *J. Food Safety, 10,* 165-174.

[129] Hood, S. K. & Zottola, E. A. (1997). Isolation and identification of adherent gram-negative microorganisms from four meat-processing facilities. *J. Food Protect, 60,* 1135-1138.

[130] Guobjornsdottir, B., Einarsson, H. and Thorkelsson, G. (2005). Microbial adhesion to processing lines for fish fillets and cooked shrimp: Influence of stainless steel surface finish and presence of gram negative bacteria on the attachment of *Listeria monocytogenes*. *Food Technol Biotechnol*, *43*, 55-61.

[131] Farber, J.M. & Peterkin, P.I. (1991). *Listeria monocytogenes*: A food borne pathogen. *Microbiol Rev, 55*, 476–571.

[132] Kumar, C.G. & Singh, R.S. (1994). *Yersinia enterocolitica*, as an emerging foodborne pathogen—a review. *Indian J. Dairy Sci*, *47*, 537–544.

[133] Stern, N.J., & Kazmi, S.U., 1989. *Campylobacter jejuni*. In: Doyle, M.P. editor. *Foodborne Bacterial Pathogens*. Marcel Dekker, New York, USA, pp. 71–110.

[134] Dewanti, R. & Wong, A.C.L. (1995). Influence of culture conditions on biofilm formation by *Escherichia coli* O157:H7. *Int. J. Food Microbiol*, *26*, 147–164.

[135] Mafu, A.A., Roy, D., Goulet, J., Hagny, P. (1990). Attachment of *Listeria monocytogenes* to stainless steel, glass, polypropylene and rubber surfaces after short contact times. *J. Food Prot*, *53*, 742–746.

[136] Blackman, I.C. & Frank, J.F. (1996). Growth of *Listeria monocytogenes* as a biofilm on various food-processing surfaces. *J. Food Prot*, *59*, 827–831.

[137] Herald, P.J. & Zottola, E.A. (1988). Attachment of *Listeria monocytogenes* to stainless steel surfaces at various temperatures and pH values. *J. Food Prot*, *53*, 1549– 1562.

[138] Austin, J.W. & Bergeron, G. (1995). Development of bacterial biofilms in dairy processing lines. *J. Dairy Res*, *62*, 509–519.

[139] Ridgeway, H.F. & Olson, B.H. (1981). Scanning electron microscopic evidence for bacterial colonization of a drinking water distribution system. *Appl. Environ. Microbiol*, *41*, 274–278.

[140] LeChevalier, M.W., Bancock, T.M., Lee, R.G. (1987). Examination and characterization of distribution system biofilms. *Appl. Environ. Microbiol*, *53*, 2714–2724.

[141] Block, J.C. (1992). Biofilms in drinking water distribution systems. In: Melo, L.F., Bott, T.R., Fletcher, M., Capdeville, B. (Eds.), *Biofilms-Science and Technology*, Kluwer Academic Press, Dordrecht, The Netherlands; pp. 469–485.

[142] Marshall, K.C. (1992). Biofilms: an overview of bacterial adhesion, activity and control at surfaces. *Am Soc Microbiol News*, *58*, 202–207.

[143] Fraser, D.W., Tsai, T.R., Orenstein, W., Parkin, W.E., Beecham, H.J., Sharrar, R.G., Harris, J., Mallison, G.F., Martin, S.M., McDade, J.E., Shepard, C.C., Brachman, P.S. (1977). Legionnaires' disease: description of an epidemic of pneumonia. *N Engl J Med*, *297*, 1189-1197.

[144] Fields, B. S. (1996). The molecular ecology of legionellae. *Trends Microbiol*, *4*, 286-290.

[145] Nguyen, T.M., Ilef, D., Jarraud, S., Rouil, L., Campese, C., Che, D., Haeghebaert, S., Ganiayre, F., Marcel, F., Etienne, J., Desenclos, J.C. (2006). A community-wide outbreak of legionnaires disease linked to industrial cooling towers--how far can contaminated aerosols spread? *J. Infect Dis*, *193*, 102-111.

[146] Liu, Y., Zhang, W., Sileika, T., Warta, R., Cianciotto, N.P., Packman, A. (2009) Role of bacterial adhesion in the microbial ecology of biofilms in cooling tower systems. *Biofouling*, 25, 241–253.

[147] Costerton, J.W., Cheng, K. J., Geesey G.G., Ladd, T.I., Nickel, J.C., Dasgupta, M., Marrie, T.J. (1987). Bacterial biofilms in nature and disease. *Annu. Rev. Microbiol*, *41*, 435–464.

[148] Murga, R., Forster, T.S., Brown, E., Pruckler, J.M., Fields, B.S., Donlan, R.M. (2001). The role of biofilms in the survival of *Legionella pneumophila* in a model potable water system. *Microbiology*, *147*, 3121-3126.

[149] Guerrieri E., Bondi M., Sabia C., de Niederhäusern S., Borella P., Messi P. (2008) Effect of bacterial interference on biofilm development by *Legionella* pneumophila. *Curr Microbiol*, *57*, 532-536.

[150] Declerck, P., Behets, J, van Hoef, V., Ollevier, F. (2007). Replication of *Legionella pneumophila* in floating biofilms. *Curr. Microbiol*, *55*, 435-440.

[151] Guerrieri E., Bondi M., Borella P., Messi P. (2007). Influence of aquatic microorganisms on *Legionella pneumophila* survival. *New Microbiologica 30*, 247-251.

[152] Yildiz, F.H. (2007). Processes controlling the transmission of bacterial pathogens in the environment. *Res. Microbiol*, *158*, 195-202.

[153] Becker, P., Hufnagl, W., Peters, G., Herrmann, M. (2001). Detection of differential gene expression in biofilm-forming versus planktonic populations of *Staphylococcus aureus* using micro-representational-difference analysis. *Appl. Environ. Microbiol*, *67*, 2958–2965.

[154] Gaylarde, C.C., & Morton, L.H.G. (1999) Deteriogenic biofilms on buildings and their control: a review. *Biofouling*, *14*, 59–74.

[155] Meesters, K.P.H., Van Groenestijn, J.W., Gerritse, J. (2003) Biofouling reduction in recirculating cooling systems through biofiltration of process water. *Water Res.*, *37*, 525–53.

[156] Lee, Y.K., Lim, C.Y., Teng, W.L., Ouwehand, A.C., Tuomola, E.M., Salminen S. (2000). Quantitative approach in the study of adhesion of lactic acid bacteria to intestinal cells and their competition with enterobacteria. *Appl. Environl. Microbiol, 66*, 3692–3697.

[157] Savage, D.C. (1977). Microbial ecology of the gastrointestinal tract. *Annu Rev Microbiol, 31*, 107–133.

[158] Fuller, R. (1989). Probiotics in man and animals. *J. Appl. Bacteriol, 66*, 365–378.

[159] Collins, E.B. & Hardt, P. (1980). Inhibition of *Candida albicans* by *Lactobacillus acidophilus*. *J. Dairy Sci, 63*, 830–832.

[160] Anand, S.K., Srinivasan, R.A., Rao, L.K. (1984). Antibacterial activity associated with *Bifidobacterium bifidum*. *Cult. Dairy Prod J., 19*, 6–8.

[161] Vanbelle, M., Teller, E., Focant, M. (1989). Probiotics in animal nutrition: a review. *Arch. Anim. Nutr., 7*, 543–567.

[162] Hawkins, S.M. (1993). Bifidobacteria in dairy products. *Cult. Dairy Prod. J. 28*, 16–20.

[163] Johannsen, M.L., Molin, G., Jeppson, B., Nobaek, S., Ahrne, S., Bengmark, S. (1993). Administration of different *Lactobacillus* strains in fermented oatmeal soup: in vitro colonization of human intestinal mucosa and effect on the indigenous flora. *Appl Environ Microbiol, 59*, 15–20.

[164] Masaaki M. (2006). Beneficial biofilm formation by industrial bacteria *Bacillus subtilis* and related species. *J Biosc Bioeng, 101*, 1-8.

[165] Powell, K.A. & Jutsum, A.R. (1993). Technical and commercial aspects of biological control products, *Pestic Sci, 37*, 315–321.

[166] Chattopadhyay, A., Bhatnagar, N.B. and Bhatnagar, R. (2004). Bacterial insecticidal toxins. *Crit. Rev. Microbiol, 30*, 33–54.

[167] Licina, G.J. (1988) Sourcebook for microbiologically influenced corrosion in nuclear power plants *RP2812-2*. Electric Power Research Institute, Palo Alto, CA.

[168] Zuo, R. & Wood, T.K. (2004). Inhibiting mild steel corrosion from sulfate-reducing and iron-oxidizing bacteria using gramicidin-*S*-producing biofilms. *Appl. Microbiol. Biotechnol, 65*, 747–753.

[169] Taras, M., Hakansson, K. and Guieysse, B. (2005) Continuous acetonitrile degradation in packed-bed bioreactor. *Appl. Microbiol Biotechnol, 66*, 567-574.

[170] Hall, E.R. (1987). Biofilm reactors in anaerobic wastewater treatment. *Biotech. Adv, 5*, 257-269.

[171] Meyer, A. & Wallis, F.M. (1997). Development of microbial biofilms on various surfaces for the treatment of heavy metal containing effluents. *Biotechnol. Tech, 11*, 859-863.

[172] Fuchs, S., Haritopoulou, T., Wilhelmi, M. (1996). Biofilms in freshwater ecosystems and their use as a pollutant monitor. *Water Sci. Technol, 37*, 137–140.

[173] Kanekar, P. & Sarnaik, S. (1991). An activated sludge process to reduce the pollution load of a dye-industry waste. *Environ. Pollution, 70*, 27–33.

[174] Kanekar, P. and Sarnaik, S. (1995). Microbial process for the treatment of phenol bearing dye-industry effluent in a fixed film bioreactor. *J. Environ Sci. Health, A30*, 1817–1826.

[175] Pedersen, A.R., Moller, S., Molin, S., Arvin, E. (1997). Activity of toluene-degrading *Pseudomonas putida* in the early growth phase of a biofilm for waste gas treatment. *Biotechnol Bioeng, 54*, 131–141.

[176] Sarnaik, S. & Kanekar, P. (1995). Bioremediation of colour of methyl violet and phenol from a dye-industry waste effluent using *Pseudomonas* spp. isolated from factory soil. *J Appl Bacteriol*, *79*, 459–469.

[177] Nigam, P., McMullan, G., Banat, I.M., Marchant, R. (1996). Decolourisation of effluent from the textile industry by a microbial consortium. *Biotechnol Lett*, *18*, 117–120.

[178] Beg, S.A., Hassan, M.M., Chaudhry, M.A.S. (1995). Multi-substrate analysis of carbon oxidation and nitrification in an upflow packed-bed biofilm reactor. *J. Chem. Technol Biotechnol.* *64*, 367–378.

[179] Cecen, F. & Orak, E. (1996). Nitrification of fertilizer waste water in a biofilm reactor. *J Chem Technol Biotechnol*, *65*, 229–238.

[180] Demirci, A. & Pometto III, A.L. (1995). Repeated batch fermentation in biofilm reactors with plastic-composite supports for lactic acid production. *Appl. Environ. Microbiol*, *43*, 585–590.

[181] Pakula, R. & Freeman, A. (1996). A new continuous biofilm bioreactor for immobilized oil-degrading filamentous fungi. *Biotechnol. Bioeng*, *49*, 20–25.

[182] Bryers, J.D. Biofilms in biotechnology. In: Characklis, W.G., Marshall, K.C. Editors. *Biofilms*, Wiley-Intersciences, New York, USA; 1990; pp. 733–773.

[183] Macaskie, L.E., Empson, R.M., Lin, F.,Tollet, M.R. (1995). Enzymatically-mediated uranium accumulation and uranium recovery using a *Citrobacter* sp. immobilized as a biofilm within a plug-flow reactor. *J. Chem Technol Biotechnol*, *63*, 1–16.

[184] Jianlong, W. (2000). Production of citric acid by immobilized *Aspergillus niger* using a rotating biological contactor (RBC). *Bioresour Technol*, *75*, 245-247.

[185] Jayasinghearachchi, H.S., Seneviratne, G. (2006). Fungal solubilization of rock phosphate is enhanced by forming fungale rhizobial biofilms. *Soil Biol. Biochem*, *38*, 405-408.

[186] Patterson, T.F. (2005). Advances and challenges in management of invasive mycoses. *Lancet*, *366*, 1013–1025.

[187] Soni, N. & Wagstaff, A. (2005). Fungal infection, *Curr Anaesth Crit Care*, *16*, 231–241.

[188] Danese P. N. (2002). Antibiofilm. *Approaches Chem Biol*, *9*, 873-880.

[189] Bauer & Robinson 2002 Disruption of bacterial quorum sensing by other organisms. *Curr Opin Biotechnol*; *13*; 234-237.

[190] Hentzer, M., Riedel, K., Rasmussen, T.B., Heydorn, A., Andersen, J.B., Parsek, M.R., Rice, S.A., Eberl, L., Molin, S., Hoiby, N., et al., (2002). Inhibition of quorum sensing in *Pseudomonas aeruginosa* biofilm bacteria by a halogenated furanone compound. *Microbiology*, *148*, 87-102.

[191] Ren, D., Sims, J.J., Wood, T.K. (2001). Inhibition of biofilm formation and swarming of *Escherichia coli* by (5Z)-4-bromo-5- (bromomethylene)-3-butyl-2(5H)-furanone. *Environ Microbiol*, *3*, 731-736.

[192] Donlan, R.M. (2009). Preventing biofilms of clinically relevant organisms using bacteriophage. *Trends Microbiol*, *17*, 66-72.

[193] Meylheuc, T., Methivier, C., Renault, M., Herry, J.M., Pradier, C.M., Bellon-Fontaine M.N. (2006). Adsorption on stainless steel surfaces of biosurfactants produced by gram-negative and gram-positive bacteria: consequence on the bioadhesive behavior of *Listeria monocytogenes*. *Colloids Surfaces B Biointerfaces*, *52*, 128-37.

[194] Guerrieri, E., de Niederhäusern, S., Messi, P., Sabia, C., Iseppi, R., Anacarso, I., Bondi, M. (2008). Use of lactic acid bacteria (LAB) biofilms for the control of *Listeria monocytogenes* in a small-scale model. *Food Control.*, *20*, 861-865.

In: Biofilms in Bioengineering
Editors: Manuel Simões and Filipe Mergulhão

ISBN: 978-1-62948-161-6

Chapter II

Methods for Biofilm Characterization

***Sofia Andersson*[*] *and Gunaratna Kuttuva Rajarao*[†]**
Department of Environmental Microbiology, School of Biotechnology,
Royal Institute of Technology (KTH), Sweden

Abstract

Biofilms consist of microbial cells and extracellular polymeric substances including exopolysaccharides, proteins, nucleic acids and lipids. Together with ions and trapped particles these components make up a very sophisticated and complex architecture. A deeper understanding of the structure-function relationship of the biofilm matrix can provide us with the necessary tools for development of methods for biofilm removal and disinfection, biofilm bio-production or for optimization of environmental bioprocesses such as wastewater treatment or bioremediation. A profound knowledge about biofilms is also essential for development within the medical field where microbial attachment is commonly the first step of bacterial infections. The tools for the study of biofilms are in constant development. Methods frequently used for biofilm studies include techniques for controlled, *in vitro* biofilm growth, quantification of biofilm mass or thickness, assessment of microbial activities within biofilms, visualization of biofilms and their components by labelling and microscopy, extraction and characterization of extracellular polymeric substances, genotyping and assessment of quorum sensing molecules. This chapter aims to give an overview of commonly used methods for biofilm studies. However, the ambition is not to cover all available techniques, since the multitude of methods makes this virtually impossible, instead we want to provide the reader with basic knowledge which can be used to further explore the subject.

Keywords: Biofilm characterization; Extracellular polymeric substances; Methods; Microscopy

[*] sofiaa@kth.se.
[†] gunar@biotech.kth.se.

Introduction

In recent years, experimental methods used to study biofilms have advanced greatly. A wide range of techniques: microscopic, microbiological, molecular-biological, chemical and physical, are nowadays available for the exploration of different aspects of biofilm morphology, physiology and genetics [1]. The aim of studying biofilms is to obtain information about their components, architecture or the biological and chemical processes taking place within the biofilm, in order to determine the relationship between microbial genetics, structure and function [2]. Since biofilms consist of microbial cells, Extracellular Polymeric Substances (EPS) and trapped or sorbed particles and substances, the analytical techniques often focus on these components.

Biofilms exist as mixed or pure cultures, in batch or continuous systems, in laboratory or natural environments, and may induce beneficial of detrimental effects to human beings. Studies focus on the macroscale or the microscale features or activities of the biofilms. The techniques used to study biofilms have to be compliant with this diversity. This chapter aims to summarize some of the most widely used analytical techniques in current biofilm research.

Biofilm Quantification

In the early ages of biofilm research, much focus was placed on analysis of the amount and thickness of biofilms [1]. These parameters are still of interest in current research, often in combination with analysis of the biofilm components, function and structure. Biofilm mass can be measured as Wet Weight (WW), Dry Weight (DW) or Total Volatile Solids (TVS) by recording the weight of the carrier material before and after biofilm growth or by measurements on detached biofilm [3, 4]. The mass is then usually expressed as gram of WW/DW/TVS per biofilm area. Biofilm mass measurements can be indirectly used to assess other properties of the biofilm such as the mechanical stability. Mechanical stability of biofilms can be assessed by determining the percent of WW mass loss due to exposure to flows with increasing Reynolds number [3].

A semi-quantitative way to measure the biofilm amount is to stain a dried biofilm sample with crystal violet dye, which binds to cell surfaces and negatively charged EPS components, followed by rinsing off unbound dye and dissolving the biofilm bound dye in ethanol before colorimetric analysis of the coloured ethanol [5]. The above mentioned methods are destructive and the same biofilm cannot be followed over time. A non-destructive, fast and low-cost alternative have been developed to overcome this shortcoming. This technique is based on opacity measurement by flatbed scanning of live biofilms grown on a transparent surface. Slides or discs can be removed from the experimental reactor (rotating annular reactor or rotating disc reactor), scanned and put back into the reactor again. Results correlated well with results from crystal violet staining, DW and thickness measurement [6].

Biofilm thickness can be measured in different ways, by use of three dimensional imaging using Confocal Laser Scanning Microscopy (CLSM), phase contrast microscopy – digital imaging acquisition of biofilm profiles [6] or by calculations based on biofilm WW [7].

Other ways to quantify biofilms include cell counts by microscopic or microbiological methods. Direct total counts can be done microscopically after staining of the cells with a fluorescent dye [3]. Viable counts can also be made microscopically by use of live/dead straining [4] or by plate count using detached, diluted biofilm samples [8].

Microbial Activity

The function of biofilms can be further investigated by analysing the activity of cells in the biofilm. Activity assays are important also when inhibition of biofilm formation or disinfection efficiency of antimicrobial or chemical compounds is determined or when the biofilm is used for degradation purposes. A simple and widely used method to assess the latter is through calculation of the substrate removal rate by analysing the substrate concentration in the bulk phase over time [9]. The removal rate of organic matter, nutrients or a specific contaminant per bulk volume, biofilm mass or biofilm support area provides estimation of general microbial activity or of specific metabolic activities. However, this method does not distinguish between microbial degradation and other removal mechanisms such as adsorption or precipitation; thus the whole system is evaluated, not exclusively the microbial activity.

To get a measurement of the total microbial activity in a biofilm, microbial respiratory activity is often determined. Microbial respiratory activity analysis can also be used as an indirect measurement of viable cells in biofilms. Two commonly used methods to asses this are staining by the tetrazolium compound XTT and measurement of oxygen uptake rate. Live bacteria convert XTT to water soluble orange formazan which can be measured spectrophotometrically at 450 nm. The measurement can be directly related to number viable bacteria cells [10] or to biofilm mass as determined by the crystal violet assay, expressed as OD490/OD570 [11]. The oxygen uptake rate (mg $O_2/g_{biofilm}$, min) is generally analysed by measuring the dissolved oxygen concentration in the bulk phase over time. Biofilms are submerged in oxygen-saturated liquid supplemented with glucose [7]. As the glucose is consumed, the oxygen concentration in the bulk phase decreases due to microbial respiration. The oxygen uptake rate together with microelectrode measurements of oxygen concentration profiles in the biofilm was used to calculate the oxygen diffusion coefficient in a heterotrophic biofilm grown in a tubular reactor [7].

A new method for microbial activity assessment in biofilms based on heat production due to metabolic activities, detected by a miniaturized calorimeter was developed by Lerchner and colleagues [12]. The method displayed good agreement of the measured metabolic heat values with oxygen uptake rate measurements and CLSM cell counts as well as an excellent fit to microbial respiratory models.

Microsensors or microelectrodes allow *in situ* study of mass transfer and microbial processes within live biofilms. Microsensors for measurements of vertical concentration profiles of O_2, N_2O, pH, NH_4^+, NO_3^-, S_2^-, H_2S, NO_2^-, CH_4 and CO_2 are available. Such measurements, used to study substrate transport phenomena, were elemental for the creation of the now widely accepted model of the biofilm matrix as a heterogeneous structure [13]. Using ion-selective microelectrodes for NH_4^+ and NO_3^-, the mass boundary layer and chemical ion fluxes could be measured in a nitrifying biofilm. The ammonia concentration rapidly decreased from the surface to a depth of 2 mm due to chemical reactions carried out by ammonia oxidizing bacteria [14].

Visualization

Microscopes constitute the most fundamental tool available in microbiology. The combination of microscopy with various labelling techniques and digital imaging acquisition/analysis software is immensely useful for the study of biofilms [15]. Visualization techniques can be used on biofilms cultured *in vitro* or on natural biofilm samples collected from the environment, medical devices or bioreactors.

Labelling of Biofilm Components

Virtually all components of the biofilm can be individually labelled using a combination of oligonucleotide probe hybridization, fluorescent dye staining and lectin specific binding. The distribution and composition of microbial cells, polysaccharides, proteins, lipids and extracellular DNA in the biofilm matrix can be visualized consecutively or simultaneously. The most widely used techniques employ fluorescently labelled marker molecules or fluorescent dyes in combination with CLSM or EPI-fluorescent microscopy. Fluorochromes, or fluorophores, comprise a functional group in a molecule which are excited at a specific wavelength and emit light at another, but equally specific, wavelength. For simultaneous visualization of multiple biofilm components, the excitation and emission wavelengths of the fluorochromes should not overlap. Thus, highly specific fluorochromes with minimum spectral peak interference should be used [16]. A comprehensive table of different types of fluorochromes is published on the Center for Cytometry and Molecular Imaging at Salc Institute for Biological Studies web-page (http://flowcyt.salk.edu/fluo.html).

The cells in a biofilm can be visualized in various ways. Non-specific staining of nucleic acids using DNA binding dyes is simple and fast. By using DNA binding dyes with the ability to penetrate intact and/or damaged cell membranes, discrimination between live and dead cells can be made. The use of Fluorescent *In Situ* Hybridization (FISH) allows labelling of a specific microbial species or group of microorganisms based on the small ribosomal subunit phylogenetic marker gene (16S rRNA gene). By design of oligonucleotide probes targeting specific regions of the ribosome, the occurrence and spatial distribution of a selected species, genus, family or domain can be visualized. Since an active cell has a larger number of ribosomes, they will emit stronger fluorescent light then dormant cells. In order to preserve the cell integrity and make the cell membranes permeable, the samples must be fixed with formaldehyde and dehydrated before hybridization. Use of multiple probes tagged with different fluorochromes allows detailed studies of interspecies interactions, synergism and antagonism. Delatolla and colleagues [17] used FISH-CLSM to study the size and distribution of bacterial colonies, the areal coverage of bacterial cells and the number of ammonia and nitrite oxidizing bacteria present throughout a nitrifying biofilm, while still attached to its substratum. FISH-CLSM is a suitable technique for examination of the microbial composition of natural biofilms, but it is also useful for monitoring *in vitro* grown multi-species biofilms.

The major concern when it comes to FISH is that specific probes only can be designed targeting well characterized microorganisms with known gene sequence.

In addition to the above mentioned concern, staining and FISH protocols require interruption of biofilm growth and the subsequent preparatory procedures inevitably and irreversibly affect the cell viability and the biofilm structure. To visualize biofilm cells in real time, without distortion of *in situ* perspective, fluorescent proteins reporter genes have been extensively employed, the most famous being the green fluorescent protein (*gfp*). Constructed cassettes, holding the gene of one fluorescent protein are inserted into the host cell chromosome by conjugation [18]. Combined with CLSM and a flow-cell growth system, the intracellular synthesis of fluorescent protein enables continuous monitoring of cell growth and distribution. Breugelmans and colleagues [19] used red and green fluorescent proteins to study the time course of synergistic interactions and structure-function relationships in a biofilm made up of three herbicide degrading bacterial strains.

The distribution and composition of EPS in the biofilm matrix can be visualized by the use of dyes, fluorescent derivatives or lectins binding to the different macromolecules which constitute the biofilm matrix. The polysaccharide fraction of EPS is commonly visualized using fluorescent dyes, such as Calcofluor white which binds to all β-1,4 and β-1,3 polysaccharides, non-fluorescent dyes, such as Alcian Blue which binds by electrostatic forces to acidic polysaccharides, or fluorescently labelled lectins. Lectins are proteins or glycoproteins, usually of plant origin, capable of reversible and highly specific binding to functional groups or parts of carbohydrates. By use of fifteen different lectins, tagged by tree distinctive fluorochromes, Lawrence et al [20] showed that extracellular polysaccharides produced by a consortium of river water microorganisms underwent enzymatic alteration, cross-linking or variable polymerization during export from the cell into the biofilm matrix. Lectin-binding analysis has proven to be a valuable tool for examination of the role of EPS in fully hydrated environmental biofilms [21]. The other fractions of EPS, proteins, lipids and extracellular DNA can be labelled using fluorescent dyes such as fluoresceine isothiocyanate (FITC), Nile red and 4',6-diamidino-2-phenylindole (DAPI) respectively. Propidium iodide, ethidium bromide, and 1,3-dichloro-7-hydroxy-9,9-dimethyl-2(9H)-acridinone (DDAO), all fluorescent when bound to DNA but unable to penetrate live bacteria, have been suggested for specific visualization of extracellular DNA [22]. Table 1 presents a summary of techniques, commonly used for labelling of biofilm components.

The labelling techniques discussed in this section, together with a number of other available protocols, can preferably be combined to provide information about all components of the biofilm. A technique where Fluorescent *In Situ* Hybridization was combined with lectin binding analysis was described by Böckelmann et al [23]. Fixation and dehydration of the samples was followed by hybridization of oligonucleotide probes before binding of the fluorescently labelled lectins and CLSM analysis. The technique was used to show the structure of polysaccharides in river snow at the same time as the microbial diversity could be observed. The method enables linking cellular and glycoconjugate identity in complex microbial communities. Chen et al., [16] proposed a scheme where six different fluorochromes were used to concurrently label total cells (Syto 63), dead cells (Sytox Blue), β-1,4/β-1,3 polysaccharides (Calcofluor white), α-mannopyranosyl/α-glucopyranosyl sugars (Con A conjugates), proteins/amino-sugars (FITC) and lipids/hydrophobic sites (Nile red). During the following microscopic investigation using CLSM each fluorochrome was excited by a specific wavelength and the emitted light was recorded through a bandpass emission filter, providing micrographs of the same motif displaying each component autonomously. By overlaying the micrographs, a complete picture of the sample could be attained.

Microscopic Techniques

Numerous microscopic techniques are available today including variations of optical and electron based systems. Basic light or phase contrast microscopy can only be used on detached samples or biofilms grown on thin transparent substrates since the technique relies on the passage of a light beam through the specimen. More appropriate techniques for the study of biofilms include EPI-fluorescent microscopy, CLSM, Transmission Electron Microscopy (TEM), Scanning Electron microscopy (SEM), Environmental Scanning Electron Microscopy (ESEM), Raman Microscopy (RM) and Atomic Force Microscopy (AFM).

CLSM in combination with fluorescent labelling of the sample is perhaps the most widely used method for biofilm examination to date (see above). In conventional wide-field fluorescence microscopy, like EPI-fluorescent microscopy, a dyed specimen is illuminated with light of an appropriate wavelength from an incoherent mercury or xenon arc-discharge lamp and an image is formed from the resulting fluorescent light, which can be directly seen through the eye-piece. The confocal microscope, on the other hand, employs point-by-point illumination of the specimen by a laser beam and the image is then assembled by a computer. This technique allows high resolution in-focus images, due to the lack of secondary fluorescence (Figure 1). It also allows production of three dimensional images by creating pictures of multiple vertical sections followed by generation of a stack in the computer [1]. EPI-fluorescent microscopes cannot examine the depth of biofilms like CLSM and are thus only suitable for examination of very thin biofilms. Both types of microscopy require staining of non-fluorescent specimens (Table 1).

Confocal RM combines the three-dimensional optical resolution of confocal microscopy and the sensitivity to molecular vibrations which characterizes Raman spectroscopy. Hence, the technique is capable of providing a three-dimensional image of the chemical composition of the biofilm with micrometer resolution. No pre-treatment is required and measurement can be done directly in the aqueous phase. This saves time and reduces the risk of altering the biofilm architecture or creation of artefacts. Although the complex biochemical composition of biofilms results in Raman spectra where signals from different polymers are super-positioned, it is possible to identify changes in structure and chemistry. Wagner and colleagues [30] used RM combined with CLSM-lectin binding assay to study the influence of hydrodynamics and culture time on EPS composition in biofilms from natural bacterial consortia. They showed that RM could detect time dependant changes in EPS composition which were not possible to identify using CLSM.

Electron microscopic methods for visualisation of biofilm surface structures at high-resolution, have the capacity to provide details of about 1 to 5 nm in size with a magnification of up to 2,000,000 times. In TEM, the electron beam is transmitted through an ultra-thin specimen (< 100 nm thick) mounted on a conducting sample support mesh. Parts of the electrons pass unaffected and parts are scattered out of the beam. The interactions between electrons and specimen are transferred into a picture which can be captured by a CCD camera. To ensure that biological samples will withstand the vacuum in the specimen chamber, they have to be fixated and dehydrated before being embedded in plastic or frozen (cryo-embedding) and sliced in ultrathin sections by a microtome. To increase the contrast of different components of the specimen, staining is performed using heavy metal compounds. Reese and Guggenheim [31] developed a new technique for fixation and dehydration of biofilm samples and contrasting of extracellular polysaccharides. The method provided images of biofilms with an intact matrix in which six types of bacteria and exopolysaccharides could be differentially contrasted.

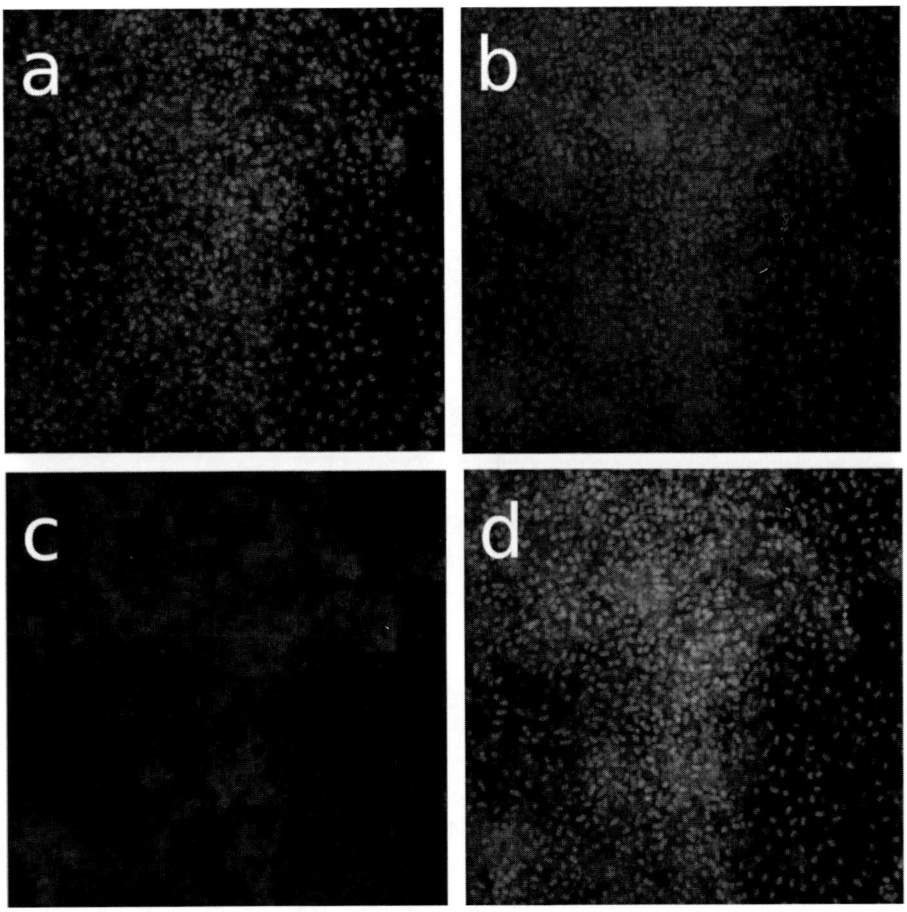

Figure 1. CLSM images of one vertical section of a dual species biofilm. The two types of bacteria were labelled using FISH. *Brachymonas denitrificans* (b) were visualized using the probe OTU6-178 (Cy 3 – red) and *Comamonas denitrificans* were vizualized using the probe EUB 338 (Fluos – green) targeting all eubacteria (a)) and the exopolysaccharides (c) were labelled with Calcofluor white (blue). Image (d) is an overlay of all tree images.

Unlike the TEM, where a complete image of the specimen is created, the electron beam of a SEM does not at any time carry a complete image of the specimen. Instead the specimen surface is scanned and an image is then created in a computer. SEM can be used on samples up to several cm thick and rely on surface processes rather than transmission, and can therefore create three dimensional images which are useful in biofilm research for structural characterization (Figure 2). In addition to fixation and dehydration of the specimen, coating with a conducting material, commonly gold, is also required in order to create an electrically conducting surface. The pre-treatment in combination with the vacuum in the specimen chamber may compromise the structure of the biofilm. It has been observed that biofilm samples can cause structural misconception using SEM since EPS have a tendency to dry up and be visualised as fibrous threads instead of a hydrated matrix [32].

Table 1. Summary of some common biofilm labelling techniques

Biofilm component	Marker molecule	Targets	Specificity	Reference
Cells	Fluorescently labelled oligonucleotide probes (FISH[1])	Hybridize to ribosomal RNA (16S rRNA)	Microorganisms (general to species specific)	[9, 23]
	DAPI *etc.* (fluorescent stains)	DNA	Binds strongly to T-A rich regions in DNA	[24]
	SYTO 9/Propidium Iodide *or*Syto 63/Sytox Blue *etc.* (Live/Dead)	DNA	SYTO 9/Syto 63 penetrates intact cell membranes, propidium iodide/Sytox blue only penetrate damaged cell membranes	[16, 25]
	Fluorescent protein (*gfp* and others)	Intracellular production of fluorescent protein	Insertion of reporter gene into host cell chromosome, resulting in fluorescent live cells.	[19]
Polysaccharides	Fluorescently labelled lectins	Polysaccharides (specific parts)	Sugar-binding proteins with high specificity for specific moieties of polysaccharides	[16, 20, 21, 26]
	Calcofluor white (CFW, fluorescent stain)	β-1,4/β-1,3 poly-saccharides (cellulose and chitin)	CFW reactive polysaccharides	[27]
	Alcian Blue (non-fluorescent stain)	Acidic Polysaccharides	Binds by electrostatic forces to negatively charged polysaccharides	[26]
Proteins	FITC (fluorescent stain)	Protein/amino Acids	Reactive towards amine and sulfhydryl groups on proteins.	[16]
	Fluorescently labelled antibodies	Specific types of Proteins	Specific antibody-antigen binding	[28]
	Congo red (fluorescent stain)	Fibrillated proteins	binding to the β-sheets of the proteins	[28]
Lipids	Nile red (fluorescent stain)	Lipid, hydrophobic sites	Emits fluorescent light in presence of lipids or hydrophobic proteins	[29]

[1] Fluorescent *In Situ* Hybridization (FISH).

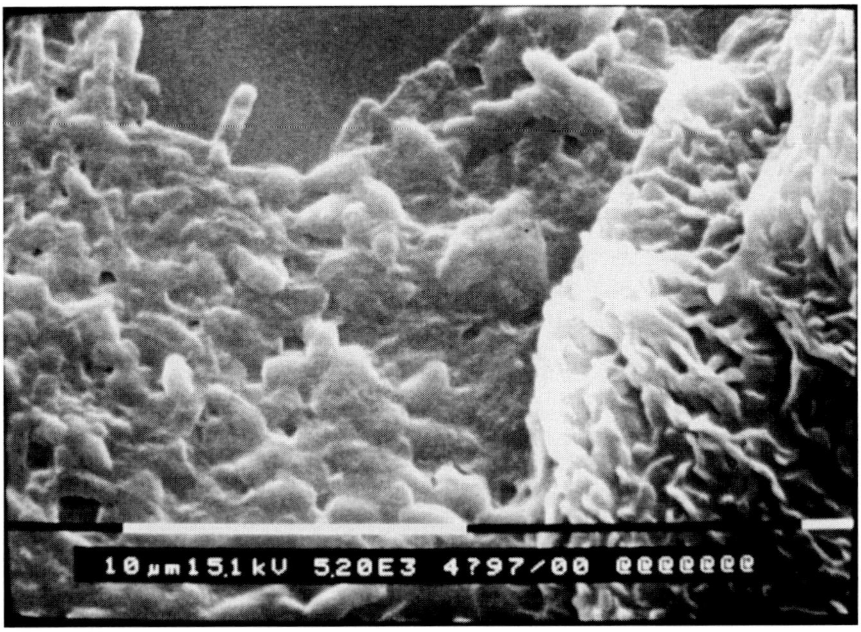

Figure 2. SEM micrograph of 5 days old multispecies biofilm grown on glass slides in a batch culture (picture by Andersson and Kuttuva Rajarao).

A method which circumvents parts of the pre-treatment is nowadays available in the form of ESEM. ESEM enables imaging of wet systems with no or little prior specimen preparation. This is made possible by (i) keeping the specimen chamber at low pressure instead of vacuum and (ii) development of the detector. ESEM was successfully used by Delatolla and colleagues [17] to determine the surface coverage and thickness of biofilm on polystyrene beads and by Planchon et al., [33] to study the structure of pure strain biofilms formed by *Staphylococcus xylosus* strains.

AFM is a powerful tool with high resolution in visualising cell morphology in nanometre or sub-nanometre scale. The surface topography of the sample is recorded with direct depth information by allowing a probe to scan the specimen surface [34]. The main advantage using AFM, compared to electron microscopy, is that sample preparation is not required. Jonas and co-workers studied the importance of curli, cellulose and biofilm associated protein (BapA) on biofilm formation [35]. They observed that curli was more important for cell aggregation than cellulose, and that BapA do not have an impact on the formation and morphology of a biofilm. In another study, bacterial adhesion based on cohesive energy per unit volume of biofilm was measured *in situ* using TEM, showing that calcium increased the cohesiveness of biofilms [36]. Recently the possibilities of AFM combined with other microscopic techniques have gained attention for the study of biofilm structure. Schmid and colleagues [37] used AFM combined with CLSM to study biofilms. The fluorescent labelling and the nanometre-scale spatial resolution enabled imaging of nanometre-sized extracellular structures, such as EPS, alginate bundles, flagella, and pili. The non-destructive techniques AFM and RM in combination allowed high resolution visualization of a biofilm with simultaneous analysis of its components [38].

EPS Extraction

EPS are vital for biofilm formation, function and integrity; furthermore, some microbial exopolysaccharides have commercial value. This has led to intensified ambitions to characterize the EPS produced by single microbial strains, controlled consortia of microorganisms or natural mixed populations. Characterization can be done *in situ* using staining and microscopic techniques [16], as described in previous sections of this chapter, or *in vitro* by extraction of EPS from the biofilm followed by separation, purification and analysis of the different fractions of EPS [39, 40, 41].

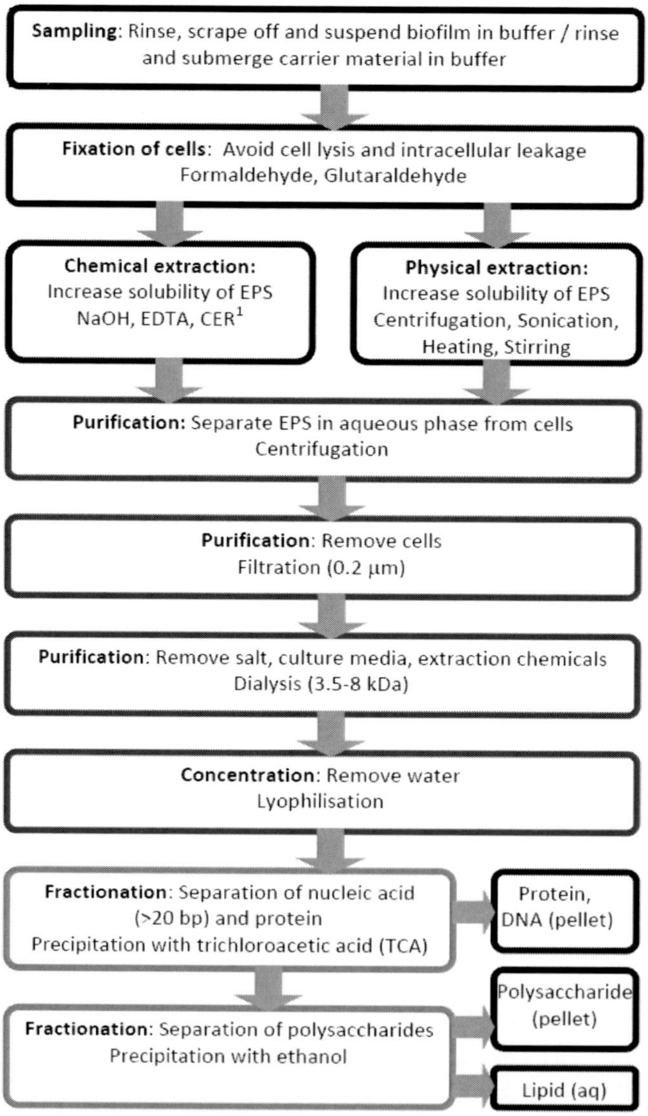

Figure 3. Schematic drawing of methods commonly used for EPS extraction, purification and separation of EPS fractions. [1]Cation exchange resin (CER) [39, 40, 41, 44, 45].

EPS extraction methods can be boldly grouped into physical and chemical ones. Physical methods include centrifugation, stirring, sonication, heating and cation exchange resin while chemical methods comprise the use of e.g. aldehydes, sodium hydroxide (NaOH) and ethylenediaminetetraacetic acid (EDTA) [39, 40, 41]. The extraction is usually followed by several purification steps including centrifugation, filtration and dialysis.

The characterization of EPS components depends on the extraction method used. Extraction protocols are generally a compromise between high yield EPS extraction and minimum cell lysis. In general, chemical methods result in higher EPS yields than physical methods; however, drawbacks such as reactions with the EPS or contamination of the product must be taken into account [39]. For example, low protein yield has been observed in samples treated with formaldehyde, since it cross-links proteins and makes their detection difficult using colorimetric methods. Cell lysis during extraction can be detected by monitoring the presence of intracellular enzymes, such as glucose-6-phosphate 1-dehydrogenase [42]. Most protocols easily extract water soluble EPS components while many hydrophobic components, such as cellulose, are lost [42]. For further analysis of a specific fraction of the EPS, precipitation of proteins, nucleic acids or polysaccharides can be made. Figure 3 provides an overview of common methods for extraction, purification and separation of EPS fractions. Each step can be used separately or in combination with one or several of the other treatments.

Analysis of Exopolysaccharides

The usually largest and also most well characterized fraction of microbial EPS is the polysaccharide fraction. The simplest way to estimate the amount of polysaccharides in extracted EPS is to determine the overall carbohydrate content by colorimetric analysis using e.g. the phenol-sulfuric acid method [46]. The size distribution of the polysaccharides can be analyzed with Size Exclusion Chromatography (SEC). Results from SEC can sometimes reveal if more than one polysaccharide type is present. Hydrolysis of polysaccharide glycosidic bonds with acid at elevated temperatures or enzyme degradation yield mono- or oligosugar molecules which can be identified and quantified using High-Performance Anion Exchange Chromatography (HPAEC) or, after reduction and acetylation, Gas Chromatography coupled to a Mass Spectrophotometer (GC-MS) [47]. These analyses provide the molar ratio between the sugar residues [48]. GC-MS can also be used for sugar linkage analysis if the polysaccharide hydroxyl groups are methylated prior to hydrolysis [49]. The knowledge of the mono-sugar molar rates and relative abundance of different glycosidic bonds can sometimes be sufficient to infer the actual polysaccharide structure. The monosaccharide composition and the stereo-chemical configuration of bacterial exopolysaccharides with repeated structural motifs of 2-8 sugar residues can also be determined by Nuclear Magnetic Resonance (NMR) spectroscopy. Olsson et al [50] determined the structure of the O-antigen polysaccharide from *Escherichia coli* O152 by performing a number of different spectroscopic NMR analysis. However, bacterial exopolysaccharides are often highly heterogeneous and a complete determination of the structure is not always possible.

Analysis of EPS-Derived Protein, Lipid and DNA

For proteins, colorimetric and spectrophotometric methods are widely used to determine the overall content in crude EPS [51, 52, 53]. More detailed analysis of the protein fraction can be done by Sodium Dodecyl Sulphate-PolyAcrylamide Gel Electrophoresis (SDS-PAGE) which separates the proteins based on molecular weight [54]. This provides unique fingerprints from each sample which are suitable for comparative studies [55]. To further characterize the proteins separated by SDS-PAGE, the bands can be cut and purified followed by MS based partial sequencing of tryptic peptides [45]. The size distribution of EPS-proteins can also be determined using SEC [56]. The quantification of enzymes in the EPS can be determined by measurement of the enzyme activity [45, 57].

The overall lipid content in EPS is commonly analyzed using calorimetric methods [58]. More detailed studies of the lipids in extracted EPS are rare, however, Conrad and colleagues [59] separated and quantified the content of neutral lipids, phospholipids, glycolipids and lipopolysaccharides by lipid extraction using chloroform/methanol/phosphate buffer followed by liquid chromatography with a silicic acid matrix. They further determined the fatty acid content of each lipid-fraction by GC-MS.

Analysis of nucleic acids can be done with the colorimetric diphenylamine method [60] or by UV-spectrophotometric measurement at 260nm, the absorption maximum of the nitrogenous bases. By reading the absorbance at 230 (absorbance maximum for phenols and sugars), 260 and 280nm (absorbance maximum for proteins) and calculation of the ratios, the purity of the nucleic acid fraction of EPS can be estimated [61]. The use of a nanodrop instrument, instead of a conventional spectrophotometer allows sample volumes as small as 1-2μL for a quantitative analysis. Another both semi-quantitative and qualitative method for nucleic acid analysis is agarose gel electrophoresis which separates nucleic acid fragments with respect to electrical charge density, i.e. size [62, 63].

Genotyping

Molecular genotyping by enterobacterial repetitive intergenic consensus-based PCR was performed to study the relationship between phenotype and genotype in biofilm formation and showed some correlation in different biofilm-forming abilities [64]. Biofilm formation on the surface of reverse osmosis membranes was studied by 16S rRNA gene targeted Degenerating Gradient Gel Electrophoresis (DGGE) and phylogenetic analysis of individual clones was reported [65]. Clone/DGGE technique was used for the microbial community analysis of biofilms from a laboratory reactor [66]. Microbial community analysis by Amplified Ribosomal DNA Restriction Analysis (ARDRA) was reported by Massol-Deya [67]. The slime secretion and biofilm formation of *Staphylococcus epidermis* strains from lung cancer patients were tested and it was found that the genotype *ica*+, *aap*+ was strong biofilm producer phenotype. However, *ica* -independent biofilm formation was also observed [68]. Seo et al [69] reported that Microbial Surface Component Recognizing Adhesive Matrix Molecules (MSCRAMMs) genes were investigated by PCR method and found that *eno* gene is mainly associated with biofilm formation.

Quorum Sensing

Quorum Sensing (QS) is a term to describe the circumstances where the accumulation of signalling molecules that helps a single cell to sense the number of bacteria (cell density) thereby assist to co-ordinate their behaviour. The secondary metabolites produced trigger the transcription factors followed by gene expression and coordination between bacterial communities. A sensitive analytical tool (LC-Electrospray ionization-MS-MS) for quantitative analysis of Activated Methyl Cycle (AMC) which serves as a metabolite pool for the formation of QS signal molecules was reported [70]. CD-NMR spectroscopy was used to analyze the QS signaling peptides for structure-activity analysis [71]. QS is well studied using microfluidic and microfabrication techniques [72, 73]. Particle video microscopy was used to analyse the particle droplets contained in process reactor vessels or pipelines and the effect of enzymatic treatments on biofilm formation was studied [74]. Recently, Mitchell and co-workers reviewed the advanced technologies developed for understanding the chemo-communication of bacterial cells for various applications [75].

References

[1] Denkhaus E, Meisen S, Telgheder U, and Wingender J, (2007), Chemical and physical methods for characterisation of biofilms. *Microchimica Acta. 158(1)*, 1-27.

[2] Davey ME and O'Toole GA (2000) Microbial Biofilms, from Ecology to Molecular Genetics. *Microbiology and Molecular Biology Reviews 64(4)*, 847-867

[3] Simões M, Cleto S, Pereira MO and Vieira MJ (2007a) Influence of biofilm composition on the resistance to detachment. *Water Science & Technology 55(8–9)*, 473–480

[4] Simões M., Simões L. C., Vieira M. J. (2009) Species association increases biofilm resistance to chemical and mechanical treatments. *Water Research* 43,229-237

[5] Djordjevic D, Wiedmann M, and McLandsborough LA, (2002), Microtiter plate assay for assessment of Listeria monocytogenes biofilm formation. *Appl Environ Microbiol.* 68(6), p. 2950-2958.

[6] Alnnasouri M, Dagot C and Pons M-N (2011) Comparison of four methods to assess biofilm development. *Water Science & Technology 63(3)*, 432-439

[7] Horn H and Morgenroth E (2006) Transport of oxygen, sodium chloride, and sodium nitrate in biofilms. *Chemical Engineering Science 61*, 1347 – 1356

[8] Stewart PS, Griebe T, Srinivasan R, Chen CI, Yu FP (1994) Comparison of Respiratory Activity and Culturability during Monochloramine Disinfection of Binary Population Biofilms. *Applied and Environmental Microbiology 60(5)*, 1690-1692

[9] Andersson S, Dalhammar G and Kuttuva Rajarao G. (2011) Influence of microbial interactions and EPS/polysaccharide composition on nutrient removal activity in biofilms formed by strains found in wastewater treatment systems. *Microbiological Research. 166(6)*, 449-457

[10] Stevens MG and Olsen SC (1993) Comparative analysis of using MTT and XTT in colorimetric assays for quantitating bovine neutrophil bactericidal activity. *Journal of Immunological Methods* 157(1-2), 225-231

[11] Simões L. C., Simões M, Vieira M. J. (2007) Biofilm Interactions between Distinct Bacterial Genera Isolated from Drinking Water. *Appl Environ Microbiol 73(19),* 6192-6200

[12] Lerchner J, Wolf A, Buchholz F, Mertens F, Neu TR, Harms H, Maskow T (2008) Miniaturized calorimetry - A new method for real-time biofilm activity analysis. *Journal of Microbiological Methods* 74, 74–81

[13] de Beer D and Schramm A (1999) Micro-environments and mass transfer phenomena in biofilms studied with microsensors. *Wat. Sci. Tech. 39(7)*, 173-178

[14] Chae K-J, Rameshwar T, Jang A, Kim SH and Kim IS (2008) Analysis of the nitrifying bacterial community in BioCube sponge media using fluorescent in situ hybridization (FISH) and microelectrodes. *Journal of Environmental Management 88(4),* 1426-1435

[15] Stewart PS and Franklin MJ, (2008), Physiological heterogeneity in biofilms. *Nat Rev Microbiol. 6(3),* 199-210.

[16] Chen MY, Lee DJ, Tay JH and Show KY (2007), Staining of extracellular polymeric substances and cells in bioaggregates. *Appl Microbiol Biotechnol 75*,467–474

[17] Delatolla R, Tufenkji N, Comeau Y, Lamarre D, Gadbois A, Berka D (2009) *In situ* characterization of nitrifying biofilm, Minimizing biomass loss and preserving perspective. *Water Research 43*, 1775-1787

[18] Tolker-Nielsen T, Brinch UC, Ragas PC, Andersen JB, Jacobsen CS, Molin S (2000) Development and Dynamics of *Pseudomonas sp.* Biofilms. *Journal of Bacteriology 182(22)*, 6482–6489

[19] Breugelmans P., Barken K. B., Tolker-Nielsen T., Hofkens J., Dejonghe W., Springael D. (2008) Architecture and spatial organization in a triple-species bacterial biofilm synergistically degrading the phenylurea herbicide linuron. *FEMS Microbiol Ecol 64*,271–282

[20] Lawrence JR, Swerhone GDW, Kuhlicke U and Neu TR (2007) In situ evidence for microdomains in the polymer matrix of bacterial microcolonies. *Can. J. Microbiol. 53,* 450-458

[21] Neu TR, Swerhone GDW and Lawrence JR (2001) Assessment of lectin-binding analysis for *in situ* detection of glycoconjugates in biofilm systems. *Microbiology 147*, 299-313

[22] Allesen-Holm M, Barken KB, Yang L, Klausen M, Webb JS, Kjelleberg S, Molin S, Givskov M, and Tolker-Nielsen T (2006) A characterization of DNA release in *Pseudomonas aeruginosa* cultures and biofilms. *Molecular Microbiology 59(4),* 1114–1128

[23] Böckelmann U, Manz W, Neu TR, Szewzyk U (2002) Investigation of lotic microbial aggregates by a combined technique of fluorescent *in situ* hybridization and lectin-binding-analysis. *Journal of Microbiological Methods 49*, 75–87

[24] Tanious FA., Veal JM., Buczak H. Ratmeyer LS., and Wilson WD. (1991) DAPI (4',6-Diamidino-2-phenylindole) Binds Differently to DNA and RNA, Minor-Groove Binding at AT Sites and Intercalation at AU Sites. *Biochemistry 31(12),* 3103-3112

[25] Zhang X, Tani A, Kawai F, Kimbara K (2010) Rapid and multiple in situ identification and analyses of physiological status of specific bacteria based on fluorescent in situ hybridization. *Journal of Bioscience and Bioengineering 110(6),* 716–719

[26] Kolari M (2003) Attachment mechanisms and properties of bacterial biofilms on non-living surfaces. Doctoral Thesis. Department of Applied Chemistry and Microbiology, University of Helsinki, Finland. ISBN 952-10-0344-8

[27] McLennan MK, Ringoir, DD, Frirdich E, Svensson SL, Wells DH, Jarrell H, Szymanski CM and Gaynor ECm (2008). *Campylobacter jejuni* biofilms up-regulated in the absence of the stringent response utilize a calcoflour white-reactive polysaccharide, *Journal of Bacteriology, 190(3),* 1097-1107

[28] Larsen P, Nielsen JL, Dueholm MS, Wetzel R, Otzen D and Nielsen PH (2007) Amyloid adhesins are abundant in natural biofilms. *Environmental Microbiology 9(12),* 3077-3090

[29] Greenspan P., Mayer E. P., Fowler S.D. (1985) "Nile Red" A Selective Fluorescent Stain for Intracellular Lipid Droplets. *The Journal of Cell Biology 100,* 965-973

[30] Wagner M, Ivleva NP, Haisch C, Niessner R and Horn H (2009) Combined use of confocal laser scanning microscopy (CLSM) and Raman microscopy (RM), Investigations on EPS – Matrix. *Water Research* 43,63-76

[31] Reese S and Guggenheim B (2007) A Novel TEM Contrasting Technique for Extracellular Polysaccharides in *In Vitro* Biofilms. *Microscopy Research and Technique 70*, 816–822

[32] Wimpenny J, Manz W, and Szewzyk U, (2000), Heterogeneity in biofilms. *FEMS Microbiol Rev. 24(5),* 661-671.

[33] Planchon S, Gaillard-Martinie B, Dordet-Frisoni E, Bellon-Fontaine MN, Leroy S, Labadie J, Hébraud M, Talon R (2006) Formation of biofilm by *Staphylococcus xylosus. International Journal of Food Microbiology 109,* 88–96

[34] Dufrêne YF (2002) Atomic Force Microscopy, a Powerful Tool in Microbiology. *Journal of Bacteriology 184(19),* 5205-5213

[35] Jonas K, Tomenius H, Kader A, Normark S, Römling U, Belova LM and Melefors O (2007) Roles of curli, cellulose and BapA in *Salmonella* biofilm morphology studied by atomic force microscopy. *BMC Microbiol, 7*:70

[36] Ahimou F, Semmens MJ, Novak PJ and Haugstad G 2007. Biofilm cohesiveness measurement using a novel atomic force microscopy methodology, *Appl Env Microbiol, 73(9),* 2897-2904

[37] Schmid T, Burkhard J, Yeo B-S, Zhang W and Zenobi R (2008) Towards chemical analysis of nanostructures in biofilms I, imaging of biological nanostructures. Anal Bioanal Chem *391*, 1899–1905

[38] Pradhan, N, Pradhan, S.K, Nayak, B.B, Mukherjee, P.S, Sukla, L.B, Mishra, B.K. (2008) Micro-Raman analysis and AFM imaging of *Acidithiobacillus ferrooxidans* biofilm grown on uranium ore. *Res. Microbiol. 159*, 557-561.

[39] Comte S, Guibaud G, and Baudu M, (2006), Relations between extraction protocols for activated sludge extracellular polymeric substances (EPS) and EPS complexation properties, Part I. Comparison of the efficiency of eight EPS extraction methods. *Enzyme Microb Technol. 38(1-2)*, 237-245

[40] Comte S, Guibaud G, and Baudu M, (2007), Effect of extraction method on EPS from activated sludge, An HPSEC investigation. *Journal of Hazardous Materials. 140(1-2),* 129-137.

[41] Liu H and Fang HH, (2002), Extraction of extracellular polymeric substances (EPS) of sludges. *J Biotechnol. 95(3),* 249-256.

[42] Flemming HC and Wingender J (2010) The biofilm matrix. *Nature Reviews 8*, 623-633

[43] Oliveira R, Marques F, and Azeredo J, (1999), Purification of polysaccharides from a biofilm matrix by selective precipitation of proteins. *Biotechnology Techniques. 13(6),* 391-393.

[44] Ras M, Girbal-Neuhauser E, Paul E, Spérandio M, Lefebvre D (2008), Protein extraction from activated sludge, an analytical approach. *Water Res. 42(8-9),* 1867-1878.

[45] DuBois M, Gilles DA, Hamilton JK, Rebers PA, Smith F, (1956), Colorimetric Method for Determination of Sugars and Related Substances. *Anal. Chem. 28(3),* 350-356.

[46] Andersson S, Gunaratna KR, Land CJ, Dalhammar G (2008) Biofilm formation and interactions of bacterial strains found in wastewater treatment systems. *FEMS microbiology letters 283(1)*, 83-90

[47] Sutherland IW, (1990), *Cambridge studies in biotechnology 9: Biotechnology of Microbial Exopolysaccharides.* (1 ed). New York, Cambridge University Press.

[48] McConville MJ, Homans S W, Thomas-Oates J E, Dell A and Bacic A (1990), Structures of the glycoinositolphospholipids from Leishmania major. A family of novel galactofuranose-containing glycolipids. *J Biol Chem. 265(13),* 7385-7394.

[49] Olsson U, Lycknert K, Stenutz R, Weintraub A and Widemalm G (2005) Structural analysis of the O-antigen polysaccharide from *Escherichia coli* O152. *Carbohydrate Research 340*, 167–171

[50] Bradford MM (1976) A rapid and sensitive method for the quantitation of microgram quantities of protein utilizing the principle of protein-dye binding. *Anal Biochem. 72*, 248-254.

[51] Lowry, O.H., Rosebrough, N.J., Farn, A., Randall, R., (1951) Protein measurement with the folin phenol reagent. J. Biol. Chem. *193*, 265–275.

[52] Frølund B, Griebe T, Nielsen PH (1995) Enzymatic activity in the activated sludge floc matrix. *Appl Microbiol Biotechnol 43*,755–61.

[53] Laemmli UK, (1970), Cleavage of structural proteins during the assembly of the head of bacteriophage T4. *Nature. 227(5259),* 680-685.

[54] Denecke M, (2006), Protein extraction from activated sludge. *Water Sci Technol. 54(1),* 175-181.

[55] Garnier C, Görner T, Lartiges B S, Abdelouhab S, de Donato P (2005), Characterization of activated sludge exopolymers from various origins, a combined size-exclusion chromatography and infrared microscopy study. *Water Res. 39(13),* 3044-3054.

[56] Yu G-H, He P-J, Shao L-M, Zhu Y-S (2008) Extracellular proteins, polysaccharides and enzymes impact on sludge aerobic digestion after ultrasonic pretreatment. *Water Research 42*, 1925-1934

[57] Frings CS, Fendley TW, Dunn RT, and Queen CA, (1972), Improved determination of total serum lipids by the sulfo-phospho-vanillin reaction. *Clin Chem. 18(7),* 673-674.

[58] Conrad A, Kontro M, Keinänen MM, Cadoret A, Faure P, Mansuy-Huault L, Block J-C (2003) Fatty Acids of Lipid Fractions in Extracellular Polymeric Substances of Activated Sludge Flocs. *Lipids 38(10),* 1093-1105

[59] Deriaz RE, Stacey M, Teece EG, and Wiggins LF, (1949), Deoxy-sugars. Part I. The dische reaction for 2-deoxypentoses. *Journal of the Chemical Society*, 1222 - 1232.

[60] Shokere LA, Holden MJ, and Ronald Jenkins G, (2009), Comparison of fluorometric and spectrophotometric DNA quantification for real-time quantitative PCR of degraded DNA. *Food Control. 20(4),* 391-401.

[61] Morimoto H, Ferchmin PA, and Bennett EL, (1974), Spectrophotometric analysis of RNA and DNA using cetyltrimethylammonium bromide. *Anal Biochem. 62(2),* 436-448.

[62] Sambrook J and Russell D, (2001), *Molecular cloning, A laboratory manual* (3ed). New York, Cold Spring Harbor Laboratory Press.

[63] Chen C, Liao X, Jiang H, Zhu H, Yue L, Li S, Fang B, Liu Y 2010. Characteristics of Escherichia coli biofilm production, genetic typing, drug resistance pattern and gene expression under aminoglycoside pressures, *Environ Toxicol Pharmacol, 30,* 5-10

[64] Bereschenko LA, Stams AJM, Heilig GHJ, Euverink GJW, Nederlof MM and van Loosdrecht MCM. (2007). Investigation of microbial communities on reverse osmosis membranes used for process water production, *Wat Sci Technol, 55,* 181-190

[65] Burr BD, Clark SJ, Spear CR and Camper AK (2006) Denaturing Gradient Gel Electrophoresis Can Rapidly Display the Bacterial Diversity Contained in 16S rDNA Clone Libraries, *Microbial Ecol, 51,* 479-486

[66] Massol-Deya A, Weller R, Rios-Hernandez L. Zhou J-Z, Hickey RF and Tiedje JM, (1997) Succession and Convergence of Biofilm Communities in Fixed-Film Reactors Treating Aromatic Hydrocarbons in Groundwater, *App Environ Microbiol, 63,* 270-276

[67] Los R, Sawicki R, Juda M, StankevicM, Rybojad P, Sawicki M, Malm A and Ginalska G (2010) A comparative analysis of phenotypic and genotypic methods for the determination of the biofilm-forming abilities of *Staphylococcus epidermidis. FEMS Microbiol Lett 310,* 97-103

[68] Seo Y-S, Lee DY, Rayamahji N, Kang ML, and Yoo HS, (2008) Biofilm-forming associated genotypic and phenotypic characteristics of *Staphylococcus* spp. isolated from animals and air, *Res in Vet Sci, 85,* 433-438

[69] Halliday NM, Hardie KR, Williams P, Winzer K, Barrett DA (2010) Quantitative liquid chromatography-tandem mass spectroscopy profiling of activated methyl cycle metabolites involved in LuxS-dependent quorum sensing in *Escherichia coli, Anal Chem 403,* 20-29

[70] Tian XL, Syvitskci RT, Liu TL, Livingston N, Jakeman DL, Li Y-H (2009) A method for structure-activity analysis of quorum- sensing signaling peptides from naturally transformable Streptococci, *Biol Proced Online, 11,* 207-226

[71] Weibel DB, DiLuzio WR, Whitesides GM (2007) Microfabrication meets microbiology, *Nat Rev Microbiol, 5,* 209-217

[72] Flickinger ST, Copeland MF, Downes EM, Braasch AT, Tuson HH, Eun Y-J, Weibel DB (2010) Quorum sensing between *Pseudomonas aeruginosa* biofilms accelerates cell growth, *J Am Chem Soc, 133(15),* 5966-5975

[73] Blanco A, Torres E, Fuente E, Negro C (2011) New tool to monitor biofilm growth in industrial process water, *Ind Eng Chem Res, 50(9),* 5766-5773

[74] Mitchell RJ, Lee SK, Ghim C-M (2011) Microbial linguistics, perspectives and applications of microbial cell-to-cell-communication, *BMC Rep, 44,* 1-10

In: Biofilms in Bioengineering
Editors: Manuel Simões and Filipe Mergulhão

ISBN: 978-1-62948-161-6

Chapter III

Platforms for *In Vitro* Biofilm Studies

J. S. Teodósio, M. Simões, L. F. Melo and F. J. Mergulhão*
[1]LEPAE, Department of Chemical Engineering, Faculty of Engineering, University of Porto, Porto, Portugal

Abstract

A detailed knowledge about the steps governing biofilm formation, development and elimination is only possible if the biofilms under study can be obtained in controlled settings where key variables can be manipulated and the implications of these manipulations can be readily assayed. Although totally reproducible biofilms are nearly impossible to obtain, the development of *in vitro* platforms for biofilm studies is a major step towards a standardization of procedures and also a better control of the environmental conditions that affect their development. This chapter presents a brief overview of the available biofilm reactors and details the main features of two of the most commonly used platforms: flow cells and microtiter plates.

Keywords: Biofilms; Bioreactors; Flow cells; Microtiter plates

Introduction

Biofilms are complex communities of surface-attached microorganisms, comprised either of a single or multiple species [1-5]. Although adhesion to a surface may look unattractive at first glance due to a more difficult access to nutrients, the fact is that in nature probably 99% or more of all bacteria exist in biofilms [6].

Despite some beneficial applications of biofilms like for instance in industrial production of various chemicals [7-9], wastewater treatment [10], removal of volatile compounds from waste streams [11] or even energy production [12], the fact is that detrimental biofilms are far more notorious than beneficial biofilms. Negative effects of biofilm formation are often found

* E-mail: filipem@fe.up.pt.

in industrial settings where biofouling costs can represent up to 30% of the plant operating costs [13] and in the overall expenditure of industrialized countries where estimates for fouling costs (of which biofouling possibly accounts for one third) represent 0.25% of the gross national product [13]. The biofilms enjoying the worst reputation are undoubtedly those found in the health sector [1] and in fact it has been estimated that more than 60% of all microbial infections in humans are caused by biofilms [6]. Additionally, the proliferation of biomedical devices like central venous catheters, of which more than 5 million are inserted annually in the United States alone, increases the probability of infection, resulting in up to 20,000 deaths and in an annual cost of caring for patients with associated infections of up to $2.3 billion [14]. If other hospital-acquired infections associated with artificial joints, urinary tract infections, or pulmonary infections are considered the costs are simply overwhelming.

Since beneficial and detrimental biofilms exist it is useful to develop strategies for biofilm control that enable us to promote the formation of beneficial biofilms and delay the formation of detrimental biofilms or promote their destruction. This requires intensive studies on the mechanisms of biofilm formation and resistance and prompted the need to develop *in vitro* platforms for biofilm studies. These platforms are artificial biofilm model systems that are easy to control and reproducible enabling a more detailed study of this phenomena. *In vitro* biofilm formation systems range from static mono-cultures formed on membranes to mixed culture biofilms growing under dynamic conditions. Although it is reasonable to assume that static biofilm formation is hampered by diffusion limitations on the transport of nutrients and oxygen to the biofilm structure, one must bear in mind that one of the key issues to be addressed on the selection of a certain platform for biofilm studies is the similarity of the flow regime to the conditions that are being simulated. This chapter presents the most common platforms that can be used to study dynamic biofilms. These systems comprise flow cells systems, and other different types of bioreactors but also microtiter plates incubated on shaking systems. Advantages and disadvantages of some biofilm–forming devices are showed in Table 1. The two most widely used systems, flow cells and microtiter plates are described in greater detail including validation procedures and examples of the experimental results that can be obtained with them.

Table 1. Advantages and disadvantages of some biofilm – forming devices (adapted from [45])

Biofilm device	Advantages	Disadvantages	References
Small-scale Flow cell	The flat surface may enable direct microscopic observation without disturbing the hydrodynamic conditions	Low throughput	[46]
Perfused reactor	Allows biofilm growth rate control	Low throughput	[18]
Microtiter plate	Generally high throughput. Allows direct microscopy (insertion of coupons in each well) with low throughput	Difficulties on the direct microscopic analysis; lack of reproducibility on the subsequent quantification methods	[47]
Calgary device	High throughput. Allows direct microscopy and quantification of biofilms	Introduction of an extra surface in the microtiter plate; Biofilms formed "upside-down"	[48]

Commonly Used Platforms

CDC-Type Reactor

The CDC reactor was designed at the Center for Disease Control and Prevention in Atlanta, USA. It is basically a cylindrical glass vessel with coupon holders suspended on a plastic lid in the same manner as pH or dissolved oxygen electrodes are mounted on the top of bench-scale bioreactors. The coupon holders can contain a variable number of coupons (3 were available in the original model) which serve as a substrate for biofilm formation. Mixing is achieved by using a magnetic stir bar which is rotated by a magnetic stir plate. One of the main advantages of this type of reactor is the fact that various materials can be used for manufacturing the coupons and therefore cell adhesion can be simulated in industrial equipment (using stainless steel coupons), water pipes (with polyvinyl chloride (PVC)), the oral cavity (using enamel), etc. Another advantage is the possibility of aseptically removing one coupon holder at a time and retrieving the biofilm containing coupon thus enabling sampling in a time course experiment. The coupons can be used for a series of chemical/biochemical analyses and are also amenable to observation by confocal microscopy. This reactor type has been used for the development of standardized biofilms of many species of organisms specially for *Pseudomonas aeruginosa* [15]. Several applications of this reactor were developed for instance in the evaluation of effectiveness of disinfectant substances, the effect of fluid dynamics on biofilm formation, simulation of treatment procedures for drinking water in pipes and also on the analysis of extracellular polymers formed on a biofilm [15]. Despite its broad applicability one must not forget that although the fluid dynamics present in this reactor may eventually emulate those found in circular water pipes at a narrow interval of flow rates, other contexts, like for instance in biomedical applications may not be correctly simulated has the fluid velocities and shear stresses may be too different from those found inside this reactor.

Rotating Annular Reactor

Annular reactors are biofilm growing systems that simulate the hydrodynamic conditions found in water distribution pipelines [16]. These reactors provide retrievable coupon surfaces were the biofilms grow and enable a vast array of chemical and biochemical analyses as well as microscopy observation of the biofilms. Although a couple of design variations exist most models are composed of two concentric cylinders, one static external and one rotating internal. The speed of rotation and the diameter of the cylinders can be changed to create different shear stress [17]. Coupons are flush with the walls of the reactor and are therefore subjected to the same hydrodynamic conditions. The reactor is usually operated in the same manner as a regular chemostat in which nutrients are continuously being introduced at a desired flow rate and the effluent is expelled by overflow. One great advantage of this reactor is that the rotation speed of the cylinders can be set independently of the feed flow rate. Thus, the dilution rate becomes independent of the shear stress and thus both parameters can be adjusted separately [16]. Another advantage is the reproducibility of the formed biofilms that can be obtained in continuous flow conditions. As in all biofilm growing systems containing coupons, these can be made from relevant materials (PVC, steel, plastics) or can be coated with biologically relevant substrates or materials that are used in implants [17]. This reactor has been successfully used to simulate biofilms that occur in drinking water pipe systems, river ecosystems, hulls of ships and also to assess the effect of nutrient concentration or the efficiency of antimicrobial agents in biofilm eradication [15].

Perfused Reactors

These reactors enable the growth rate control of adherent microbial populations [18]. The reactor contains a perfusion membrane which is inoculated with the microorganism of interest and then the culture medium is introduced in the reactor passing through the membrane. Thus, the young or weakly loosely adhered cells are removed as the medium crosses the membrane and can be quantified to monitor biofilm growth or cell survival. When the cell numbers in the perfused medium reach as constant value the system has reached a steady state [15]. Although the design features of this reactor force the culture medium to perfuse the biofilm from its underside, which is not the typical situation in biofilms formed on inanimate surfaces, this scenario may be representative of bacterial surface interactions in soft tissues [18]. Additionally, perfusion of the membrane with antibiotics can also emulate antibiotic therapy. In this system, the residual adherent population is enriched in older dividing cells. This may also simulate the *in vivo* situation were new cells which are less adherent leave the biofilm and colonize new surfaces thereby spreading the infection [18]. This reactor has been successfully used to evaluate the efficiency of biofilm control strategies using several microbial species [15].

Calgary Device

This device consists of a polystyrene lid with 96 pegs that can be fit into a standard 96-well microtiter plate. Its main advantages are the high throughput that enables the simultaneous testing of different experimental conditions, the fact that these devices are not prone to contamination (as they are manipulated in a laminar flow hood) and also they are not prone to leakage. Biofilm cells can be recovered from the pegs by gentle sonication for example in saline solution [17]. Although it has been recognized [19] that not all bacteria are amenable to this method of growth (particularly mucoidal isolates), this system has been successfully used to test antimicrobial susceptibility in various strains showing that reproducible biofilms can be obtained [19]. Some modifications of this technique have been reported including the growth of biofilms on lids containing 384 pins [17]. Besides the throughput, another major advantage of this system is that pegs can be detached from the lids to allow control measurements and microscopy. Major disadvantages are the fact that biofilms are formed in complex fluid dynamic conditions and the amount of biofilm that is formed is not sufficient for most chemical/biochemical analysis. Additionally, when comparing different strains, the optimization of biofilm formation conditions is required prior to high-throughput screening [17].

Flow Cells

Biofilms can be formed under laminar flow for instance in some biomedical contexts but in industrial settings they are commonly formed under turbulent conditions. In both cases, the velocity field of the fluid in contact with the microbial layer will affect biofilm structure and behaviour [20-25]. Thus, hydrodynamic conditions will determine the rate of transport of cells, oxygen and nutrients to the surface, as well as the magnitude of the shear forces acting on a developing biofilm [26]. It has been shown that higher fluid velocities associated with a higher turbulence conditions promote the formation of thinner and more stable biofilms [21, 27]. On the other hand it has been shown [28] that shear stresses of 6 to 8 Pa are sufficient to prevent adhesion of *Pseudomonas fluorescens* to stainless steel and that wall shear stresses of 12 Pa could remove attached cells.

Table 2. Methods used to analyze biofilms formed on large-scale flow cells

Device	Characteristic	Method	References
Flow cells	Biomass density	Optical density measurement (610 nm)	[30]
	Glucose concentration	Dinitrosalicylic colorimetric method (DNS)	[49]
	Epifluorescence microscopy analysis – total cells	4',6-diamidino-2-phenylindole – DAPI	[26, 50]
	Microscopy analysis – viable cells	Live/DeadR (L/D); BacLight™ bacterial viability kit (L-7012)	[26]
	Biofilm wet weight	Direct weight quantification	[30]
	Biofilm dry weight – total volatile solids (TVS)	American Standard methods (Method number 2540 A-D)	[26]
	Biofilm thickness	Direct thickness measurement	[30]
	Biofilm extraction of extracellular polymeric substances (EPS)	Dowex™ resin	[51]
	Biofilm protein assay	Bicinchoninic Acid Protein Assay Kit (BCA)	[26]
	Biofilm polysaccharide assay	Phenol – Sulphuric acid method	[52]

Flow cells have been used for more than 30 years for the study of dynamic biofilms. Although they exist in a variety of shapes and sizes, these systems can be divided into two main classes; those that contain removable coupons and those that are essentially built for microscopic observation. Flow cells containing a large number of coupons (usually between 8 and 20) will be referred as large-scale flow cells whereas those that are primarily destined for microscopic observation will be designated by small-scale flow cells. This classification can sometimes be ambiguous because there are flow cell models which are built for microscopic observation and that contain one or two coupons. These systems are more closely related to the small-scale flow cells since the number of coupons is too small to enable a time-course experiment with invasive analysis and the amount of biofilm formed on the coupon surface is also not sufficient for downstream chemical/biochemical analysis. A summary of the analytical tools and methodologies used on flow cell operation is shown in Table 2.

System Setup

Besides the flow cell, several components must be assembled in order to obtain a functional biofilm growing system. The list of components includes pumps, flow meters and other accessories although the main component of the system besides the flow cell is a reservoir containing the bacterial culture. The simplest system configuration is that of a batch culture that is circulated through the flow cell with the effluent being drained to a waste reservoir. This "once-through" configuration enables initial attachment studies particularly if the flow rate is very low due to the volume limitations of the batch culture reservoir. As an alternative, a closed loop system may be operated where the batch culture is introduced into the flow cell and the effluent drips back into the culture flask. This system may be adequate to demonstrate the adhesion potential of the cells only if very quick assays are performed. One must bear in mind that since the nutrient concentration is decreasing over time, biofilm growth is limited. Also, the attachment rate is influenced by the planktonic cell concentration which is constantly changing during the exponential phase of growth. In order to study biofilm formation using equilibrium conditions it is more convenient to use the effluent from a chemostat to feed the flow cell. This system provides a constant concentration of exponentially growing cells with the added advantage that the composition of the effluent stream is also constant. Although this setup is convenient for an extended experimental time it has the drawback of being limited to relatively low flow rates due to the time and expense entailed with media preparation. Furthermore, the flow rate in the flow cell is fixed by the dilution rate used on the chemostat which reduces the range of hydrodynamic conditions that can be studied. An alternative configuration can be used where the chemostat is part of a recirculating loop that is continuously fed by a nutrient stream. This configuration can be used to achieve the high flow rates that are common in industrial processes [29]. In this case, the flow velocity in the flow cell is independent of the dilution rate of the system and higher shear rates can be achieved. In essence, these systems are "chemostats with irregular geometries" [16]. In order to prevent gradient formation along the system the recirculation flow rates must be high and the volume of the recycle loops must be minimized thus decreasing residence times. Ideally, the residence time in the entire recycle loop and flow cell should not exceed a few minutes so that the whole system can be assumed to be completely mixed [16].

Small-Scale Flow Cells

There are several commercially available models of small-scale flow cells that can be used with video capture systems to enable real-time observation of microbial adhesion, division of biofilm cells and production of extracellular polymeric substances. Some models are autoclavable, reusable and compatible with confocal and epifluorescence microscopes [15]. The most common geometries for small-scale flow cells are the flat plate and the glass capillary flow cells. Flat plate flow cells are usually made on a polycarbonate base plate (although anodized aluminium or even stainless steel are also used for increased mechanical resistance) and usually contain one or two square or rectangular glass viewing ports. An illustrative photography of a small-scale flow cell can be observed in Figure 1. Models with two biofilm growth channels enable direct comparison of biofilms formed under two different conditions (e.g. a control and a specific treatment) in parallel. In this case each channel can be compared microscopically by simply moving the microscope stage. Biofilms can be observed through the glass viewing port with immersion microscope objectives. Some of the flat plate models include recesses to fit coupons that can be constructed form different materials to simulate cell adhesion to different surfaces. The amount of biofilm that is produced under these conditions is usually too low to enable further chemical/biochemical analysis. Since flat plate flow cells are usually restricted to low flow rates and to laminar flow applications, square glass capillary flow cells can be used as an alternative. These capillary flow cells enable non-destructive real-time analysis of biofilm formation under controlled hydrodynamic conditions. The flow regimes that can be used cover the laminar region but higher fluid velocities can be used so that high shear turbulent flow can be obtained [16]. Thinner glass square capillary tube (0.15 mm wall) is used for low flow rates enabling higher resolution observation and thicker flow cells (0.5 mm) are used for higher flow rates although the thickness of the glass imposes some constraints on microscopic observation. In most situations these flow cells can be mounted in parallel in a polycarbonate or aluminium support which allows a quick comparison between different experimental conditions. For turbulent flow regimes one must make sure that the flow cell is long enough to allow for full development of the flow in the viewing region. Avoiding sudden expansions or contractions between the connective tubing and the flow cells minimizes this entry effect [16, 28].

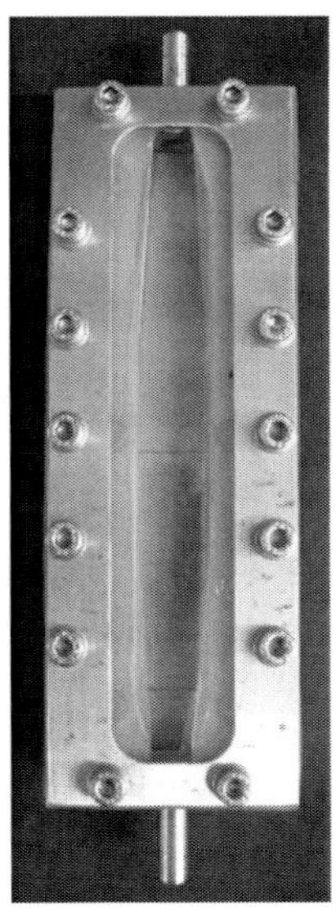

Figure 1. Small-scale flow cell reactor for biofilm formation with optical access from the top.

Large-Scale Flow Cells

Large-scale flow cells can be identified by two very important features besides their size. They usually contain a relatively high number of coupons (between 8 and 20) or similar removable biofilm forming surfaces that enable destructive analysis in a long time-course assay and they can be operated at very high flow rates in high turbulence and high shear regimes. Most of these large-scale flow cells are based on the design introduced by Jim Robins which was developed more than 30 years ago and later modified by McCoy creating what is commonly described as the Modified Robbins Device. This flow cell is basically a square channel pipe with coupons fixed to sampling plugs that can be unscrewed from the walls [16]. In order to accurately simulate biofilm formation on a specific surface, several materials like glass, silicone rubber, PVC, stainless steel or even dental enamel can be used to make the coupon [28]. Other designs include flow cells with a half-pipe geometry that more closely resemble the circular section of the tubes found in industrial piping systems [21]. Large-scale flow cells can be operated either in laminar or turbulent regime [21] but it is important to guarantee that in turbulent operation the flow cell has an entry section which is long enough to allow complete flow development before the sampling zone. Additionally, the last coupon must be placed at a far enough distance from the exit of the flow cell so that the constriction will not affect the flow at the end of sampling zone. Validation of these flow cells for a determined flow rate can be done using particle image velocity (PIV) techniques [28] and simulation of the flow by computational fluid dynamics [30]. Besides the validation of the flow characteristics within the flow cell, the whole system must be validated. In the recirculation system described before where the flow cell is considered to be part of a chemostat with irregular geometry, two important validation issues must be addressed. The system is represented on Figure 2 with all its components and it is easy to verify that a lot of tubing material is used. Since silicone tubing is highly permeable to oxygen, biofilm growth can be promoted on the lumen of the tube. If significant amounts of biofilm are formed on the tubing than these may have a greater effect on the chemistry and microbiology of the effluent than the biofilm in the flow cell [16]. This problem is particularly serious during laminar flow operations where the shear stress in the tubing is not high enough to prevent cell adhesion. Another problem that may arise during operation at very low flow rates is that significant gradients may build up along the recycle tube. If these gradients are present, the chemostat kinetics are no longer applicable because the system cannot be assumed to be completely mixed. In order to prevent this situation the recycle tube should be made as short as possible and the total residence in the whole system should not exceed a few minutes [16]. Operation of these flow cells is usually more complicated when compared to some of the other biofilm growing systems but the wealth of information that can be extracted from these systems usually pays off. Figure 3 represents the results obtained during the 12-day operation of the system depicted in Figure 2. *Escherichia coli* cells where grown in complex medium containing glucose as a carbon source and the biofilms were formed at a Reynolds number of 5400. The optical density in the recirculating tank was measured to monitor cell growth and glucose consumption within the system was assayed by the dinitrosalicylic colorimetric method (DNS) method as described in [30]. Figure 3a shows that the optical density increases until day 5 and after day 6 it starts to decrease although the glucose consumption is increasing until day 11 where it starts to stabilize. This increase in glucose consumption is caused by biofilm growth which coincides with the decline of the planktonic cell concentration from day 6 (Figure 1c). Biofilm growth was monitored by wet weight analysis (pre-weighted coupons were weighted after coupon retrieval) and biofilm thickness measurements that were made using a digital micrometer [30]. Figure 3b shows that the molecular weight and thickness of the biofilm seem to stabilize from day 9 indicating that the system reached a steady state. These results are further confirmed when the biofilm from the coupon is ressuspended in saline to enable total cell counting using 4,6-diamidino-2-phenylindole (DAPI) stain [26]. The same methodology was used for planktonic cell counting in samples taken from the recirculation tank. Figure 3c shows that the total number of cells in the planktonic and sessile states is relatively constant from day 9 onwards. Coupons retrieved from these flow cells are amenable to microscopy analysis. Most common techniques for this purpose are confocal microscopy, scanning electron microscopy (SEM) and transmission electron microscopy (TEM).

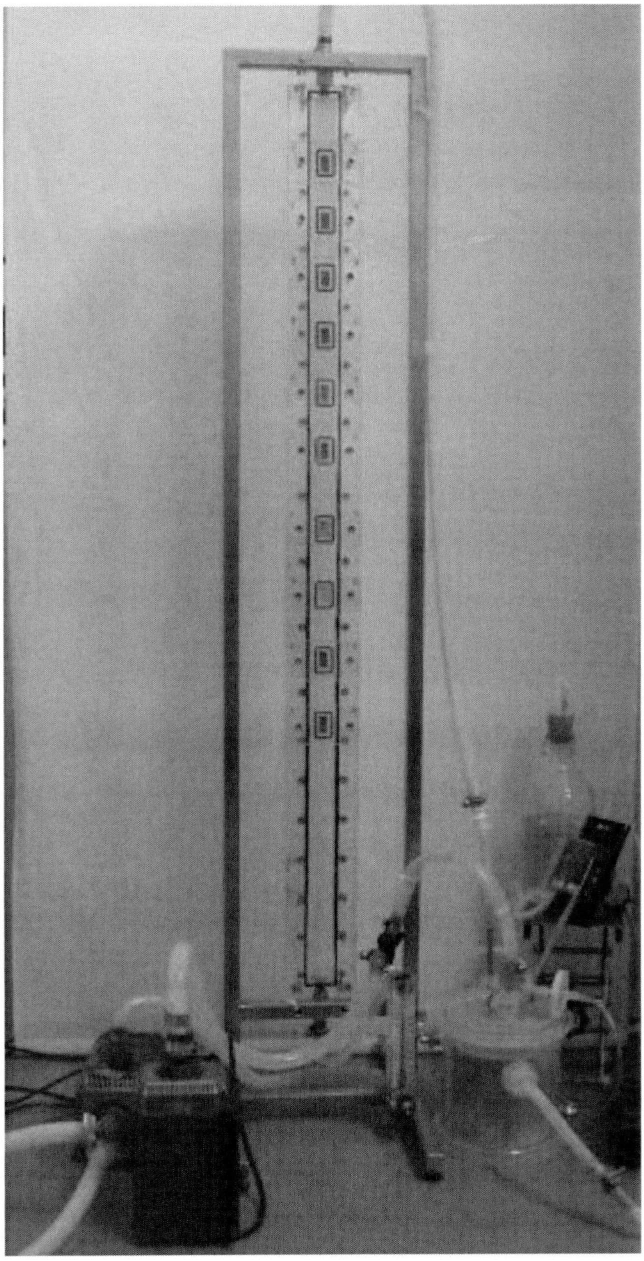

Figure 2. Large-scale flow cell reactor for biofilm formation containing removable coupons that can be made from different materials.

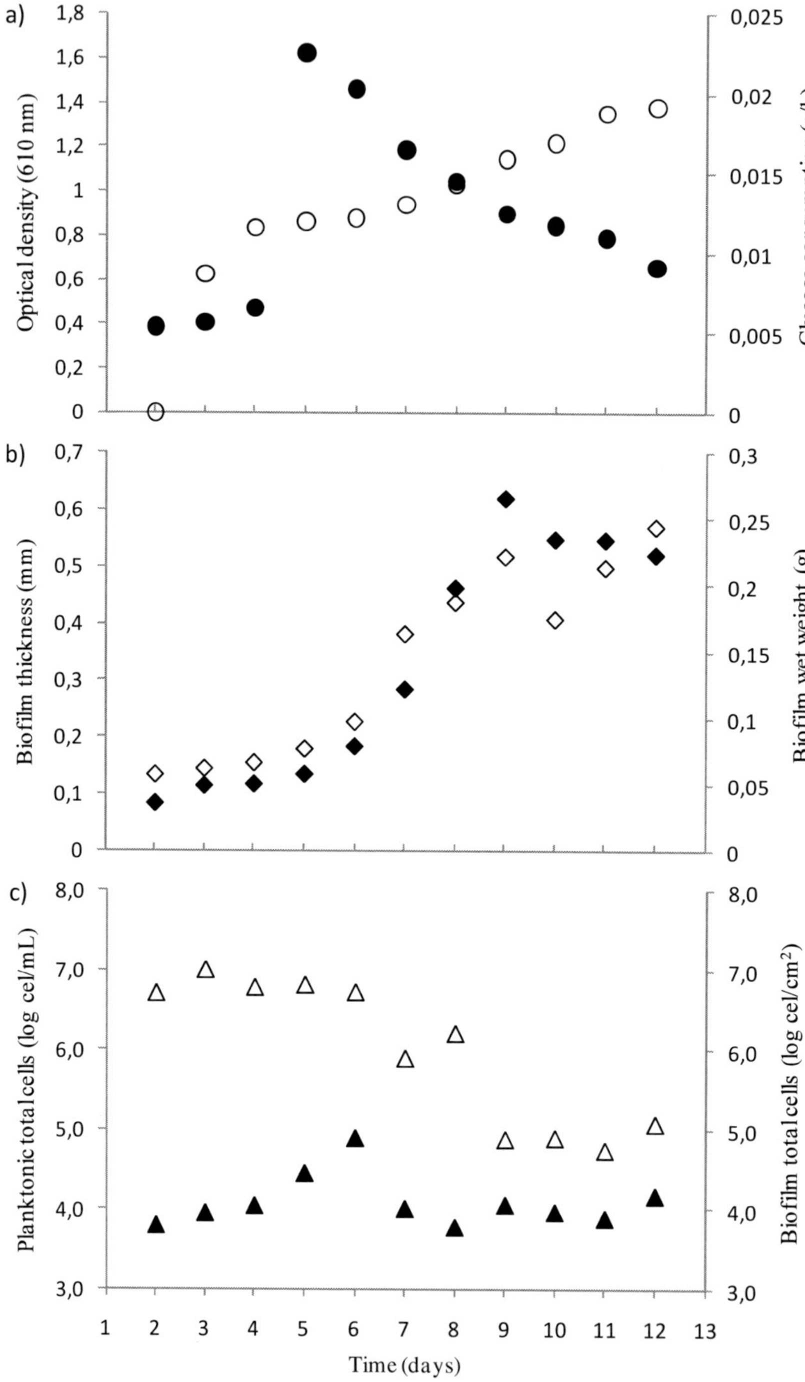

Figure 3. Experimental results obtained for planktonic and biofilm cells using a culture of *E. coli* JM109(DE3) on the large-scale flow cell reactor depicted in figure 2. During the experimental period of 12 days, the following parameters were assayed: a) closed circles (●) - optical density (610 nm), open circles (○) - glucose consumption ($g{\cdot}h^{-1}$); b) closed diamonds (♦) biofilm thickness (mm), open diamonds (◊) biofilm wet weight (g); c) closed triangles (▲) planktonic total cells (log $cells{\cdot}mL^{-1}$), open triangles (Δ) biofilm total cells (log $cells{\cdot}mL^{-1}$).

Microtiter Plates

Several bioreactor designs are currently used for bacterial growth in a laboratorial scale. Taking into account the volumetric capacity of these bioreactors, preference is given to the smallest ones due to their economic advantage: the use of minimum amounts of reagents and culture media and also the simple control of hydrodynamic and temperature conditions. In biotechnology, preference is given to shaken bioreactors, including flasks, test tubes and microtiter plates [31]. The first two are mainly used in academic and industrial environments dealing with biological processes. In contrast, the use of microtiter is relatively recent, since they were only developed in 1951. A microtiter plate is a flat plate with multiple wells arranged in a rectangular array with a 2:3 aspect ratio, resulting in 6, 12, 24, 48, 96 or 384 wells. The wells have designs similar to test tubes, and may have flat bottoms, rounded or tapered, and their volumes can range from tens of nanoliters and a few milliliters, depending on the number of wells [32]. Currently it is possible to find microtiter plates constructed in different types of polymers although the transparent polystyrene (PS) is the most widely used for the measurement of absorbance. Plates may also be white pigmented, for use in luminescence assays [33], or black, to be used in fluorescence tests [34]. Polypropylene microplates (PP) are also manufactured; they are more resistant to high temperatures and ideal for storage of chemicals for longer periods of time [35]. PVC microtiter plates may also be used instead of the previous ones; they are non-sterile, disposable and cheaper. An illustrative photography of PS microtiter plates with 6, 48 and 96 wells can be observed in Figure 4.

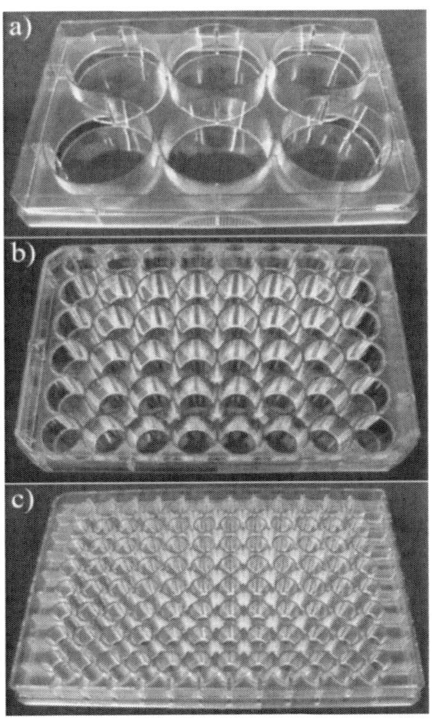

Figure 4. Illustrative photograph of Polystyrene microtiter plates used on biofilm formation: a) 6 - well microplates, b) 48 - well microplates c) 96 - well microplates.

Microtiter plates can be used in applications ranging from microbial cell culture, tissue culture, enzyme linked immunosorbent assay (ELISA) tests, high-throughput screening [36], determination of antimicrobial activity of natural compounds [37], antimicrobial susceptibility determination [38], adhesion studies [39] and biofilm inhibition quantification, among others. As mentioned before, the bioreactor type is an important factor and in the context of microtiter plates a clear advantage consists on the existence of a large number of reactors in miniature, with similar shapes and hydrodynamic conditions [32]. Additionally, these devices are easy to handle, with the use of robotic systems and multi-channel micropipettes, allowing the rapid development of biofilms in different conditions. Thus it is possible to study the adhesion of different bacterial strains and consequent biofilm formation in rapid and inexpensive assays[40].

Several techniques have been developed for the quantification of biofilm formation and its bioactivity in microtiter plates, such as the Crystal Violet technique for biomass determination [41], or Alamar Blue technique for cell viability determination [42, 43]. A summary of the analytical tools and methodologies used on microtiter plate operation is represented in Table 3.

Table 3. Methods used to analyze biofilms formed on microtiter plates (adapted from [45])

Device	Characteristic	Method	References
Microtiter plate	Biofilm biomass	Crystal violet assay	[41]
	Biofilm dry weight	Total solids determination	[53]
	Microbial physiological activity	XTT assay	[40, 42]
		Resazurin assay	[42, 44]
		Fluorescein diacetate assay	[54]
		Dimethyl methylene blue assay	[55]
	Microscopy analysis of viable cells	Live/Dead (L/D); BacLight™ bacterial viability kit (L-7012)	[26]

Microtiter plates are thus a tool for biofilm study in a laboratory scale. Research continues on the development of new strategies for removal and control of biofilms involving the study of initial adhesion and biofilm formation in PS microtiter plates [39] and also the use of biocides [44] for their control. On the other hand, studies are also being directed towards the development of biofilms with beneficial properties to use for example on the removal of pollutants by biodegradation.

Figure 5 represents examples of results obtained on biofilm formation in microtiter plates for different bacterial strains - *Listeria monocytogenes*, *Staphylococcus aureus*, *Escherichia coli* and *Pseudomonas aeruginosa*. Figure 5a) shows the results for biofilm biomass. Higher cell amounts were obtained for *Escherichia coli* followed by *Pseudomonas aeruginosa*. Residual values were obtained for *Listeria monocytogenes* and *Staphylococcus aureus*.

In Figure 5b it is possible to observe the results obtained for the respiratory activity of the biofilms formed by the different bacterial strains. A higher activity was observed for *Listeria monocytogenes* and *Staphylococcus aureus* cells.

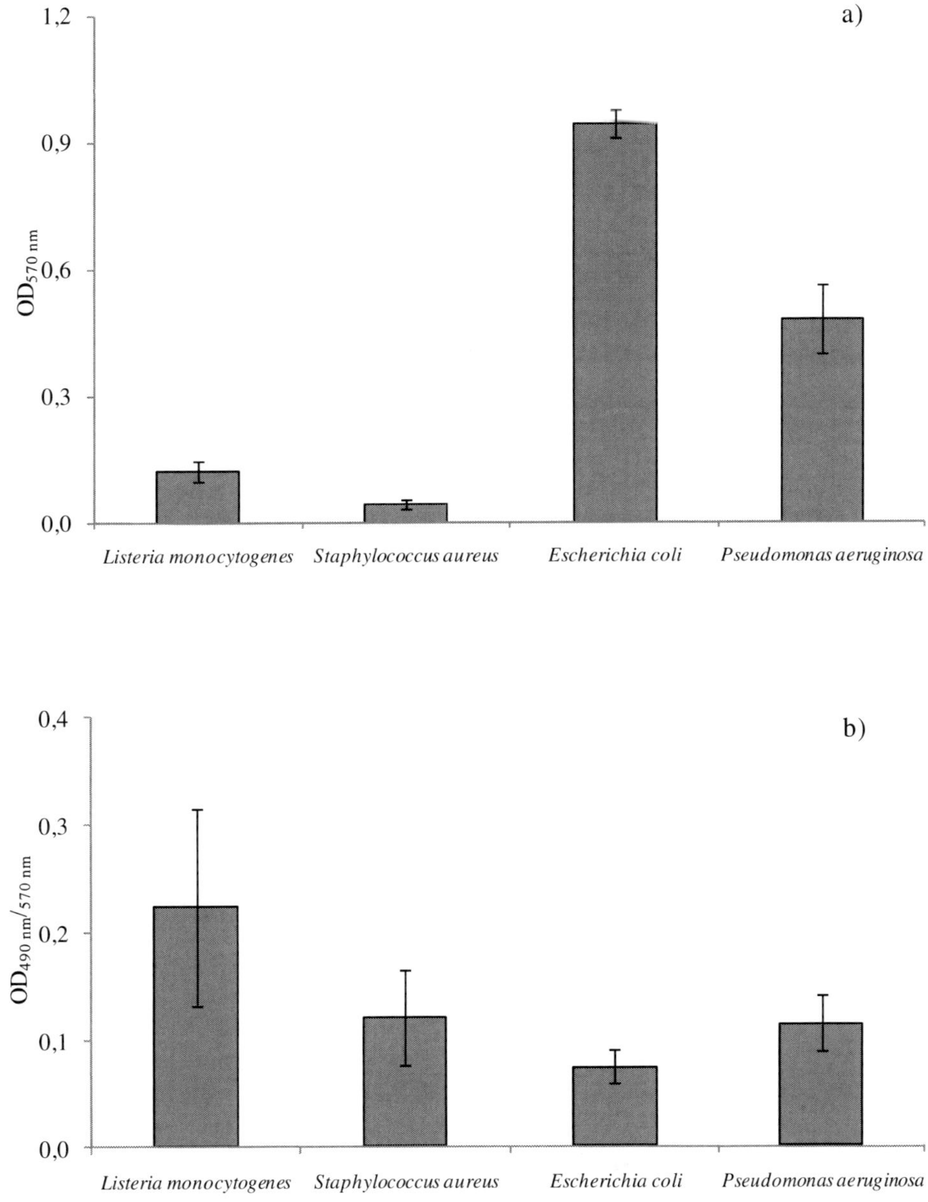

Figure 5. Results obtained on biofilm formation in microtiter plates for different bacterial strains – a) biofilm biomass (optical density 570 nm); b) microbial activity (O.D.490nm/O.D.570nm).

Conclusion

Several platforms are now available for *in vitro* biofilm formation and each of them comes with its own unique advantages and limitations. Traditionally, biofilm formation systems were designed by biofilm researchers and were custom-built to serve the researcher's needs. As a consequence of this fact a great diversity in the design of these systems was created. This enabled a variety of studies in quite different conditions but has also made comparison of the results obtained by different groups far more complicated. The development of commercial biofilm formation platforms leads to system standardization which facilitates the validation of the results by independent groups. Several key issues like throughput, reproducibility, possibility of contamination or the cost per assay affect the choice of a particular system. One must however bear in mind that the best *in vitro* biofilm growing system is the one that emulates with greater perfection the particular environment that needs to be simulated. Thus two very important characteristics have to be maintained in simulation, the flow regime must be as close as possible to the "real life" situation and the biofilm formed by the system must also resemble the "real-life" biofilm.

Acknowledgments

The authors acknowledge the financial support provided by the Portuguese Foundation for Science and Technology (Projects PTDC/EBB-BIO/102863/2008 and PTDC/EBB-BIO/104940/2008).

References

[1] Bryers, J. D., (2008). Medical Biofilms. *Biotechnology and Bioengineering, 100*, 1-18.

[2] Davey, M. E. & O'Toole G, A., (2000). Microbial biofilms: from ecology to molecular genetics. *Microbiology and Molecular Biology Reviews, 64*, 847-867.

[3] O'Toole, G., Kaplan, H. B. & Kolter, R., (2000). Biofilm formation as microbial development. *Annual Reviews of Microbiology, 54*, 49-79.

[4] Stoodley, P., Sauer, K., Davies, D. G. & Costerton, J. W., (2002). Biofilms as complex differentiated communities. *Annual Reviews of Microbiology, 56*, 187-209.

[5] Watnick, P. & Kolter, R., (2000). Biofilm, city of microbes. *Journal of Bacteriology, 182*, 2675-2679.

[6] Shunmugaperumal, T., (2010). Inrtoduction and overview of biofilm. In Shunmugaperumal, T. (Eds.), *Biofilm Eradication and Prevention* (1st edition, pp. 3-35). Hoboken, New Jersey: John Wiley & Sons, Inc.

[7] Qureshi, N., Annous, B. A., Ezeji, T. C., Karcher, P. & Maddox, I. S., (2005). Biofilm reactors for industrial bioconversion processes: employing potential of enhanced reaction rates. *Microbial Cell Factories, 4*.

[8] Qureshi, N., Karcher, P., Cotta, M. & Blaschek, H. P., (2004). High-productivity continuous biofilm reactor for butanol production - Effect of acetate, butyrate, and corn steep liquor on bioreactor performance. *Applied Biochemistry and Biotechnology, 113*, 713-721.

[9] Tay, A. & Yang, S. T., (2002). Production of L(+)-lactic acid from glucose and starch by immobilized cells of *Rhizopus oryzae* in a rotating fibrous bed bioreactor. *Biotechnology and Bioengineering, 80*, 1-12.

[10] Rabah, F. K. J. & Dahab, M. F., (2004). Nitrate removal characteristics of high performance fluidized-bed biofilm reactors. *Water Research, 38*, 3719-3728.

[11] Manolov, T., Kristina, H. & Benoit, G., (2005). Continuous acetonitrile degradation in a packed-bed bioreactor. *Applied Microbiology and Biotechnology, 66*, 567-574.

[12] Wang, Z. W. & Chen, S. L., (2009). Potential of biofilm-based biofuel production. *Applied Microbiology and Biotechnology, 83*, 1-18.

[13] Melo, L. F. & Flemming, H. C., (2010). Mechanistic aspects of heat exchanger and membrane biofouling and prevention. In Amjad, Z. (Eds.), *The Science and Technology of Industrial Water Treatment* (1st edition, pp. 365-380). Boca Raton, USA: Taylor and Francis Group.

[14] Shunmugaperumal, T., (2010). Rationale for Biofilm Eradication from Modern Medical Devices. In Shunmugaperumal, T. (Eds.), *Biofilm Eradication and Prevention* (1st edition, pp. 36-72). Hoboken, New Jersey: John Wiley & Sons, Inc.

[15] Pavarina, A. C., Dovigo, L. N., Sanitá, P. V., Machado, A. L., Giampaolo, E. T. & Vergani, C. E., (2011). Dynamic models for in vitro biofilm formation In Bailey, W. C. (Eds.), *Biofilms:Formation, Development and Properties.* (1st edition, pp. 125-161). Nova Sicence Puslishers, Inc.

[16] Stoodley, P. & Warwood, B. K., (2003). Use of flow cells an annular reactors to study biofilms. In Lens, P., O'Flaherty, V., Moran, A. P., Stoodley, P. & Mahony, T. (Eds.), *Biofilms in Medicine, Industry and Environmental Biotechnology: Characteristics, Analysis and Control* (1st edition, pp. 197-213). Cornwall. UK: IWA Publishing.

[17] Peterson, S. B., Irie, Y., Borlee, B. R., Murakami, K., Harrison, J. J., Colvin, K. M. & Parsek, M. R., (2011). Different Methods for Culturing Biofilms In Vitro In Bjarnsholt, T., Jensen, P. Ø., Moser, C. & Høiby, N. (Eds.), *Biofilm infections* (1st edition, pp. 251-266). New York: Springer.

[18] Gilbert, P., Allison, D. G., Evans, D. J., Handley, P. S. & Brown, M. R. W., (1989). Growth-rate control of adherent bacterial populations. *Applied and Environmental Microbiology, 55*, 1308-1311.

[19] Harrison, J. J., Ceri, H., Yerly, J., Rabiei, M., Hu, Y. P., Martinuzzi, R. & Turner, R. J., (2007). Metal ions may suppress or enhance cellular differentiation in *Candida albicans* and *Candida tropicalis* biofilms. *Applied and Environmental Microbiology, 73*, 4940-4949.

[20] Hall-Stoodley, L. & Stoodley, P., (2002). Developmental regulation of microbial biofilms. *Current Opinion in Biotechnology, 13*, 228-233.

[21] Pereira, M. O., Kuehn, M., Wuertz, S., Neu, T. & Melo, L. F., (2002). Effect of flow regime on the architecture of a *Pseudomonas fluorescens* biofilm. *Biotechnology and Bioengineering, 78*, 164-171.

[22] Purevdorj, B., Costerton, J. W. & Stoodley, P., (2002). Influence of hydrodynamics and cell signaling on the structure and behavior of *Pseudomonas aeruginosa* biofilms. *Applied and Environmental Microbiology, 68*, 4457-4464.

[23] Stoodley, P., Dodds, I., Boyle, J. D. & Lappin-Scott, H. M., (1999). Influence of hydrodynamics and nutrients on biofilm structure. *Journal of Applied Microbiology, 85*, 19S-28S.

[24] Tsai, Y. P., (2005). Impact of flow velocity on the dynamic behaviour of biofilm bacteria. *Biofouling, 21*, 267-277.

[25] Vieira, M. J., Melo, L. F. & Pinheiro, M. M., (1993). Biofilm formation: Hydrodynamic effects on internal diffusion and structure. *Biofouling, 7*, 67-80.

[26] Simoes, M., Pereira, M. O., Sillankorva, S., Azeredo, J. & Vieira, M. J., (2007). The effect of hydrodynamic conditions on the phenotype of *Pseudomonas fluorescens* biofilms. *Biofouling, 23*, 249-258.

[27] Liu, Y. & Tay, J. H., (2001). Metabolic response of biofilm to shear stress in fixed-film culture. *Journal of Applied Microbiology, 90*, 337-342.

[28] Bakker, D. P., van der Mats, A., Verkerke, G. J., Busscher, H. J. & van der Mei, H. C., (2003). Comparison of velocity profiles for different flow chamber designs used in studies of microbial adhesion to surfaces. *Applied and Environmental Microbiology, 69*, 6280-6287.

[29] Melo, L. F. & Vieira, M. J., (1999). Physical stability and biological activity of biofilms under turbulent flow and low substrate concentration. *Bioprocess Engineering, 20*, 363-368.

[30] Teodosio, J. S., Simoes, M., Melo, L. F. & Mergulhao, F. J., (2011). Flow cell hydrodynamics and their effects on *E. coli* biofilm formation under different nutrient conditions and turbulent flow. *Biofouling, 27*, 1-11.

[31] Funke, M., Diederichs, S., Kensy, F., Mueller, C. & Buechs, J., (2009). The Baffled Microtiter Plate: Increased Oxygen Transfer and Improved Online Monitoring in Small Scale Fermentations. *Biotechnology and Bioengineering, 103*, 1118-1128.

[32] Kumar, S., Wittmann, C. & Heinzle, E., (2004). Minibioreactors. *Biotechnology Letters, 26*, 1-10.

[33] Tauriainen, S., Karp, M., Chang, W. & Virta, M., (1998). Luminescent bacterial sensor for cadmium and lead. *Biosensors & Bioelectronics, 13*, 931-938.

[34] Tani, C., Inoue, K., Tani, Y., Harun-ur-Rashid, M., Azuma, N., Ueda, S., Yoshida, K. & Maeda, I., (2009). Sensitive fluorescent microplate bioassay using recombinant *Escherichia coli* with multiple promoter-reporter units in tandem for detection of arsenic. *Journal of Bioscience and Bioengineering, 108*, 414-420.

[35] Fukazawa, T., Yamazaki, Y. & Miyamoto, Y., (2010). Reduction of non-specific adsorption of drugs to plastic containers used in bioassays or analyses. *Journal of Pharmacological and Toxicological Methods, 61*, 329-333.

[36] Nakayama, G. R., (1998). Microplate assays for high-throughput screening. *Current opinion in drug discovery & development, 1*, 85-91.

[37] Jagani, S., Chelikani, R. & Kim, D.-S., (2009). Effects of phenol and natural phenolic compounds on biofilm formation by *Pseudomonas aeruginosa*. *Biofouling, 25*, 321-324.

[38] Nateche, F., Martin, A., Baraka, S., Palomino, J. C., Khaled, S. & Portaels, F., (2006). Application of the resazurin microtitre assay for detection of multidrug resistance in *Mycobacterium tuberculosis* in Algiers. *Journal of Medical Microbiology, 55*, 857-860.

[39] Simoes, L. C., Simoes, M. & Vieira, M. J., (2010). Influence of the Diversity of Bacterial Isolates from Drinking Water on Resistance of Biofilms to Disinfection. *Applied and Environmental Microbiology, 76*, 6673-6679.

[40] Stepanovic, S., Vukovic, D., Hola, V., Di Bonaventura, G., Djukic, S., Cirkovic, I. & Ruzicka, F., (2007). Quantification of biofilm in microtiter plates: overview of testing conditions and practical recommendations for assessment of biofilm production by *Staphylococci*. *Apmis, 115*, 891-899.

[41] Stepanovic, S., Vukovic, D., Dakic, I., Savic, B. & Svabic-Vlahovic, M., (2000). A modified microtiter-plate test for quantification of *Staphylococcal* biofilm formation. *Journal of Microbiological Methods, 40*, 175-179.

[42] Pettit, R. K., Weber, C. A., Kean, M. J., Hoffmann, H., Pettit, G. R., Tan, R., Franks, K. S. & Horton, M. L., (2005). Microplate alamar blue assay for *Staphylococcus epidermidis* biofilm susceptibility testing. *Antimicrobial Agents and Chemotherapy, 49*, 2612-2617.

[43] Sarker, S. D., Nahar, L. & Kumarasamy, Y., (2007). Microtitre plate-based antibacterial assay incorporating resazurin as an indicator of cell growth, and its application in the in vitro antibacterial screening of phytochemicals. *Methods, 42*, 321-324.

[44] Pitts, B., Hamilton, M. A., Zelver, N. & Stewart, P. S., (2003). A microtiter-plate screening method for biofilm disinfection and removal. *Journal of Microbiological Methods, 54*, 269-276.

[45] Azevedo, N. F., Lopes, S. P., Keevil, C. W., Pereira, M. O. & Vieira, M. J., (2009). Time to "go large" on biofilm research: advantages of an omics approach. *Biotechnology Letters, 31*, 477-485.

[46] Braganca, S. M., Azevedo, N. F., Simoes, L. C., Keevil, C. W. & Vieira, M. J., (2007). Use of fluorescent in situ hybridisation for the visualisation of *Helicobacter pylori* in real drinking water biofilms. *Water Science and Technology, 55*, 387-393.

[47] Azevedo, N. F., Pacheco, A. P., Keevil, C. W. & Vieira, M. J., (2006). Adhesion of water stressed Helicobacter pylori to abiotic surfaces. *Journal of Applied Microbiology, 101*, 718-724.

[48] Harrison, J. J., Turner, R. J. & Ceri, H., (2005). High-throughput metal susceptibility testing of microbial biofilms. *Bmc Microbiology, 5*.

[49] Miller, G. L., (1959). Use of dinitrosalicylic acid reagent for determination of reducing sugar. *Analytical Chemistry, 31*, 426-428.

[50] Saby, S., Sibille, I., Mathieu, L., Paquin, J. L. & Block, J. C., (1997). Influence of water chlorination on the counting of bacteria with DAPI (4',6-diamidino-2-phenylindole). *Applied and Environmental Microbiology, 63*, 1564-1569.

[51] Frolund, B., Palmgren, R., Keiding, K. & Nielsen, P. H., (1996). Extraction of extracellular polymers from activated sludge using a cation exchange resin. *Water Research, 30*, 1749-1758.

[52] Dubois, M., Gilles, K. A., Hamilton, J. K., Rebers, P. A. & Smith, F., (1956). Colorimetric method for determination of sugars and related substances. *Analytical Chemistry, 28*, 350-356.

[53] Cerca, N., Martins, S., Pier, G. B., Oliveira, R. & Azeredo, J., (2005). The relationship between inhibition of bacterial adhesion to a solid surface by sub-MICs of antibiotics and subsequent development of a biofilm. *Research in Microbiology, 156*, 650-655.

[54] Honraet, K., Goetghebeur, E. & Nelis, H. J., (2005). Comparison of three assays for the quantification of *Candida* biomass in suspension and CDC reactor grown biofilms. *Journal of Microbiological Methods, 63*, 287-295.

[55] Tote, K., Vanden Berghe, D., Maes, L. & Cos, P., (2008). A new colorimetric microtitre model for the detection of *Staphylococcus aureus* biofilms. *Letters in Applied Microbiology, 46*, 249-254.

In: Biofilms in Bioengineering
Editors: Manuel Simões and Filipe Mergulhão

ISBN: 978-1-62948-161-6

Chapter IV

Biofilm Monitoring Strategies

Ana Pereira[*]
LEPAE, Department of Chemical Engineering, Faculty of Engineering, University of Porto, Porto, Portugal

Abstract

Biofouling is a major problem for systems that involves fluid/ surface interfaces. The deleterious effects of biofilms associated with the impossibility to eradicate them, requires new solutions. One of these solutions is monitoring biofouling build-up/ removal, to provide online, real-time information about the deposit layers attached to the monitored surface. This information, may ultimately lead to, a process optimization based on the implementation of pro-active, and informed counter-measures, and on the assessment of their efficacy.

Monitoring in real-time and non-destructively ~~of~~ the build-up or removal of the biofouling layers over time, can be accomplished in several different ways, using similar or different physical principles and relating them to different characteristics of the deposits.

This chapter discusses the challenges of monitoring strategies and briefly describes several monitoring techniques that have been developed over the last decades. Furthermore, this chapter was designed in a way that can help the reader to easily identify the techniques that can be more suited for a given process.

The discussion about the industrial applicability of most of these techniques is also an important issue that will be presented along the text. In spite of the interesting number of techniques described herein, there is still a gap between what is done in laboratorial settings and what is already implemented in industry.

Keywords: Monitoring, Biofouling, Anti-biofouling strategies, Early warning systems

* Enkrott Química, S.A. (www.enkrottquimica.com), Sintra, Portugal.
Email: aalexp@live.com.pt.

Introduction

Biofilm is a complex and fascinating world! This book, and other reference literature [1-4] wisely demonstrates this complexity and shows completely different areas of research around this subject. Many of this research is focused on understanding microorganisms, the interaction between them, biofilms nature, formation/removal mechanisms, kinetics and their impact on different fields where they are a 'must have' or a 'do not even think about having it'.

Biofilms are dynamic, adaptable to new conditions and affect not only the bulk fluid but mostly the fluid surface interface. The existence of such microbial communities which adhere to surfaces and protect themselves by an EPS matrix (extracellular polymeric substances) is a well-known problem to many processes. The contamination of medical instruments/implants [5] or the contact lens colonization with bacteria [6] are examples of the attention and concern that biofouling requires in daily life. Special focus is given to fouling within industry, since a large number of industrial processes favor the adhesion and growth of such unwanted deposits. Cooling and drinking water systems, heat exchangers, food and beverage processing plants and membrane systems are some of the industrial operations typically affected by fouling formation. The type of fouling (more information on this point can be found in [7] and [8]) and its effect on the overall economics of process is highly dependent on the specificities of each process, but it is often responsible for a reduction of the thermal (increase of heat transfer resistance) and hydrodynamic performance (increase of pressure drop) of production lines.

In parallel with the build-up of the deposits, the removal of the fouling layers from the equipment surfaces (cleaning procedure) is also a critical point. This point is not only associated with the operational constraints during cleaning but also regarding chemicals and water consumption and discharge to the environment and with health related problems (associated with hygienic of the operating lines).

The impact of biofilm depends on the specificities of each process. Some of the figures that can illustrate the impact of fouling (transversely to different processes and industries) in the economy of processes, on the environment and on human-health are:

1) Müller-Steinhagen et al., [9] points out that the economic penalties associated with fouling within heat exchangers represents around 0.25% of the gross national product of industrialized countries, like the USA;
2) the problems associated with improper cleaning and disinfection in the food and beverage processing plants, may compromise safety and quality of the final product [10]. For instances, the microbial contamination of the final product, by microorganisms like *Listeria spp.*, *Bacillus spp.*, *Salmonella* spp., *Clostridium botulinu,* may result in product spoilage and foodborne illnesses [11].
3) within the dairy industry, cleaning procedures are carried out on a daily basis, due to the severity and high extent of the fouling layers [12]. These semi-empirically cleaning procedures [13] are associated with consumption of large amounts of water and cleaning agents (e.g. biocides, alkaline and acid solutions) which have a great impact on the thermal and energy consumption associated with their recirculation, on

the wastewater treatment and with the fact that production is stopped during cleaning.

4) filtration processes (mainly reverse osmosis systems) are well positioned to fulfill the worldwide necessities of water (since they are capable of treating almost all kind of water, such as seawater, ground and surface water) and a technique widely applied in pharmaceutical and medical processes [14]. Filtration are very costly processes, mainly due to biofouling build-up on the membranes, which: decrease of the product quality and quantity of the final product; need for extensive cleaning procedures; high energy and necessity to replace membranes; increases production down-time. Flemming [15] has estimated that biofouling reduces the working life of a desalinization process within a nanofiltration plant from 3 to 1 year.
5) biofilm is a host for pathogenic microrganisms, which enter and survive within the biofilm structure. Some of the most well-known are *Legionella pneumophila*, *Pseudomonas aeruginosa* and *Helicobater pylori* which are often found in drinking and cooling water systems. A portion of biofilm can detach and these pathogens are released to the bulk water of the system, and this may lead to serious health-related problems [16].
6) the increase of the energy demands, resulting from the inefficient energy transfer caused by biofouling problems lead the increase of greenhouse gas emission (in the cases where energy depends on combustion). Similar conclusions have been taken, for example by Watkinson [17] which found that the effect of fouling in crude oil systems has a significant impact on fuel consumption and on greenhouse gas emissions. A growing concern, not only because of its environmental impact, but also because of the world environmental policies which demand for controlled greenhouse gas emissions and water sustainability.

Prevention and Monitoring a Step Towards Pro-Activity

The studies focused on planktonic and sessile microorganisms are consensual on demonstrating that microorganisms, and therefore biofilms, cannot be totally eradicated. No system is indifferent to biofouling formation!

Several efforts have been made regarding biofouling mitigation, which encompasses the theoretical and experimental research carried by academia and the know-how from industry. These efforts result on the devise and combination of new and old strategies to alleviate the biofouling problem.

The proposals for fouling mitigation can have different backgrounds, although are inevitably based on the study of the mechanisms behind biofilm formation [8] and their removal [18], [19]. These works include for example the study of the impact of flow velocities, temperature, nutrient availability on the formation and removal of biofilms, such as the ones formed by *Pseudomonas fluorescens* [20], [21].

The chemical strategy is still one of the most used resources to control biofouling build-up, and in spite of its popularity is facing a drawback regarding the environmental problems it causes [22]. The issues of microorganism resistance or the impact of the chemicals in the environment, in association with the increasingly strict regulation, demand for the complete understanding of the functioning mechanism of biocides e.g. [16], to optimize the biocide dosage by studying the best working concentration, frequency and duration of the injection [18], the design of more 'environmental friendly' chemical substances [23] or the use of 'smart' techniques to dose such chemicals, like the use of drug delivery systems [24].

In parallel with an optimized chemical treatment there are operational good practices that can reduce biofilm adhesion and/ or promote their removal. Some of these approaches rely on a better design of equipment with the elimination of hydrodynamic stagnation zones within the bulk circulation, with the implementation of new surface materials [25], [26], [27], [28], among many others.

With the evolution of informatics a new input has also been given to computational and numeric techniques that combine the fundamental and applied results in order to predict biofouling impact [29].

Monitoring the build-up and/or the removal of microbiological deposits from surfaces is a complementary approach for biofouling minimization and for the optimization of the cleaning procedures. The traditional way to evaluate the existence and removal of a biofouling layer in a given system is based on indirect methods such as process product quality or performance indicators [30], which are often complemented by collecting samples for off-line analysis (chemical, physical or microbiological). This 'way of doing things' is based on very poor, limited and retrospective data [10].

The idea of an eco-friendly set of tools that embraces preventive (adjusted) strategies (which include, among many others, the strategies previously mentioned) and also reliable techniques that gather on-line, real-time, non-destructively and in-situ information about the biofouling formation is gaining more importance. The implementation of such strategy aims to gather different kinds of information that will lead to timely and supported counter-measures. The ultimate goal of the monitoring approach is to enable the decision maker to answer to the following questions pro-actively:

a) When should the plant/equipment be cleaned?
b) How should the plant be cleaned?
c) When is the cleaning end-point reached?
d) How efficient is the preventive and cleaning program?

Figure 1 demonstrates how a monitoring strategy can be a part of an integrated anti-fouling strategy.

Different monitoring techniques can provide different kinds of information (as will be discussed in section 1.3) according to their specific design [30]. As seen in Figure 1, the information about the extent (amount) and nature of the deposit allows to constantly evaluate the counter-measures, and to actuate on them according to the response of the system. This acts positively on the process economics, environmental and quality of the final product.

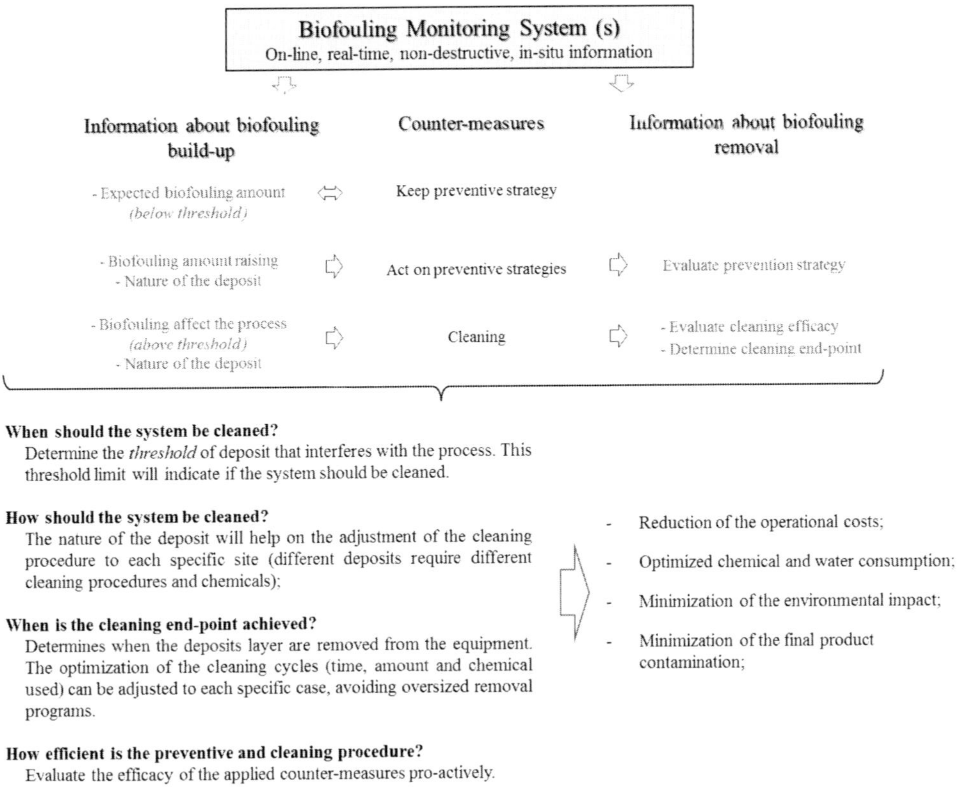

Figure 1. Monitoring strategy within an integrated anti-fouling approach.

What To Know About a Monitoring Technique?

The primary goal of a monitoring technique is to assess biofouling formation and/or its removal over time. The ultimate goal is to generate valuable information that can lead to informed and proactive actions concerning the monitored process.

It is now considered that a good monitoring technique should fulfill a set of requirements, such as: i) provide online, real-time information about the deposits build-up/removal; ii) generate different levels of information like the amount – mass, thickness, density, the nature of the deposit – organic or inorganic and the state of the microorganisms – live or dead; iii) analysis should be non-intrusive and non-destructive; iv) integrate the information over a large area, in order to avoid biasing problems; v) be simple to use, robust and reliable; vi) be non-expensive (low cost/benefit ratio) [31], [32].

It is quite difficult to find a technique that meets all of the requirements previously listed. The actual trend is to combine the use of more than one technique and assess all the information that a process needs to be correctly monitored. Figure 2 summarizes the questions that should be asked about a monitoring technique. This can help not only to characterize a given monitoring technique but also to define the basic requirements that it should have in order to provide the information needed. The final questions should be related to the ultimate goal of a monitoring strategy – to evaluate the impact and importance of the provided information: can it increase pro-activeness and can it be used to access the efficacy of control and cleaning strategies?

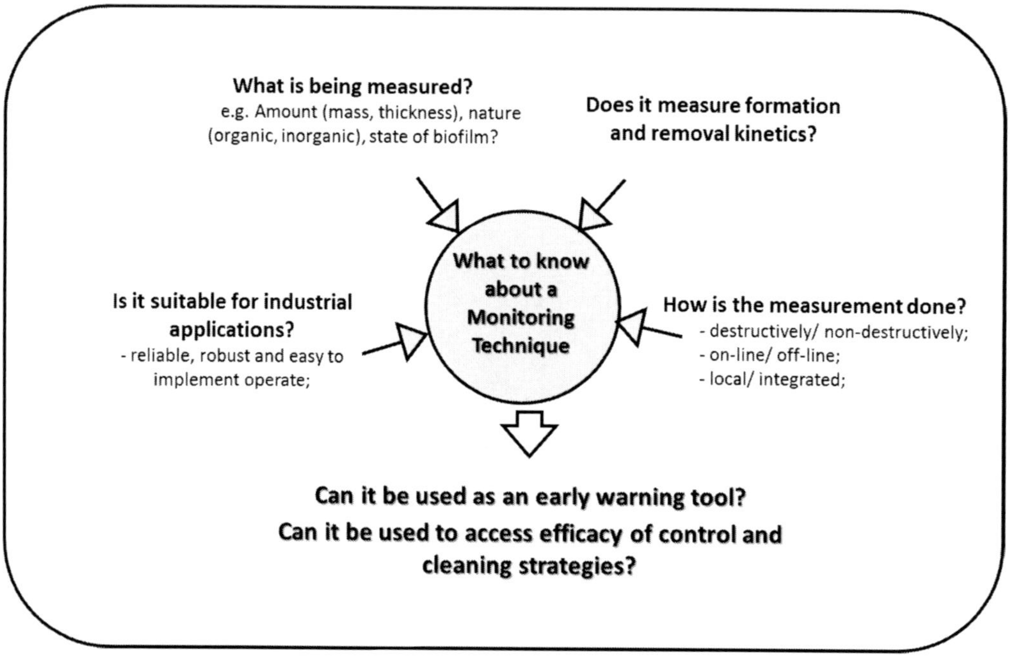

Figure 2. What to know about a monitoring technique.

In addition to those characteristics there are also some important issues that should be taken into account when searching for or implementing a biofouling monitoring technique. Technique(s) should be able to provide both extent and location of the biofouling layers. The implementation of monitoring devices should be done in places that are representative of the biofouling that occur on the overall plant and/or on the critical points of the circuit or process.

The discussions about monitoring techniques should be focused, not only on the technique itself but also on the needs that each industry/process/research has. In many cases, industries suffer similar biofouling related problems, however, due to their operational or processual specificities they demand completely different approaches to the same output. In many cases, this will lead to different monitoring technique sets.

The interesting thing about monitoring techniques is that even using the same physical principle, different outputs are observed according to the way they are 'looking' for the biofouling phenomenon. Indeed, there is not only knowledge but a high amount of creativity behind the design and conception of each monitoring device.

Biofouling Monitoring Techniques

The information previously exposed indicates that monitoring techniques may be classified following different criteria, such as: physical signal that they are measuring, the level of information they provide, enumerated by application where they are mostly used. There are some reviews [32], [33] which address in more or less detail the features of monitoring techniques, their physical principles, and how they show or seem to have an industrial potential application.

The present chapter is committed not only to the description of the monitoring techniques themselves, but also to provide a simple way to enable the reader to easily find the monitoring techniques that can provide specific required information – Figure 3. In order to accomplish this goal, sorting will be based on the type of information each technique is able to provide.

The traditional monitoring and the characterization of the biofouling layers techniques were also included on the monitoring tool screening provided in Figure 3. These techniques do not fulfill the requirements of a monitoring technique listed on section 1.3, but since they are widely used within industry they will be briefly described. The monitoring techniques are firstly divided according to their capacity to follow deposits adhesion/removal kinetics or to provide a characterization of the attached layers. Within the techniques that are able to provide kinetic information they are grouped according to their non-intrusive characteristic with an integrated measurement or intrusive with localized measurement character. Furthermore, Figure 3 also categorizes monitoring techniques according to their common physical principle and provides information about their potential for industrial application. Along the next paragraphs the reader will find a brief description of the working principle of each of the listed techniques and the main applications where they have been tested.

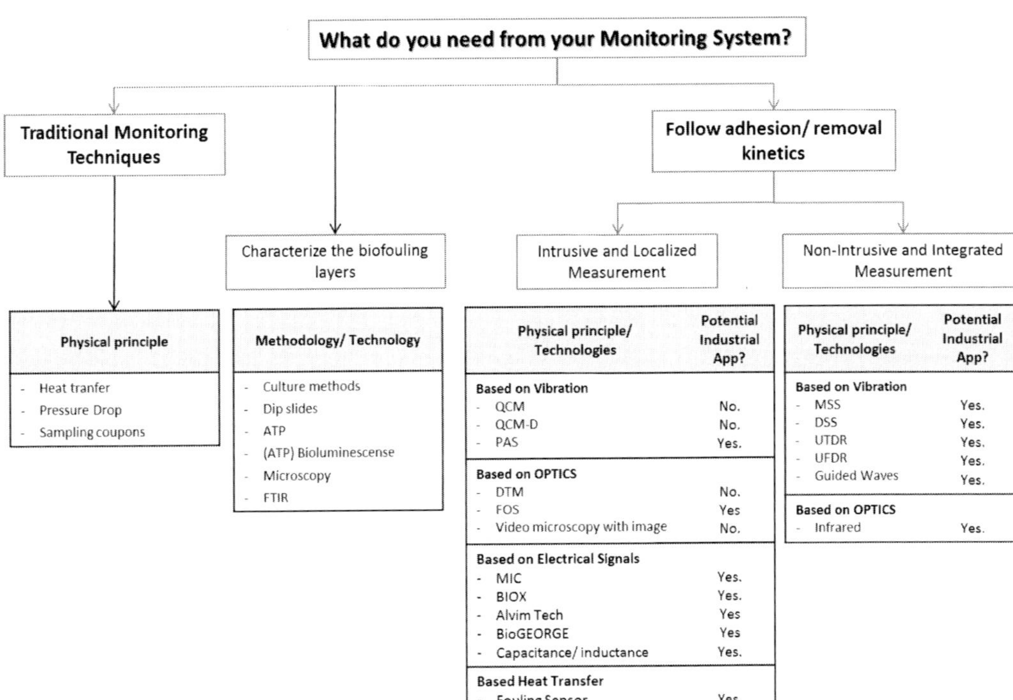

Figure 3. List of the most well-known monitoring techniques, with indication about their industrial application potential. These techniques are divided in four categories: a) the traditional monitoring techniques; b) the ones that can characterize the biofouling layers, c) the ones which provide the kinetics of formation/removal, with intrusive and localized measurements or d) non-intrusive with integrated information. The acronyms listed in this figure will be defined within the next paragraphs.

Traditional Monitoring Techniques – Performance Indicators

In spite of the innumerous technologies developed in recent years, most of the industrial practices for biofouling monitoring still rely on traditional approaches. The mostly used techniques take advantage of the fact that biofouling formation/removal affects the pressure and temperature profiles along the processing plant. Measurements of the heat transfer and pressure drop are quite easy to implement since they need traditional process instrumentation, respectively temperature sensors (e.g. thermocouples) and differential pressure sensors. The *Heat Transfer* rate calculation is based on an enthalpy balance which accounts for the inlet and outlet temperatures and flow rates of the fluids. The *Pressure Drop* represents the pressure difference between the inlet and the outlet of the equipment or the processing line. Different impact is observed during biofouling formation: enthalpy declines over time, while pressure drop increases. These two process indicators, in spite of providing real-time, integrated information about the biofouling layers, are quite insensitive to the formation of the initial layers. Accordingly, the information generated by those techniques only lead to reactive measures.

The heat transfer and pressure drop measurements are widely used in industrial applications but also in laboratorial research, regarding not only biofilm growth but also cleaning quantification [34], [35].

Another technique that might be considered as traditional is the implementation of *Sampling Coupons* that are placed in bypass of the equipment or processing lines, and that are periodically removed for biofouling examination (for example, through some of the methods reported on point 1.4.2). The fluid velocity is typically pre-adjusted in order to match the one of the process.

Characterization of the Biofouling Layers

There are many systems where the characterization of the biofouling layers is mandatory and critical to the performance of the process. The obvious choices to illustrate this are the sanitization procedures within food and beverage system and the medical implants, which if incorrectly implemented can lead to serious health problems (among many others). In spite of the efforts done in the last decades, regarding the improvement of biofilm microbial characterization (accuracy, reduction of time needed for identification, number of microbial species quantified) [36], these techniques are still intrusive and provide retrospective and localized information about the deposit layer.

The decision regarding which microbiological technique should be applied in a given process is complicated due to the variety of available methods that differ significantly in technical complexity, training required, ease of implementation and cost. Typically, a sample of the deposit is collected or swabbed from the process, which is then characterized through the implementation of different techniques. Several parameters can be determined from the biofouling layers, such as chemical composition and microbiological characterization.

The good ratio between the information provided and the cost of the **culture methods** (such as quantification of colony forming units) makes these techniques quite appealing for routine application. The quantification of colony forming unit is often applied in food and beverage processing plants in order to evaluate the sanitization efficacy and the quality of the final product, to control the microbiological population in cooling water systems and the standard procedure for drinking water regulation [37].

There are several methods based on culture methods [38], like the dip-slide analysis (which is a simplified version of the viable plate count methodology) which are employed to evaluate the microbial activity in liquid systems. The dip slide contains a paddle with one or two culture media which is immersed into the liquid or placed in contact with the surface and then set to incubate into its sterilized transparent plastic tube. The output result is established by comparing the microbial colonies observed in the culture medium with the chart to interpret the dip slide results (typically this is provided with the test kit).

On the other hand these are intrusive techniques that provide an off-line, retrospective analysis and in some cases the time delay between the sampling procedure and the output results can be very high. Furthermore, in many cases their implementation requires a skilled specialized operator besides the fact that they are time-consuming [39].

ATP measurement is a well-established technique, based on the quantification of the molecule responsible for energy carrier in all living cells – the adenosine triphosphate (ATP). ATP quantification is based on the conversion of such energy into light, which is typically accomplished by using the luciferin (substract)-luciferase (enzyme) as catalyzer. The overall functioning of the method encompasses the disruption of the cells for ATP release, and the ATP reaction with the luciferin-luciferase to produce light. The amount of light produced in such reaction is measured in a luminometer and directly related to the amount of ATP present in the tested sample. This technique is typically provided as a compact and portable kit which includes the reagents and the luminometer. Atp bioluminescence is widely used to access total surface cleanliness [40] and more specifically to detect different food soils in the food industry [41].

Aiming towards a deeper characterization of the biofouling layers there are several methods that can be employed, mostly based on the measurement of optical signals. Most of these methods cannot be industrially applied because of their complexity, degree of specialization required and high cost.

Measurement of optical signals is based on the fact that matter can emit, absorb or scatter light (in different wavelengths – visible, infrared or ultraviolet). The ATP bioluminescence, previously described is another example of an optical measurement. Given the information that is able to provide it was considered as a microbiological methodology. Several techniques use bioluminescence for biofouling characterization, both by taking advantage of the weak light signals that some organisms can naturally emit or by genetically 'introducing' it into the organisms. Kadurugamuwa et al., [42] monitored the effect of biocides and antibiotics on the removal of biofilms by bioluminescence with microorganisms that where genetically manipulated to emit light. Weinrich et al., [43] developed a method based on the measurement of the bioluminescence of *vibrio harveyi* (a marine organism that naturally emits light signals). This method aims to determine the assimilable organic carbon from that organism, in order to study the impact of the desalination pretreatment on the increase or decrease of biodegradable matter.

FTIR (Fourier Transform Infrared) spectroscopy is one of the most important analytical techniques for biofouling characterization. This technique encompasses the excitation of the sample with infrared radiation, the use of an interferometer and the output - the amount of incident radiation that is absorbed at a specific energy (vibration frequency) is determined by exploiting the Fourier transform mathematical procedure. FTIR techniques have been used to characterize the biofouling layers collected from industrial applications. However, due to its complexity, it is not suited to be used directly within industrial processes. Schmitt & flemming [44] reviews different ftir techniques as a way to characterize biofouling layers and investigation methodologies for biofilms, pointing out how these techniques can be industrially applied in the future. Suci et al., [45] used FTIR techniques combined with attenuated total reflectance (ATR) to study how antimicrobial agents behave within the biofilm. The ATR mode is a variant of the FTIR technique that overcomes the problem associated with the fact that water strongly absorbs the energy in the range where biological molecules have high resonance vibrations, thus improving the sensitivity of the technique. The sample is in contact with a high reflective prism, where the infrared beam is forced to pass through. An evanescence wave is created each time the wave is reflected at the interface sample – ATR prism, thus providing an IR spectrum of the sample. Delille et al., [46], studied the effect of dissolved organic carbon on pseudomonas fluorescens biofilms from tap water using this technique.

Monitoring Adhesion/ Removal Kinetics

The techniques within this group can provide information about the biofouling layers online, real-time and non-destructively. The way to accomplish the monitoring goal is what distinguishes the different techniques. Generically, a relation between the physical parameter behind each monitoring technique and the biofouling layer is established [33]. The scientific/ technological basis behind each monitoring system leads to completely different sensors that measure different physical parameters like vibrations, light, heat transfer, pressure drop, among others.

The physical parameters variation is typically based on a passive response of the biofouling layers [32]. These parameters are expected to change with the presence/amount/thickness/structure/etc of the deposit, and so, related to the biofilm event that is being monitored.

This set of techniques work as early warnings and can be a part of the pro-active approach previously mentioned.

Intrusive and Localized Measurement

One of the most well-known techniques based on acoustic measurements is the **QCM** (Quartz Crystal Microbalance). The QCM is an AT-cut quartz crystal oscillator which vibrates (mechanical change occurs) when a voltage is applied to its electrodes, due to the piezoelectric effect. The electrodes are typically made of gold but can be coated with some specific material. The system vibrates at a very precise frequency making it a very sensitive mass sensor. The adhesion/ adsorption of material to the crystal surface produces a change on the resonance frequency which is related to the amount of deposit. Different relations between the amount of deposit and the resonance frequency of the system can be found according to the characteristics of the deposits/ measuring media: a) Nivens et al., [47], based on the work of Sauerbrey, reported that a decrease of frequency was observed when the amount of attached mass increased, for a thin, rigid, uniform deposit (when the deposit has acoustic characteristics similar to the crystal); b) Kanazawa et al., [48] found that the change in the resonant frequency of the QCM was proportional to the square root of the liquid density-viscosity product, depending not only on the mass attached to the surface but also on its rheological properties. QCM has been widely applied for biofouling monitoring, mainly on laboratorial studies. Nivens et al., [47] reports a detection limit of $3x10^5$ cells/cm^2 regarding the adhesion of *P. cepacia* bacteria in spite of the problems found with temperature and flow velocity. Helle et al., [49] successfully monitored the biofilm adhesion to quartz polished surfaces, but the flow and the nutrients in the bulk seemed to affect the frequency values. More information about the QCM potential, applications and related literature can be found in [50].

The *QCM-D* (Quartz Crystal Microbalance with Dissipation) is a technique which overcomes the QCM drawback associated with rheological effect observed on the QCM technique, and being so, suitable for biofilm formation [51]. The main advantage of the QCM-D compared to the QCM is the simultaneous measurement of two different characteristics of the adherent layer: the resonant frequency shift and the dissipation variation (energy loss). The resonant frequency and the dissipation parameters measure respectively, the mass attached to the sensor surface and the viscoelastic characteristics of such adhered layer. This technique has been widely used on studies of proteins. The studies regarding biofouling characterization are very recent, and the results obtained so far are quite promising. For example, Poitras & Tufenkji (2009) monitored in real-time the presence of the pathogen *E. coli* O157:H7 by immobilizing polyclonal antibodies for that bacteria on the QCM-D surface, with a best sensitivity of $3x10^5$ cells/ ml. Dixon, (2008) reviews most of the work done with QCM-D, regarding the characterization of biological materials (DNA, lipids, proteins, biomolecules, etc) and the interactions between them.

There are many commercially available devices based on the QCM principle, like the QSENSE (www.Q-SENSE.com) or QCMagic (www.elbatech.com). These manufacturers have long lists of application areas from molecular biology to nanoparticle studies and refer the high sensitivity of the probes. Due to it intrinsic characteristics this technique is still more applied in laboratorial studies than in industry.

The *Photoacoustic Spectroscopy* (PAS) technique detects the layers attached to a solid surface through the combination of light and sound propagation analysis. When submitted to a certain electromagnetic radiation (e.g. laser light) the sample absorbs part of it and expands, when energy is converted into heat. This expansion produces an acoustic wave, which propagates along the sample and that can be detected by a piezoelectric transducer [54]. The output information of the system is based on the analysis of the propagation velocities (time delays) of the emitted pulse and of the acoustic response detected by the piezoelectric element. PAS has been used to study biofilms and to detect the formation/ removal of such layers. Schmid et al., [55,56] have made some fundamental work regarding the structure and the stability of biofilms to soluble and colloidal substances. By comparing the output acoustic wave produced with excitation at different wavelengths, Schmid et al., [55] were able to differentiate between biofilm constituents. Furthermore, these authors [56] studied the efficacy of different anti-fouling strategies.

The information provided by PAS is undeniably very interesting for the study of *in situ* biofilms regarding their characterization and composition. However, its sophistication, cost and operational complexity, limits its industrial implementation and targets its use to the research area.

Optical

Measurements of optical/ light signals to assess the formation/ removal of biofouling layers are used by some monitoring techniques, such as the FOS (Fiber Optical Sensor) and the DTM (Differential Turbidity Measurement Device).

The *DTM* is based on the fact that biofouling formation changes turbidity measurement. In this technique, turbidity is measured in two identical 'windows': one that is continuously cleaned and the other where biofouling is being formed. The differential analysis of the two signals provides the information of the biofouling layer formed on the equipment window. Klahre & Flemming [31] used the DTM to monitor biofouling on whitewaters from paper mill processes. The main drawback of this approach is the size of the measuring device tubes, which are very small for processes with high levels of suspended solids and with high potential for biofouling formation. Besides that, after a given thickness of deposit is reached, the turbidity signal gets saturated. The problem regarding the low threshold thickness of deposit that changes the optical signal is also observed for the FOS.

The *FOS* measures the change of intensity of backscattered light caused by the adhesion of the biofouling layers to the tip of a probe (an optical fiber). It was industrially applied to study the biofouling formation in brewery pipes [57].

Blanco et al., [58] reported the use of a *video microscopy* associated *with image analysis* to study the biofilm formation from the recycling paper mill process water and to determine the effect of specific enzymes on the growth of such biofilms. In spite of the successful laboratorial approach, the complexity of the system seems too difficult for industrial implementation in the near future.

Electrochemical Signal

The use of electric signals is the physical principle that probably gathers more outputs regarding monitoring techniques. Videla & Herrera [59] reviews the principles, features and research regarding microbiologically influenced corrosion *(MIC*) and highlights the unique features of microorganisms in such MIC processes, mainly associated to the biofilm formation on the metal surface. Most of the well-known monitoring techniques within this group rely on the fact that the metabolic activity of microorganisms affects the electrochemical characteristics of the electrodes (typically two) that are immersed on the bulk fluid. The more sensitive techniques are based on the cathodic depolarization theory which, in its fundamental approach, considers that sulfate-reducing bacteria reduces the electron acceptor by removing hydrogen at the cathode, coupled with the reduction of sulfate to sulfide, with iron release on the anode. Based on such approach, several monitoring methods have been devised, like the BIOX or the ALVIM tech. The *BIOX* has been used to optimize antifouling treatments on industrial cooling systems, and to monitor biofilm growth and microbial corrosion [60].The BIOX measures the galvanic current between a stainless steel and a sacrificial anode (less noble alloy, such as iron). The output of this measurement is the determination of the ohmic drop on a resistor placed between the two alloys. Biofilm formation causes an increase of the ohmic drop.

A similar method was used by Faimali et al., [61] which found proportionality between the measurements of the density or potential current of the electrochemical signal (between a pair of electrodes) and the amount of bacteria from aquatic biofilms per unit of area. Pavanello et al., [62] reported the use of the *ALVIM* device to monitor biofilm formation and to optimize anti-fouling treatments in industrial environments. It is based on the fact that biofilm growth on active-passive alloys (electrodes) when exposed to aerated waters changes the cathodic depolarization between such electrodes. This paper refers a set of studies that have used cathodic depolarization over the last 20 years to study biofilm growth in sea and fresh water. ALVIM technology is currently under commercialization with the focus on biofilm early detection and in antifouling water treatment monitoring (more information can be found in: www.alvimcleantech.com).

Another monitoring system based on the measurement of electrochemical signals is the *BIoGEORGE*, which has shown to be able to follow biofilm activity as well as to continuously measure the biocide effectiveness. It consists of two electrodes, each composed by a series of electrically isolated metallic discs. One of the electrodes is polarized for a short period creating a current (both applied and generated) between the electrodes [63]. The BioGeorge device was used to optimize chemical dosing on water treatment programs [64]. The variation of these currents from their baseline (this baseline corresponds to the clean state of the probe) is then related to the biofouling amount. More information can be found at: www.structint.com/what-we-do/nuclear-plant-services/products-and-software/biogeorge.

Measurement of capacitance/impedance signals takes advantage of the fact that microorganisms, due to the dielectric properties of the cells plasma, can behave as capacitors when exposed to electric fields. Measurement of electrical capacitance between two parallel electrodes separated by a non-conducting material can give information about the deposits formed/ removed from the electrode surfaces. Maurício et al., [65] found an inverse relation between the thickness of a biofilm deposit and the electric capacitance. Muñoz-Berbel et al., [66] and Dheilly et al., [67] reported the use of electrochemical impedance respectively to study biofilm formation and detachment on drinking water systems and to continuously monitor the initial bacterial adhesion and the biofilm growth of *Pseudomonas aeruginosa* and *Bacillus subtilis*.

The techniques based on the measurement of electrochemical signals are able to detect biofilm formation and removal at early stages. They have been proven to be successful in industrial applications despite of their localized and intrusive measurements.

Heat Transfer

The *Fouling Sensor* is a technique based on the thermal pulse analysis, which was patented by INRA. It combines the measurement of the local conductivity and of the heat transfer changes between the probe and the fluid due to biofouling formation. A periodic electrical pulse is injected and the temperature increases temporarily on the probe. The amount of deposit attached to the monitoring system is determined by measuring the temperature before and after the thermal pulse – an increase on the deposit thickness, decreases the heat transfer between the probe and the fluid. This technique has been tested within the food and beverage industries by Neosens (more information can be found at: www.neo-sens.com).

Non-Intrusive and Integrated Measurement

There are many techniques for biofouling monitoring based on the propagation and measurement of vibration parameters, providing a non-intrusive and integrated output.

Several strategies are found within this group, but they are mainly based on the implementation of transducer(s) on the outer surface of the monitored system.

Vibration

The uses of low-frequency (0-20 kHz) acoustic techniques have also been exploited to quantitative and qualitatively monitor fouling over time. The Mechatronic Surface Sensor (*MSS*) is based on the resonance frequency of the system. It uses two transducers, which are placed on the outer surface of the monitoring tubing, one piezoelectric actuator that excites the surface and makes it vibrate, and a sensor (placed a few centimeters apart) that captures the systems response to the vibration induced by the actuator. The amplitude (after Fast Fourier Transform application) and the damping factor (or a related characteristic) of the system vibration are respectively related to the amount and structure of the deposit. The MSS technique has been used in laboratory settings to follow biofilm formation and to distinguish between these deposits and silicate fouling [68]. Furthermore, Pereira et al., [69] found that the MSS was able to provide information about the build-up and removal of different milk components (minerals and proteins) and to distinguish them.

More recently, the Surface Sensor technology was industrially redesigned in order to adjust the technique to a new geometry and to increase its sensitivity to monitor fouling and cleaning phenomenon in cooling water systems, the so called Diveil Surface Sensor (*DSS*). The Surface Sensor technology is currently under commercialization by Enkrott Química (www.enkrottquimica.com). This new approach relies on the assessment of fouling build-up/ removal on a tubular geometry and includes an external heating source to provide a heat flux between the bulk fluid and the surface wall (similarly to what occurs on a heat exchanger). The goal is to mimic the hydrodynamic and the thermal conditions of the most problematic zone of the operating plant regarding biofouling formation. This configuration was also confirmed by the Faculty of Engineering of the University of Porto, in order to extract, from the vibrational response of the system, not only the amount but also the structure of the monitored deposit. The preliminary results seem to be promising and suggest that from the relation between the Amplitude and the ratio between the Amplitude and the Damping Factor, it is possible to verify structural changes within the attached layers [70], as well as variation in the amount of such deposits.

Merheb et al., [71] used an acoustic technique also based on low range-frequencies to detect fouling in plate heat exchangers. Three sensors were placed on the lower area of the heat exchanger plate (in different zones). These sensing elements capture the vibration produced by the exciter which is attached to the upper section of the plate. The integration of the output waves over time and the arrival time of such wave were determined for each of the sensors. These parameters were found to decrease with the fouling build-up.

Ultrasound techniques are used for Non-Destructive Acoustic Tests which are successfully implemented in a variety of areas like medicine, food or aerospace industry [72]. The fundamentals of ultrasound techniques are quite similar when applied to biofouling monitoring and they can assume different configurations depending on the approach implemented. They are mainly based on the measurement of the echo (pulse echo technique) or on the vibration along the system/ tubing section (transmission technique) which are then related to the biofouling amount. Withers [72] reviews the principles behind these techniques, and relates the output signal of a pulse echo methodology to biofouling severity (rather than with biofouling amount). The pulse echo uses only one transducer attached to the outer surface of the monitored piping, which is able to emit an ultrasound pulse and capture the echoed signals. The echo occurs each time the wave crosses different media/boundaries, such as wall/deposit/bulk fluid. In theory, the time the echo takes to return to the transducer can be related to the thickness of the deposit, however, the high impact of the biofilm on the output of the signal precludes the implementation of these approaches. This issue was overcame with the use of a different methodology, based on the determination of the damping or on the attenuation of the reflected signal, parameters that are highly affected by the presence of the biofilm and which can provide information about the presence of a deposit.

The pulse-echo technique is also the basis for the Ultrasonic Time Domain Reflectometry (*UTDR*), which was exploited by several authors for biofouling quantification, mainly in membrane technology. Measurements of the amplitude and the arrival time of the reflected wave are related to the amount of deposit. Mairal et al., [73] and Li et al., [74] applied these fundamentals to detect the initials layers of calcium sulfate deposits on reverse osmosis systems and of protein fouling on ultrafiltration membranes.

The introduction of frequency domain analysis on the output signal (Ultrasonic Frequency-Domain Reflectometry - *UFDR*), enables the study of more mature biofouling deposits (thicker deposits). Kujundzic et al., (2007) used the UFDR technique to monitor mature biofilms and their removal from polymeric matrixes. More recently, Kujundzic et al., [76] applied a different statistical methodology to the collected ultrasonic spectrum of bovine serum albumin, in order to improve the sensitivity of this technique and were able to detect a concentration of 14 $\mu g/cm^2$.

These ultrasound techniques were able to detect very low concentrations of foulant and to follow the fouling formation in real-time. The main question about these acoustic techniques is that the experiments were performed on a laboratorial scale, and no information about their applicability to industrial devices (mainly to membrane systems) is known. The higher the number of boundary conditions encountered during wave travelling, the lower is the amount of reflected wave that reaches the transducer. In fact, the amount of signal that reaches the transducer is one of the main concerns in industrial implementation of ultrasound techniques, because of the higher number of existing boundary conditions.

Lohr & Rose [77] and Hay & Rose [78], exploited the use the interference between longitudinal and shear waves (the so called *Guided Waves*) within pipe tubing. An acoustic pulse is generated and transmitted along the pipe system (which includes the wall, the bulk fluid and the deposit – when it exists) and the output response (frequency and amplitude) is measured over time. Fouling was simulated as a viscous film on the tubing, and the guided waves were able to provide indication about the bulk viscosity. These authors tested the acoustic guided waves within the transmission and the pulse-echo mode. The main configuration differences between them are that in the first mode two transducers are used (one as an exciter and another as a sensor) and in the second, only one transducer is placed on the pipe surface and acts as exciter and sensor.

Optical

Optical systems can use *Infrared Radiation* that is passed through the monitoring system, perpendicularly to the water flow, from the infrared emitter source to the infrared receiver. The difference between the emitted and received radiations equals the amount absorbed by all the components of the system, which includes: emitter, both sides of a transparent glass wall, the biofilm (if it exists) on both sides of the tubing, the flowing water and the glass cell of the infrared sensor. In order to eliminate the effect of the radiation absorption of the monitoring system a baseline measurement is performed when there is no biofilm but when water is flowing. The Infrared Monitor has been widely used in laboratorial scale to determine the effect of several physico-chemical parameters like pH, flow velocity or biocide on the growth of *Pseudomonas fluorescens* biofilms [79]. According to these authors, the Infrared Monitor has been studied both under laboratory, pilot plant and full-scale approaches in side-stream. When using the Infrared Monitor there are some points that should be taken into consideration: a) the system works in bypass, b) the material of the system is glass which disables the opportunity to match with the most commonly used construction materials; c) the measurement is localized, although Tinham & Bott [79] reported the possibility to perform the integration of the biofilm signal manually.

Conclusion

Biofouling is a ubiquitous problem for aqueous systems and environments! It is accepted that the focus should be on the minimization of biofouling phenomenon rather than on its eradication, which does not seem feasible so far.

The present chapter emphasized the role of biofouling monitoring strategies integrated into an anti-fouling approach, where counter-measures are driven on an informed based about the attached layer from earlier stages of formation. An on-line, real-time, non-destructive information about the fouling severity within the process systems, can lead to the implementation of timely, pro-active and adjusted counter-measures and to the assessment (also in real-time) about the efficacy of those actions. The ultimate goal of a monitoring strategy is to provide inputs to some of the critical questions regarding industrial biofouling namely:

- when and how should the system be cleaned?
- when is the cleaning end-point achieved?
- how efficient is the preventive and/or anti-fouling approach?

The difficulty to find a monitoring technique which fulfills all of the ideal monitoring requirements together with the fact that each industrial process has its specificities, suggests that each process will require a specific combination of tools. The main questions that should be asked to and/ or about a monitoring technique have been discussed along this text. Answering these questions will help to understand if a monitoring technique is suited for a given process or, on the other hand, what type of monitoring technique does a given process requires.

Finally, some of the most well-known and important monitoring techniques have been reviewed. Screening of the techniques was based on the type of information that they are able to provide, in order to enable the reader to easily find the best technique(s) that fits to its process.

The gaps between what is done in laboratory and what is industrially implemented are still evident. In some cases, this difficulty seems to be related with the technical complexity of the monitoring technique and, on others cases, more information about the industrial tests and behavior of the monitoring techniques. Nevertheless, the demand for monitoring techniques which can be industrially applied is increasing.

References

[1] Characklis, W.G., & Cooksey, K.E., (1983). Biofilms and microbial fouling. *Advances in Applied Microbiology, 29*, p. 93–138.

[2] Hall-Stoodley, L., & Stoodley, P., (2002). Developmental regulation of microbial biofilms. *Current opinion in biotechnology*, *13*(*3*); 228–33.

[3] Chmielewski, R.A.N., & Frank, J.F., (2003). Biofilm Formation and Control in Food Processing Facilities. *Comprehensive Reviews in Food Science and Food Safety*, *2*; 22–32.

[4] Bott, T.R., (2011). *Industrial Biofouling: ocurrence and control*. Elsevier.

[5] Tunney, M.M., Dunne, N., Einarsson, G., Mcdowell, A., Kerr, A., & Patrick, S., (2007). Biofilm Formation by Bacteria Isolated from Retrieved Failed Prosthetic Hip Implants in an In Vitro Model of Hip Arthroplasty Antibiotic Prophylaxis. *Journal of Orthopaedic Research*, *25*(*1*); 2–10.

[6] McLaughlin-Borlace, L., Stapleton, F., Matheson, M., & Dart, J.K., (1998). Bacterial biofilm on contact lenses and lens storage cases in wearers with microbial keratitis. *Journal of applied microbiology*, *84*(*5*); 827–38.

[7] Epstein, N., (1981). Thinking about Heat Transfer Fouling: A 5 × 5 Matrix. *Heat Transfer Engineering*, *4*(*1*); 43–56.

[8] Melo, L.F., & Bott, T.R., (1997). Biofouling in water systems. *Experimental Thermal and Fluid Science*, *14*(*4*); 375–81.

[9] Müller-Steinhagen, H., Malayeri, M.R., & Watkinson, A.P., (2005). Fouling of Heat Exchangers-New Approaches to Solve an Old Problem. *Heat Transfer Engineering*, *26*(*1*); 1–4.

[10] Hasting, A.P.M., (2005). Improving the monitoring of fouling, cleaning and disinfection in closed process plant. In: Lelieveld HLM, Mostert MA, & Holah J, editors. *Handbook of hygiene control in the food industry*, p. 572–87.

[11] Verran, J., (2002). BIOFOULING IN FOOD PROCESSING Biofilm or biotransfer potential? *Trans IChemE*, *80*(*Part C*); 292–8.

[12] Changani, S.D., Belmar-Beiny, M.T., & Fryer, P.J., (1997). Engineering and chemical factors associated with fouling and cleaning in milk processing. *Experimental Thermal and Fluid Science*, *14*(*4*); 392–406.

[13] Fryer, P.J., Christian, G.K., & Liu, W., (2006). How hygiene happens: physics and chemistry of cleaning. *International Journal of Dairy Technology*, *59*(*2*); 76–84.

[14] Al-Juboori, R. a., & Yusaf, T., (2012). Biofouling in RO system: Mechanisms, monitoring and controlling. *Desalination*, *302*; 1–23.

[15] Flemming, H.-C., (2011). Microbial Biofouling: Unsolved Problems, Insufficient Approaches, and Possible Solutions. In: Flemming H-C, Wingender J, & Szewzyk U, editors. *Biofilm Highlights*, Berlin, Heidelberg; p. 81–110.

[16] Simões, M., Simões, L.C., & Vieira, M.J., (2010). A review of current and emergent biofilm control strategies. *LWT - Food Science and Technology*, *43*; 573–83.

[17] Watkinson, A.P., (2007). Deposition from Crude Oils in Heat Exchangers. *Heat Transfer Engineering*, *28*(*3*); 177–84.

[18] Grant, D.M., & Bott, T.R., (2005). Biocide Dosing Strategies for Biofilm Control. *Heat Transfer Engineering*, *26*(*1*); 44–50.

[19] Simões, M., Simões, L.C., Machado, I., Pereira, M.O., & Vieira, M.J., (2006). Control of Flow-Generated Biofilms with Surfactants. *Food and Bioproducts Processing*, *84*(*C4*); 338–45.

[20] Vieira, M.J., & Melo, L.F., (1999). Intrinsic kinetics of biofilms formed under turbulent flow and low substrate concentrations. *Bioprocess Engineering*, *20*(*4*); 369.

[21] Pereira, M.O., Morin, P., Vieira, M.J., & Melo, L.F., (2002). A versatile reactor for continuous monitoring of biofilm properties in laboratory and industrial conditions. *Letters in applied microbiology*, *34*(*1*); 22–6.

[22] Cloete, T.E., (2003). Biofouling control in industrial water systems: What we know and what we need to know. *Materials and Corrosion*, *54*(*7*); 520–6.

[23] Guiamet, P.S., & Gómez de Saravia, S.G., (2005). Laboratory studies of biocorrosion control using traditional and environmentally friendly biocides: an overview. *Latin American applied research*, *35*(*4*); 295–300.

[24] Ferreira, C., Rosmaninho, R., Simoes, M., Pereira, M.C., Bastos, M.M.S.M., Nunes, O.C., et al., (2010). Biofouling control using microparticles carrying a biocide. *Biofouling*, *26*(*2*); 205–12.

[25] Rosmaninho, R., & Melo, L.F., (2006). Calcium phosphate deposition from simulated milk ultrafiltrate on different stainless steel-based surfaces. *International Dairy Journal*, *16*(*1*); 81–7.

[26] Bruellhoff, K., Fiedler, J., Möller, M., Groll, J., & Brenner, R.E., (2010). Surface coating strategies to prevent biofilm formation on implant surfaces. *The International journal of artificial organs*, DWI e.V. and Institute of Technical and Macromolecular Chemistry, RWTH Aachen University, Aachen, Germany.; *33*(*9*); 646–53.

[27] Yoshinari, M., Kato, T., Matsuzaka, K., Hayakawa, T., & Shiba, K., (2010). Prevention of biofilm formation on titanium surfaces modified with conjugated molecules comprised of antimicrobial and titanium-binding peptides. *Biofouling*, *26*(*1*); 103–10.

[28] Su, X.J., Zhao, Q., Wang, S., Noormofidi, S.Z.N., Kreutzwiesner, E., & Slugovc, C., (2011). Antimicrobial efficiency of metal-polymer nanocomposite coatings. In: Malayeri MR, Muller-Steinhagen H, & Watkinson AP, editors. *Proceedings of International Conference on Heat Exchanger Fouling and Cleaning*, p. 259–63.

[29] Wallhäußer, E., Hussein, M.A., & Becker, T., (2012). Detection methods of fouling in heat exchangers in the food industry. *Food Control*, *27*(*1*); 1–10.

[30] Flemming, H.-C., (2003). Role and levels of real-time monitoring for successful anti-fouling strategies: an overview. *Water Science and Technology*, *47*(*5*); 1–5.

[31] Klahre, J., & Flemming, H.-C., (2000). Monitoring of biofouling in papermill process waters. *Water Research*, *34*(*14*); 3657–65.

[32] Janknecht, P., & Melo, L.F., (2004). Online biofilm monitoring. (*2002*); 269–83.

[33] Pereira, A., & Melo, L.F., (2009). Monitoring of biofilms in the food and beverage industries. In: Fratamico PM, Annous B, & Guenther NW, editors. *Biofilms in the food and beverage industries*, p. 131–51.

[34] Vieira, M.J., Melo, L.F., & Pinheiro, M.M., (1993). Biofilm formation: Hydrodynamic effects on internal diffusion and structure. *Biofouling*, *7*(*1*); 67–80.

[35] Chew, Y., Paterson, W., & Wilson, D., (2007). Fluid dynamic gauging: A new tool to study deposition on porous surfaces. *Journal of Membrane Science*, *296*(*1-2*); 29–41.

[36] Warner, J.C., Rothwell, S.D., & Keevil, C.W., (2008). Use of episcopic differential interference contrast microscopy to identify bacterial biofilms on salad leaves and track colonization by Salmonella Thompson. *Environmental Microbiology*, *10*(*4*); 918–25.

[37] Flemming, H., (1999). Methods for Investigation of Biofilms.

[38] Nivens, D.E., Co, B.M., & Franklin, M.J., (2009). Sampling and quantification of biofilms in food processing and other environments. In: Fratamico PM, Annous B, & Guenther NW, editors. *Biofilms in the food and beverage industries*, p. 539 – 568.

[39] Simões, M., Pereira, M.O., & Vieira, M.J., (2005). Validation of respirometry as a short-term method to assess the efficacy of biocides. *Biofouling*, *21*(*1*); 9–17.

[40] Jasson, V., Jacxsens, L., Luning, P., Rajkovic, A., & Uyttendaele, M., (2010). Alternative microbial methods: An overview and selection criteria. *Food microbiology*, *27*(*6*); 710–30.

[41] Whitehead, K.A., Smith, L.A., & Verran, J., (2008). The detection of food soils and cells on stainless steel using industrial methods: UV illumination and ATP bioluminescence. *International journal of food microbiology*, *127*(*1-2*); 121–8.

[42] Kadurugamuwa, J.L., Sin, L. V, Yu, J., Francis, K.P., Purchio, T.F., & Contag, P.R., (2004). Noninvasive Optical Imaging Method To Evaluate Postantibiotic Effects on Biofilm Infection In Vivo. *Antimicrobial Agents Chemothery*, *48*(*6*); 2283–7.

[43] Weinrich, L.A., Schneider, O.D., & LeChevallier, M.W., (2011). Bioluminescence-based method for measuring assimilable organic carbon in pretreatment water for reverse osmosis membrane desalination. *Applied and environmental microbiology*, *77*(*3*); 1148–50.

[44] Schmitt, J., & Flemming, H.-C., (1998). FTIR-spectroscopy in microbial and material analysis. *International Biodeterioration & Biodegradation*, *41*(*1*); 1–11.

[45] Suci, P.A., Vrany, J., & Mittelman, M., (1998). Investigation of interactions between antimicrobial agents and bacterial biofilms using attenuated total reflection Fourier transform infrared spectroscopy. *Biomaterials*, *19*; 327–39.

[46] Delille, A., Quilès, F., & Humbert, F., (2007). In Situ Monitoring of the Nascent Pseudomonas fluorescens Biofilm Response to Variations in the Dissolved Organic Carbon Level in Low-Nutrient Water by Attenuated Total Reflectance-Fourier Transform Infrared Spectroscopy. *Applied and Environmental Microbiology*, *73*(*18*); 5782–8.

[47] Nivens, D.E., Chambers, J.Q., Anderson, T.R., & White, D.C.L., (1993). Long-term online monitoring of microbial biofilms using quartz crystal microbalance. *Analytical Chemistry*, *65*; 65–9.

[48] Kanazawa, K.K., & Gordon, J.G., (1985). Frequency of a quartz microbalance in contact with liquid. *Analytical Chemistry*, *57*(*8*); 1770–1.

[49] Helle, H., Vuoriranta, P., Valimaki, H., Lekkala, J., & Aaltonen, V., (2000). Monitoring of biofilm growth with thickness-shear mode quartz resonators in different flow and nutrition conditions. *Sensor and Actuators B*, *71*; 47–54.

[50] Becker, B., & Cooper, M. a, (2011). A survey of the 2006-2009 quartz crystal microbalance biosensor literature. *Journal of molecular recognition*, *24*(*5*); 754–87.

[51] Rudh, M., (2003). Quartz crystal microbalance with dissipation monitoring: a new tool for studying biofilm in real time. In: Lens P, Moran A, Mahony T, Stoodly P, & O'Flaherty V, editors. *Biofilms in Medicine, Industry and Environmental Biotechnology*, p. 450–60.

[52] Poitras, C., & Tufenkji, N., (2009). A QCM-D-based biosensor for E. coli O157:H7 highlighting the relevance of the dissipation slope as a transduction signal. *Biosensors & bioelectronics*, *24*; 2137–42.

[53] Dixon, M.C., (2008). Quartz crystal microbalance with dissipation monitoring: enabling real-time characterization of biological materials and their interactions. *Journal of biomolecular techniques*, *19*(*3*); 151–8.

[54] Bageshwar, D. V, Pawar, A.S., Khanvilkar, V. V, & Kadam, V.J., (2010). Photoacoustic Spectroscopy and Its Applications – A Tutorial Review. *Euroasian Journal Analytical Chemistry*, *5*(*2*); 187–203.

[55] Schmid, T., Helmbrecht, C., Panne, U., Haisch, C., & Niessner, R., (2003). Process analysis of biofilms by photoacoustic spectroscopy. *Analytical and Bioanalytical Chemistry*, *375*; 1124–9.

[56] Schmid, T., Panne, U., Adams, J., & Niessner, R., (2004). Investigation of biocide efficacy by photoacousticbiofilm monitoring. *Water Research*, *38*; 1189–96.

[57] Tamachkiarow, A., & Flemming, H.-C., (2003). On-line monitoring of biofilm formation in a brewery water pipeline system with a fibre optical device. *Water science and technology : a journal of the International Association on Water Pollution Research*, *47*(*5*); 19–24.

[58] Blanco, A., Torres, E., Fuente, E., & Negro, C., (2011). New Tool To Monitor Biofilm Growth in Industrial Process Waters. *Industrial & Engineering Chemistry Research*, *50*(*9*); 5766–73.

[59] Videla, H.A., & Herrera, L.K., (2005). Microbiologically influenced corrosion: looking to the future. *International Microbiology*, *8*; 169–80.

[60] Mollica, A., & Cristiani, P., (2003). On-line biofilm monitoring by "BIOX" electrochemical probe. *Water science and technology : a journal of the International Association on Water Pollution Research*, CNR-ICMM, Via De Marini 6, 16149 Genova, Italy.; *47*(*5*); 45–9.

[61] Faimali, M., Chelossi, E., Pavanello, G., Benedetti, A., Vandecandelaere, I., De Vos, P., et al., (2010). Electrochemical activity and bacterial diversity of natural marine biofilm in laboratory closed-systems. *Bioelectrochemistry (Amsterdam, Netherlands)*, *78*(*1*); 30–8.

[62] Pavanello, G., Faimali, M., Pittore, M., Mollica, A., Mollica, A., & Mollica, A., (2011). Exploiting a new electrochemical sensor for biofilm monitoring and water treatment optimization. *Water research*, *45*(*4*); 1651–8.

[63] Licina, G.J., Nekoksa, G., & Howard, R.., (1994). An alectrochemical method for on-line monitoring of biofilm activity in cooling water using BioGeorge. In: Kearns JR, & Little BJ, editors. *Microbiology Influenced Corrosion Testing*, p. 118–27.

[64] Bruijs, M.C.M., Venhuis, L.P., Jenner, H.A., Daniels, D.G., Licina, G.J., & Technology, M.R., (2001). Biocide optimisation using an on-line biofilm monitor. *Journal of Power PLant Chemistry*, *3*(*7*); 400–5.

[65] Maurício, R., Dias, C.J., & Santana, F., (2006). Monitoring biofilm thickness using a non-destructive, on-line, electrical capacitance technique. *Environmental monitoring and assessment, 119(1-3)*; 599–607.

[66] Muñoz-Berbel, X., García-Aljaro, C., & Muñoz, F.J., (2008). Impedimetric approach for monitoring the formation of biofilms on metallic surfaces and the subsequent application to the detection of bacteriophages. *Electrochimica Acta, 53(19)*; 5739–44.

[67] Dheilly, A., Linossier, I., Darchen, A., Hadjiev, D., Corbel, C., & Alonso, V., (2008). Monitoring of microbial adhesion and biofilm growth using electrochemical impedancemetry. *Applied microbiology and biotechnology, 79*; 157–64.

[68] Pereira, A., Mendes, J., & Melo, L.F., (2008). Using nanovibrations to monitor biofouling. *Biotechnology and bioengineering, 99(6)*; 1407–15.

[69] Pereira, A., Rosmaninho, R., Mendes, J., & Melo, L.F., (2006). Monitoring deposit build-up using a novel Mechatronic Surface Sensor (MSS).pdf. *Trans IChemE, Part C, Food and Bioproducts processing, 84(C4)*; 1–5.

[70] Teixeira, R., Pereira, A., Mendes, J., & Melo, L.F.,. Identifying the nature of fouling layers by online monitoring of the propagation of vibrations along the deposition surface. *Heat Transfer Engineering, (submitted.*

[71] Merheb, B., Nassar, G., Nongaillard, B., Delaplace, G., & Leuliet, J.C., (2007). Design and performance of a low-frequency non-intrusive acoustic technique for monitoring fouling in plate heat exchangers. *Journal of Food Engineering, 82*; 512–27.

[72] Withers, P.M., (1996). Ultrasonic, acoustic and optical techniques for the non-invasive detection of fouling in food processing equipment. *Trends in Food Science & Technology, 7*; 293–8.

[73] Mairal, A.P., Greenberg, A.R., & Krantz, W.B., (2000). Investigation of membrane fouling and cleaning using ultrasonic time-domain reflectometry. *Desalination, 130*; 45–60.

[74] Li, J.-X., Sanderson, R.D., & Chai, G.Y., (2006). A focused ultrasonic sensor for in situ detection of protein fouling on tubular ultrafiltration membranes. *Sensors and Actuators B: Chemical, 114(1)*; 182–91.

[75] Kujundzic, E., Fonseca, A.C., Evans, E.A., Peterson, M., Greenberg, A.R., & Hernandez, M., (2007). Ultrasonic monitoring of early-stage biofilm growth on polymeric surfaces. *Journal of microbiological methods, 68(3)*; 458–67.

[76] Kujundzic, E., Greenberg, A.R., Fong, R., & Hernandez, M., (2011). Monitoring Protein Fouling on Polymeric Membranes Using Ultrasonic Frequency-Domain Reflectometry. *Membranes, 1*; 195–216.

[77] Lohr, K.R., & Rose, J.L., (2003). Ultrasonic guided wave and acoustic impact methods for pipe fouling detection. *Journal of Food Engineering, 56(4)*; 315–24.

[78] Hay, T.R., & Rose, J.L., (2003). Fouling detection in the food industry using ultrasonic guided waves. *Food Control, 14(7)*; 481–8.

[79] Tinham, P., & Bott, T.R., (2003). Biofouling assessment using an infrared monitor. *Water science and technology: a journal of the International Association on Water Pollution Research*, School of Engineering, University of Birmingham, Edgbaston, Birmingham B15 2TT, UK.; *47(5)*; 39–43.

In: Biofilms in Bioengineering
Editors: Manuel Simões and Filipe Mergulhão
ISBN: 978-1-62948-161-6

Chapter V

Biofilm Formation, Mathematical Modeling and Quorum Sensing

Qi Wang[1], Xiangrong Xu[2]*, Tianyu Zhang[3] and Hao Xu[2]
[1]University of South Carolina, Columbia, SC, US
[2]Anhui University of Technology, Ma'anshan, Anhui, China;
[3]Montana State University, Bozeman, MT, US

Abstract

This chapter gives a brief overview of the progress made in the mathematical modeling of biofilms over the last 30 years. Biofilms constitute a spectrum of dynamical microorganisms, whose interaction with the surrounding environment and thereby induced dynamics dictates the complex properties of the living organism. Modeling of biofilms began with a low dimensional continuum description first based on kinematics and translational diffusions; later, more sophisticated microscopic dynamical mechanisms are introduced leading to the anomalous diffusion and dissipation encountered by various components in biofilms. Recently, biofilm and bulk fluid (or solvent) coupling has been investigated using discrete-continuum, multi-fluid and single fluid multicomponent models to treat the entire biofilm-bulk-fluid system either as a system consisting of various components whose dynamics exhibits different time scales or as a whole. We classify the models into roughly four classes: low-dimensional continuum models, diffusion limited aggregation models, continuum-discrete models, and fully coupled biofilm-fluid models. We will address some hybrid models that combine the ideas from the above categories and new protocols combining the existing computational tools for cell dynamics coupled with the discrete-continuum biofilm model. In this chapter, we also briefly describe the progress of bacterial quorum sensing in biofilms. Bacterial quorum sensing involves in the regulation of a variety of biological functions including the causative agents of human, animal and plant Human, animal and plant pathogen virulence. The discovery of the quorum sensing process derived from the bioluminescence phenomenon of *vibrio fischeri*, afterward many scholars have studied

* Email: xuxr88@yahoo.com.

different bacterial quorum sensing. We will introduce several bacterial quorum sensing modes and their regulatory mechanism.

Keywords: Biofilms, Mathematical models, EPS, Quorum Sensing

Introduction

Biofilms exist everywhere in natural environments, industrial settings and human-made materials. Biofilm formation begins when free-living, planktonic bacteria encounter a surface. Bacteria can adhere to solid surfaces and produce a slimy, slippery coat with structured features. The development of biofilms is a multistage process. First, free floating planktonic bacterial cells land on a damp surface and later get attached to it. After a sufficiently high cell density is reached, cells begin excreting extracellular polymeric substances (EPS) and forming a EPS matrix anchored on the surface. The bacterial population inside the matrix grows as well by consuming nutritious substrates supplied by the surrounding environment. Consequently, the biofilm begins to grow. Biofilm growth is a dynamical process whose structure is largely influenced by environmental conditions including the substrate diffusion limitation and the shear stress of the bulk fluid. Biofilms generally consist of multispecies of bacteria, as well as fungi, algae, yeasts, protozoa, other microorganisms, debris, corrosion products, EPS and solvent (water). All these components are held together by the EPS to form a complex heterogeneous dynamical bio-organism.

Biofilm related problems cost the US literally billions of dollars every year in energy losses, equipment damage, product contamination and medical infections. But biofilms also offer a huge potential for bio-remediating hazardous waste sites, bio-filtering municipal and industrial water and waste water, forming bio-barriers to protect soil and ground water from contamination, as well as heap leaching in the mining industry. It is important and apparently challenging to unravel the complex dynamical properties of the biofilms, prevent damage caused by unwanted biofilm growth and to utilize biofilms for good causes.

Biofilm formation is a complicated dynamical process governed by various physical, chemical principles and biological protocols. Biofilm development depends heavily on the environmental conditions and types and properties of the bacteria inside the biofilm. Bacteria in biofilms are very resistant to antimicrobial agents due to a number of protection features among which the limited penetration of antimicrobial agents. Mathematical modeling of biofilms is crucial to attain a broader and deeper understanding of this complex microorganism. The mathematical models can not only be used to verify experimental findings, but also to make qualitative and quantitative predictions that might well serve as guidelines for experimental design. There has been extensive research on the mathematical modeling of biofilms in the last three decades and a very rich literature has been accumulated. Some of the models were proposed and applied successfully in explaining and predicting the biofilm's behavior in various conditions. Chronically, the development of the biofilm models has undergone an evolution from simple to complex, from one spatial dimensional models to multidimensional models, from single species to multispecies models, from steady-state to dynamical models, from pure growth models to models including biomass growth and biofilm–fluid interaction, from models describing a particular process to models incorporating many different processes occurring simultaneously, from qualitative to quantitative models, and from single scale models to multi-scale ones, etc.

Quorum sensing (QS) is the signal communication among the bacterial population which regulates the transcription of multiple target genes by bacterial cell density. Biofilm formation is a complex and dynamic process, regulated at different stages via diverse mechanisms. The current study found that quorum sensing regulates EPS production, biofilm formation and differentiation. The bacteria use signaling molecules known as autoinducer(AI) as the "language" of interaction. They continuously secrete signaling molecules into the extracellular during the growth process, and detect the concentration so that perceive changes in the population density; when the density reaches a certain threshold, it will initiate the expression of certain genes. The discovery of the quorum sensing process derived from the bioluminescence phenomenon of *vibrio fischeri*. Many bacteria behavior belong to quorum sensing, such as *Pseudomonas aeruginosa* virulence genes expression and *P. aureofaciens* regulation of antibiotic production, etc. Bacterial quorum sensing is widespread, but there are differences between different populations of bacteria. Here, we will introduce several bacterial quorum sensing modes and their regulatory mechanism.

In this chapter, we will give a brief review of some biofilm models developed in the past 30 years, most of which have been successfully applied to studying the biofilms' growth. We classify the models based on their dimensionality, the way in which diffusion is treated, and the complexity in terms of the incorporation of the physics, chemistry and biological effects into four main categories: one dimensional continuum models, diffusion limited aggregate models, continuum-discrete diffusion models, and biofilm–fluid coupled models. There have been several excellent review articles about biofilms from the biological, physiological, mechanical and mathematical perspectives by a group of distinguished researchers in the field [1-4]. We hope to provide a review from a different view, giving a chronological perspective on the development of various models and their mathematical structure evolution.

Biofilm Formation

Biofilm formation begins when free-living, planktonic bacteria encounter a surface. Bacteria can adhere to solid surfaces and produce a slimy, slippery coat with structured features. Usually, this biofilm contains only 15% cells and 85% 'slime.' The slimy coat protects the bacteria from natural defenses such as inflammatory cells and antibodies as well as antimicrobial treatments and therefore biofilms continue to pose significant challenges in many areas, including the medical device field. Once contained in this protective coating and adherent to a surface, bacteria can multiply rapidly. Sites for biofilm formation include all kinds of surfaces: natural materials above and below ground, metals, plastics, medical implant materials—even plant and body tissue. Wherever you find a combination of moisture, nutrients and a surface, you are likely to find biofilm. The biofilm is not simply a thick layer of extracellular material. There are channels and pathways that provide access to the interior portions of the biofilm. Once contained in his protective coating and adherent to a surface, bacteria can multiply rapidly.

These direct observations of living cells show clearly that bacterial cells can sense their proximity to a surface, and that this proximity signals elaborate species-specific behavior patterns. After a planktonic bacterial cell has sensed and 'explored' a surface, the cell may commit to the active process of adhesion and biofilm formation (Figure 1).

1) Free-floating, or *planktonic*, bacteria encounter a submerged surface and within minutes can become attached. They begin to produce slimy extracellular polymeric substances (EPS) and to colonize the surface.
2) EPS production allows the emerging biofilm community to develop a complex, three-dimensional structure that is influenced by a variety of environmental factors. Biofilm communities can develop within hours.
3) Biofilms can propagate through detachment of small or large clumps of cells, or by a type of "seeding dispersal" that releases individual cells. Either type of detachment allows bacteria to attach to a surface or to a biofilm downstream of the original community.

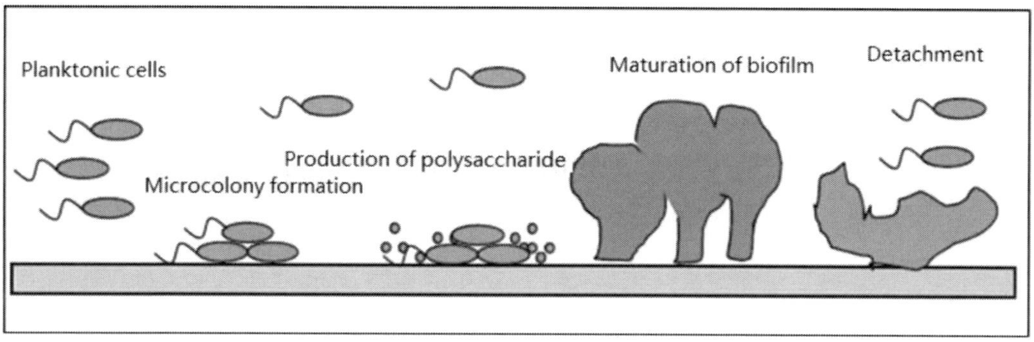

Figure 1. Biofilm formation, growth and maturation.

Mathematical Models of Biofilms

One-Dimensional Continuum Models

Mathematical models of biofilms have been used in the past three decades as tools to simulate the behavior of bacterial biofilms. The initial models described biofilms as uniform steady-state films containing a single type of organism, governed exclusively by one-dimensional (1D) mass transport and biochemical transformations. Later, stratified dynamic models able to represent multisubstrate-multispecies biofilms [9] were developed. Although these 1D models were advanced descriptions of multispecies interactions within the biofilm, they were not able to represent the characteristic structural heterogeneity that has been later elucidated through experimental observations. The new biofilm models provide sophisticated two- and three-dimensional (3D) descriptions of the baterial biofilm, and incorporate also mass transport and transformations [27], hydrodynamics [46] and population dynamics [31]. This evolution in model complexity has paralleled the advances in computational tools. Early efforts on the mathematical modeling of biofilms can be traced back to the 1980s represented by the work of [5-10]. These authors studied biofilms using continuum models primarily in one space dimension. These studies were centered mostly on the steady-state biofilm growth dynamics including the biofilms' thickness and spatial distribution of microbial species and substrate concentration. We illustrate the idea and basic mathematical structures in these models using the multispecies biofilm model introduced by Wanner and Gujer [9] as an example. The model is based on a continuum description of the biofilms' material and conservation principles; it predicts the evolution of the biofilms' thickness, dynamics and spatial distribution of microbial species and substrates in the biofilm. It also allows biomass detachment due to the shear stress and sloughing. Assuming that the biofilm contains nx different microbial species in a control volume between z and with a cross sectional area A in the support interface, a mass balance of the species i is set up for a differential volume Adz as follows (see Figure 2)

$$\frac{\partial[Adz\rho_i f_i(t,z)]}{\partial t} = Adz\mu o_i(t,z)\rho_i f_i(t,z) + Ag_i(t,z) - A[g_i(t,z) + \frac{\partial g_i(t,z)}{\partial z}dz] \quad (2.1)$$

where ρ_i is the constant density for the i-th species, f_i denotes the volume fraction of the i-th species, μo_i is the observed specific growth rate, and g_i is the mass flux of the ith species in the z-direction, namely, the mass of the species i that is displaced per unit time and unit area across a plane transverse to the z-direction.

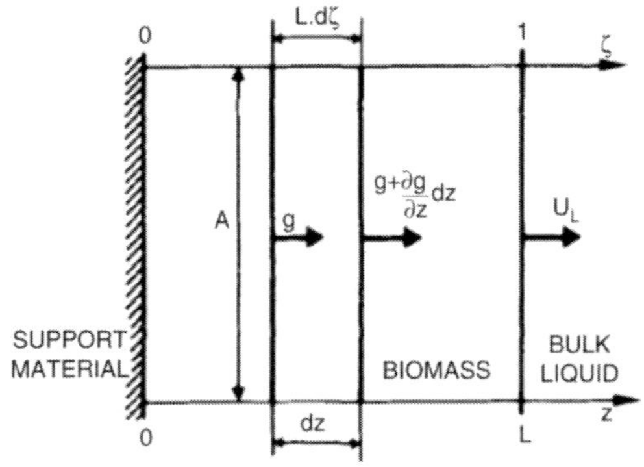

Figure 2. Flux of biomass through a differential volume element of an expanding idealized biofilm.

Dividing (2.1) by $Adz\rho_i$ yields the differential equation

$$\frac{\partial f_i}{\partial t} = \mu o_i f_i - \frac{1}{\rho_i}\frac{\partial g_i}{\partial z} \quad (2.2)$$

Let $u(t,z)$ be the velocity at which the microbial mass is displaced with respect to the film-support interface, the flux g_i can be expressed as $g_i(t,z) = u(t,z)\rho_i f_i(t,z)$. Thus, (2.2) can be written as

$$\frac{\partial f_i}{\partial t} = (\mu o_i - \frac{\partial u}{\partial z})f_i - u\frac{\partial f_i}{\partial z} \quad (2.3)$$

Summing Eq. (2.3) over all the nx microbial species and using the sum of the volume fractions equals 1, i.e., $\sum f_i = 1$, we obtain

$$\frac{\partial u(t,z)}{\partial z} = \overline{\mu o}(t,z),$$

where

$$\overline{\mu o}(t,z) = \sum_{i=1}^{nx} \mu o_i(t,z) f_i(t,z) \tag{2.4}$$

Thus the velocity μ is determined by the mean observed specific growth rate of the biomass $\mu o_i(t,z)$ by $u(t,z) = \int_0^z \overline{\mu o}(t,z')dz'$,(2.5) where $u(t,0)$ is set to zero. We remark that this microbial mass transport velocity is identical to all species, which is clearly an artifact of the model. If the biofilm grows or shrinks, its thickness $L(t)$ changes and the film–water interface moves at a velocity $u_L(t) = \frac{dL(t)}{dt}$. With $\sigma(t)$ defined as the velocity at which the biomass is exchanged between the biofilm and the bulk liquid outside the biofilm and by use of Eq. (2.5), the velocity of the film–water interface is expressed as

$$u_L(t) = u(t,L) + \sigma(t) = \int_0^L \overline{\mu o}(t,z')dz' + \sigma(t) \tag{2.5}$$

In this paper, the biomass loss due to shear stress is modeled by setting $\sigma = -\lambda L^2$ with some constant λ, and the biomass loss due to sloughing is modeled by setting σ as a σ function.

Substitution of (2.4) into (2.3) yields the mass balance equation for

$$f_i : \frac{\partial f_i}{\partial t} = [\mu o_i(t,z) - \overline{\mu o}(t,z)] f_i(t,z) - u(t,z)\frac{\partial f_i(t,z)}{\partial z} \quad i = 1,\dots,nx-1 \tag{2.6}$$

Assuming that the bulk liquid contains ns different substrates, in the same way as for the microbial species, a mass balance for the substrate i is obtained

$$\frac{\partial S_i(t,z)}{\partial t} = r_i(t,z) + \frac{\partial}{\partial z}\left(D_i \frac{\partial S_i(t,z)}{\partial z}\right), i = 1,\dots,ns \tag{2.7}$$

where S_i is the concentration, r_i the observed conversion rate, and D_i the diffusion coefficient of substrate i. The boundary conditions for the film-support interface ($z = 0$) are no-flux boundary conditions given by $\frac{df_i}{dt}(\mu o_i - \overline{\mu o})f_i, \frac{\partial S_i}{\partial z} = 0$.For the film–water interface ($z = L$), with or without an external mass transfer limitation, the boundary conditions are defined by $\frac{\partial S_i}{\partial z} = \frac{LD_{l_i}}{L_l D_i}(S_{l_i} - S_i)$ or $S_{l_i} = S_i$.where L_l is the thickness of the effective substrate transfer layer between the film–water interface and the bulk liquid, D_{l_i} is the diffusion coefficient of substrate i in the bulk fluid, S_{l_i} is the concentration of substrate i in the bulk liquid. The specific growth rates μ μo_i and the substrate conversion rates r_i are given by Monod models.

The steady-state analysis of this multispecies biofilm model predicts that for any given kinetics, the spatial distribution of a microbial species may be described by one or more layers, each being either homogeneous or mixed. Numerical solutions of the time-dependent equations were computed for 5 cases in the situation of a heterotrophic–autotrophic competition for space and for oxygen as a common substrate. The five cases are: 1. Unrestricted Growth; 2. Changes in Bulk Liquid Substrate Concentration; 3. Biomass Shear; 4. Biomass Sloughing; 5. Biofilm in a Completely Mixed Reactor with External Mass Transfer Resistance. The development of the biofilms' thickness and the removal of substrates from the bulk liquid with time are presented for each case. The model predicts how the microbial composition of the biofilm depends on the microbial kinetics, the substrate concentration as well as different reactor configurations and an alternative biomass detachment mechanism.

Based on new experimental observations, an extended mixed-culture biofilm (MCB) model was introduced by later [11]. This model includes a more flexible description of the transport of dissolved components in the biofilm and considers the diffusive transport of particulate components in the biofilm's solid matrix, changes of the biofilm's liquid phase volume fraction (porosity), and the simultaneous detachment and attachment of cells and particles at the biofilm's surface. This extended model can reproduce most of the new experimental data. This model is derived based on the general mass conservation principle and is independent of any special set-up except for limitations in one space dimension. The model can be adjusted to any specific microbial kinetics or biomass detachment mechanism, and used to predict the short and long term behavior of the biofilm's growth. Indeed the model has been successfully applied to many biofilm studies since its development, and it has helped enormously in understanding the complex bulk interactions in the multispecies biofilms. On the other hand, since it is a one-dimensional model, it significantly simplifies the interaction between the substrate and the biofilm and the spatial dynamics of the biofilm. It also lacks the ability to characterize the multi-dimensional morphology of the biofilm. Various higher dimensional models with a biofilm–flow interaction are developed later.

DLA Model for Biofilm Growth and Pattern Formation

In general the growth mechanism of the bacterial colonies is highly complex, in which the substrate concentration plays an important role. One way to model the growth pattern is through the class of models called the diffusion-limited aggregation (DLA) model. The DLA growth was first studied through computer simulations [12] and the first biological DLA example was a DLA colony of *Bacillus subtilis* [13]. More experimental and modeling work was done in [14] and [15] later. The rule of the DLA model is as follows: one chooses a seed particle as the origin of a square lattice on a plane. Another particle is released far from the origin and is allowed to move at random. When it arrives at the nearest neighboring site to the origin, it sticks to the site. Then another particle is launched and it moves until it reaches the nearest neighboring site of a cluster made of the two particles. Through the repetition of these procedures, a cluster grows with an outwardly open and randomly branched structure as shown by a computer simulation in (Figure 3(b)) [16]. Its pattern is morphologically self-similar, implying that it is a fractal.

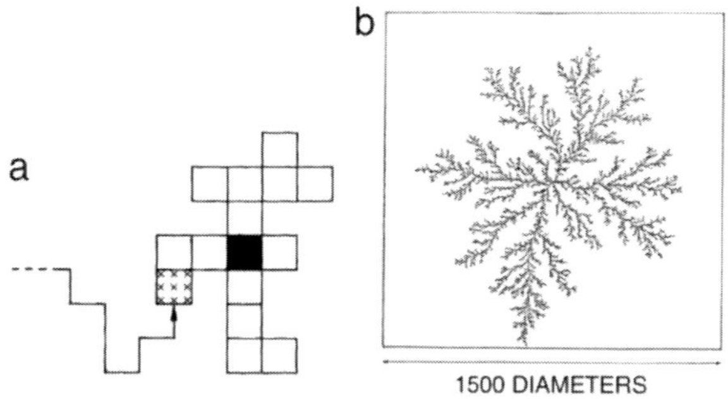

Figure 3. DLA on computer simulation: (a) the rule of the 2D DLA growth. The solid cell is the origin. The hatched cell is a new cell that has aggregated to the cluster; (b) a typical example of a 2D DLA cluster which consists of 100,000 particles.

Figure 4 [15] shows the morphology of a *B. subtilis* strain that grows through the DLA process at a very low nutrient level on an agar plate. The colony had an outwardly open and randomly branched structure, and these findings were in good agreement with the computer simulations of the DLA model. This suggests that *B. subtilis* cells grow predominantly obeying a simple physical law, diffusion, at low limit of the nutrient concentration. A study in [15] also shows that a smoother morphology was developed under a higher nutrient concentration.

Figure 4. An example of a DLA colony. The colony was photographed 21 days after inoculation. Its diameter was about 47 mm.

Discrete-Continuum/Cellular Automaton (CA) Model

One of the origins of the Cellular Automaton (CA) model is Conway's Game of Life [17]. It consists of a regular grid of cells, each in one of a finite number of states. The grid can be in any finite number of dimensions. Time is also discrete, and the state of a cell at time t is a function of the states of a finite number of cells (called its neighborhood) at time $t-1$. Every cell has the same rule for updating, based on the values in this neighborhood. Each time the rules are applied to the whole grid, a new generation is created. A clear advantage of the CA model is that it can produce a complex behavior despite its simplicity, and this makes it particularly attractive to simulate biological systems which often exhibit complex spatial structures [18-20].

CA models for biofilms incorporating the biological rules including bacterial reproduction and movement, cell to cell communication, and diffusion of the nutrient were extensively studied in [21-26]. These models can capture various biofilm growth patterns observed in experiments, and strongly suggest that the biofilm's structure is largely determined by the surrounding substrate concentration. [24] claims that the three widely accepted conceptual models of the biofilm's structure:

1) penetrated water-channel biofilm,
2) heterogeneous mosaic biofilm,
3) dense confluent biofilm, can all be reproduced by CA models at very different substrate concentrations.

We use the model in [23] as a representative of this class of models. The nutrient concentration $c(\vec{r},t)$ at the position $\vec{r}$ and the time t is governed by the diffusion-reaction equation on a triangular lattice

$$\frac{\partial c(\vec{r},t)}{\partial t} = D_c \nabla^2 c(\vec{r},t) - \sum_{active-walkers} \delta(\vec{r}-\vec{r}_i)\min(c_r, c(\vec{r},t))$$

where D_c is the diffusivity for the nutrient and $\delta(r)$ is the Kronecker symbol. The second term on the right represents the consumption of the nutrient. A mesoscopic unit of the bacterial cells called a 'walker' is characterized by its location $\vec{r}_i$ and 'internal energy' W_i which affects its activity. The walker loses the 'internal energy' at a fixed rate e and increases it through consuming nutrients at a fixed rate c_r if sufficient nutrient is available. Otherwise, it consumes the available amount. Wherever W_i drops to zero, the walker becomes stationery (sporulation). When nutrient is sufficient, W_i increases; and when it reaches some threshold t_r, the walker divides into two. The equation for the internal energy W_i is $\frac{dW_i}{dt} = \min(c_r, c(\vec{r}_i,t)) - e$ .The walkers perform an off-lattice random walk within a well-defined envelope on the triangular lattice. At each time step, the random walk is of size d at an random angle Θ uniformly chosen from $[0, 2\pi]$. The new location $\vec{r_i'}$ is given by $\vec{r_i'} = \vec{r_i} + d(\cos\Theta, \sin\Theta)$. If the step $\vec{r_i} \to \vec{r_i'}$ crosses the envelope, the step is not performed and a counter on the appropriate segment of the envelope is increased by 1. When a segment counter reaches N_c, the envelope segment is shifted by one lattice step. Here the level of N_c represents the agar concentration, as more 'collisions' are needed to push the envelope on a harder substrate.

Numerical results from [23] are shown in Figure 5a. The growth patterns are compact at high peptone levels and become fractal with decreasing food level. At a given peptone level, the patterns are more ramified as the agar concentration increases. The results capture the experimentally observed patterns shown in Figure 4 (also from [23]) very well except that there is a discrepancy at very low peptone levels. When the nutrient concentration is low, the observed growth patterns (Figure 5d) become denser again, but the simulation result (Figure 6a with small P value) does not capture this feature.

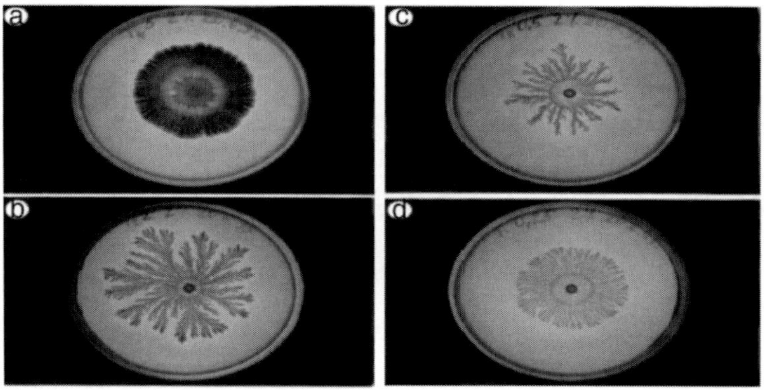

Figure 5. Observed patterns of colonies growth on a substrate with 2% agar concentration. The peptone level is 5, 2, 0.5 and 0.25 gl^{-1} for a, b, c and d, respectively.

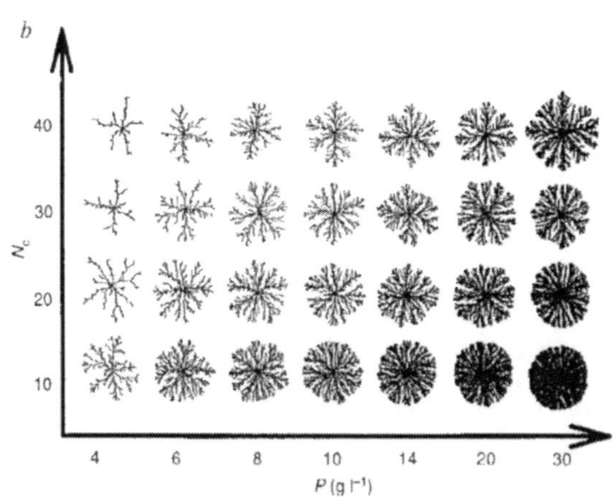

Figure 6a. Numerical simulation results of growth patterns as function of peptone level P (initial value of c) and N_c (agar concentration).

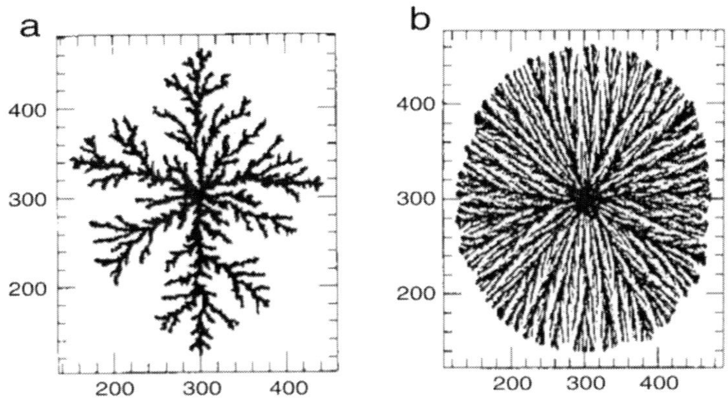

Figure 6b. Effect of chemotaxis on CA model with $P = 10gl^{-1}$, $N_c = 40$. (a) No chemotaxis, (b) With chemotaxis.

The explanation of the denser pattern at a low nutrient level is that as the environmental conditions become more hostile, a higher level of cooperation is required for a more efficient response of the colony. Non-local communication between different colonies might be necessary. A chemotactic mechanism is introduced in [23] to achieve this information exchange between colonies. Generally chemotaxis means a movement of the microorganisms in response to the concentration gradient of certain chemicals. In the modified CA model, each stationary walker produces a communication chemical s at a fixed rate s_r, and each active walker consumes the chemical at a fixed rate c_c. Thus s is governed by the following reaction-diffusion equation

$$\frac{\partial s(\vec{r},t)}{\partial t} = D_c \nabla^2 s(\vec{r},t) + \sum_{stationary-walkers} \delta(\vec{r} - \vec{r}_i) s_r - \sum_{active-walkers} \delta(\vec{r} - \vec{r}_i) \min(c_c, s(\vec{r},t)),$$

where D_s is the diffusivity of the communication chemical. The movement of the active walker changes from a pure random walk to a random walk with a bias along the gradient of the communication field (high probability to move in the direction of the signaling material). Figure 6b shows the simulation result with chemotaxis included in the model. The pattern has a denser structure with thin branches and a well-defined circular envelope, and the number of walkers (bacterial density) in the colony is much lower than in the absence of chemotaxis. It is also found that a chemotactic response to the nutrient concentration gradient does not produce this effect, thus it confirms that the communication between different colonies is crucial for this growth feature.

Their apparent simplicity and great success in capturing the biofilm's structure makes the CA models very attractive. The model described above is a close approximation to the molecular diffusion in a much coarser scale. The coarseness emphasizes the stochastic nature of the model which is the main mechanism of producing a heterogeneous biofilm's morphology.

Picioreanu et al.,[27] developed a combined differential-discrete CA biofilm model. In this model, the substrate concentration was assumed to be continuous and governed by the conventional diffusion–reaction equation. The biomass density was computed by a direct integration of the biomass balance equation, taking into account only the growth as a result of substrate consumption. The biomass spreading was modeled by CA model. But instead of performing an entirely random walk for the "walkers", the newly formed biomass finds a place by "pushing" its neighbors to an adjacent, unoccupied space. Numerical simulations evolving in real time were carried out for both 2D and 3D, and the calculated values of the global oxygen uptake rate, the concentration profiles for the oxygen, and biomass and colonies, size were both qualitatively and quantitatively in agreement with experimental data [28-30].

Noguera et al., [31] developed a three-dimensional multispecies biofilm model based on CA approach and tested it with a two-species biofilm composed of sulfate-reducing bacteria and methanogens. The model predicts different biofilm structures in the absence of sulfate, when a syntrophic association between the two organisms develops, and in the presence of sulfate, when the two organisms compete for the available hydrogen.

In [32], a 2D CA model was developed to study the effect of the thickness of the substrate's concentration and the hydrodynamic boundary layers in the biofilm's structure, motivated by the experimental results of [33,34]. The model predicts that when external mass transfer (for substrate) limitations are significant, the biofilm develops more open structures; when the substrate concentration layer is reduced and the external mass transfer is enhanced, the biofilm develops denser and smooth structures. The simulation reproduced the "mushroom" shaped biofilm described by [35].

A computer model called BacLAB based on the solute diffusion model coupled with the CA principles for the bacterial growth was developed by Hunt et al., [36,37] to study the detachment mechanisms in a biofilm. This computer model has been used to test various hypothetical detachment mechanisms in 3-D simulations [38,39]. These studies show that the detachment is a critical determinant of the biofilm's structures and of the dynamics of the biofilm's accumulation and loss.

For the models described in this section, the effect of the surrounding bulk fluid is not coupled explicitly. We next discuss models that couple the bulk fluid to the biofilm completely. More sophisticated models combining the CA and a continuous description of the bulk fluid will be described later.

Biofilms Models with Biomass and Flow Coupling

For most of the biofilm models described in previous sections, the biofilm's structure is primarily determined by the substrate concentration which is solely modeled by diffusion process. However, the heterogeneous structures including a nonuniform distribution of cells and polymers, variable biofilm thickness and surface shape, and variable density and porosity, have been observed and quantitatively measured by many researchers [40-42]. It is also found that the hydrodynamics of the bulk fluid plays an important role in shaping the structure of the biofilms through both the convection of the nutrient and the detachment of biomass by shear stress [43-45]. It is generally accepted that the formation of the biofilm's structure is the result of a combination of several simultaneously occurring biological, chemical and physical processes including cell transport to the substratum and attachment, the biofilm's generation by cellular growth and extracellular polymer production, the biofilm's detachment by shear stress, and substrate and product transport to and from the biofilm. Based on the previously established models and a better understanding of the biofilm's properties through experiment, several biofilm models coupling the bulk fluid were developed.

Hybrid Discrete-Continuum Models

In [46-48], a fully quantitative two- and three-dimensional biofilm model was developed which incorporates the flow over the irregular biofilm's surfaces, convective and diffusive mass transfer of substrate, bacterial growth and biomass spreading. In this model, the biomass growth and spreading is modeled by the discrete CA model, the mass balance of the substrate is modeled by the continuity equation and the convection–diffusion–reaction equation, respectively, and the flow field is governed by the momentum balance (Navier–Stokes) equation. The fluid flow affects the biofilm's growth by both regulating the substrate concentration at the biofilm-fluid interface, shearing the biofilm's surface and eroding the protuberances. The interaction is reciprocal since a new biofilm's shape leads to a different boundary condition and thus different flow and substrate concentration.

The governing equations describing the biofilm model given in [46] are:

- Total mass conservation for fluid phase (the continuity equation): $\nabla \cdot u = 0$.
- Momentum conservation for the fluid phase flowing around the biofilm (the Navier-Stokes equation): $\frac{\partial u}{\partial t} + u \cdot \nabla u = -\frac{1}{\rho}\nabla p + \upsilon \nabla^2 u$
- Mass conservation for one dissolved substrate (for the limiting substrate only): $\frac{\partial C_S}{\partial t} + u \cdot \nabla C_S = -R_S(C_S, C_X) + D\nabla^2 C_S$.
- A kinetic equation for biomass growth: $\frac{dC_X}{dt} = R_X(C_S, C_X)$

where u is the fluid velocity, p is the pressure, ρ is the liquid density, υ is the kinematic viscosity of the fluid, D is the substrate's diffusion coefficient, R_S and R_X are the rates of the substrate's consumption and biomass formation, C_S and C_X are the concentration of the substrate and biomass, respectively.

There is a clear separation of time scales in the biofilm's growth [47], namely, (1) the biomass growth and decay, in the order of hours or days, (2) the substrate transport, in the order of minutes, and (3) the hydrodynamic processes, in the order of seconds. Instead of solving all the equations in the smallest time-step given by the momentum equation, a process occurring at a particular time scale is solved while assuming all the other processes at different time scales are at steady state. For example, while solving the substrate mass-balance equation, the flow pattern is considered at a pseudoequilibrium for a given biofilm shape, and at the same time the biomass growth, decay, and detachment are in a frozen state. This leads to the following strategy in simulating the biofilm's development in time [46]:

1. Momentum transfer and continuity equations are solved to find the flow field variables: velocity u, pressure p, and the normal and tangential stresses τ acting on the biofilm's surface. This hydrodynamic step is needed each time the biofilm's geometry changes, either due to growth or detachment.
2. The flow velocity u calculated in the previous step (the flow is considered completely established) is used to solve the convection-diffusion-reaction equation for the mass conservation of the substrate. This equation is solved towards a pseudosteady-state substrate concentration, namely, $\frac{\partial C_S}{\partial t}$ is set to zero. In the first two steps, the equations were solved by a lattice Boltzmann algorithm [49-51]. In the absence of a convection term, a finite difference method was used to solve the substrate equation.
3. The new biomass content of each grid element, C_X, is calculated using the Monod and Herbert–Pirt equation including the substrate's concentration at a steady-state as calculated before.
4. As the biomass grows, it is redistributed in space according to the discrete CA rules used in [27]: by splitting the biomass content in the element in which it has grown above the maximum biomass density and by pushing the neighbors to reach the free space.
5. A detachment step can be performed before the redistribution. A traditional finite element method can be used to compute the deformation energy in the biofilm induced by the fluid stresses at the biofilm's surface, and places where the biofilm will break can be determined. The loss of biomass will cause a change of the external biofilm's structure so that a new hydrodynamic step is necessary.

One advantage of the discrete model of the biomass is that, once the model equations have been developed, the boundary conditions and the geometry of the system can be easily changed. For example, the carrier's surface can be spherical instead of flat, and it can be used to model the development of biofilms in particle reactors.

Numerical simulations in 2D and 3D were carried in [46]. The 3D model predicted that in a substrate limited regime the biofilm develops as a heterogeneous structure, presenting pores, channels and "mushroom"-like bacterial clusters, whereas in a substrate sufficient regime a compact, and densely-packed biofilm forms. The 2D models predicted that at high R_e numbers, the thickness of the substrate concentration boundary layers (CBL) decreases and the biofilm develops a smooth surface; whereas at low R_e numbers, the thickness of the CBL increases and yields a steep substrate concentration gradient, thus a rough biofilm with a finger-like shape forms.

A more detailed and quantitative study on the effect of diffusive and convective substrate transport on the biofilm's structure formation using this model was performed in [47] in a 2D case. The model's parameters characterizing the biofilm growth condition include: the growth ratio G, defined as G = (maximum biomass growth rate)/(maximum internal transport rate of substrate), the Reynolds number R_e, and the inoculation density s_0 (the percentage of substratum surface initially covered by the biofilm). Note that the large value of G corresponds to a substrate limited regime (high metabolic rate), and a small value of G corresponds to a substrate sufficient regime (low metabolic rate). A quantitative analysis of the simulated biofilm structure was achieved by calculating measures including the biofilm's area enlargement (α), the coefficient of surface roughness (σ), compactness (ξ), the porosity (ε) and the maximum biofilm's thickness (δ) as defined in [52]. The simulated biofilm's evolution in time is shown in Figure 7 [47]. The model again predicts that under the limited substrate regime, biofilms develop an irregular surface with a high surface roughness, high surface area, high porosity, and low degree of compactness. As the substrate concentration increases, biofilms gradually shift to more compact and smooth structure. The model in this paper does not include the biofilm's detachment effect. It predicts that the effects of convection and of a flow-driven mechanism on the biofilm's structures are less important. It also predicts that a smaller inoculation density leads to a rougher biofilm, and the substrate flux to the biofilm is greatly influenced by both the internal and external mass transfer rates, but not by the inoculation density.

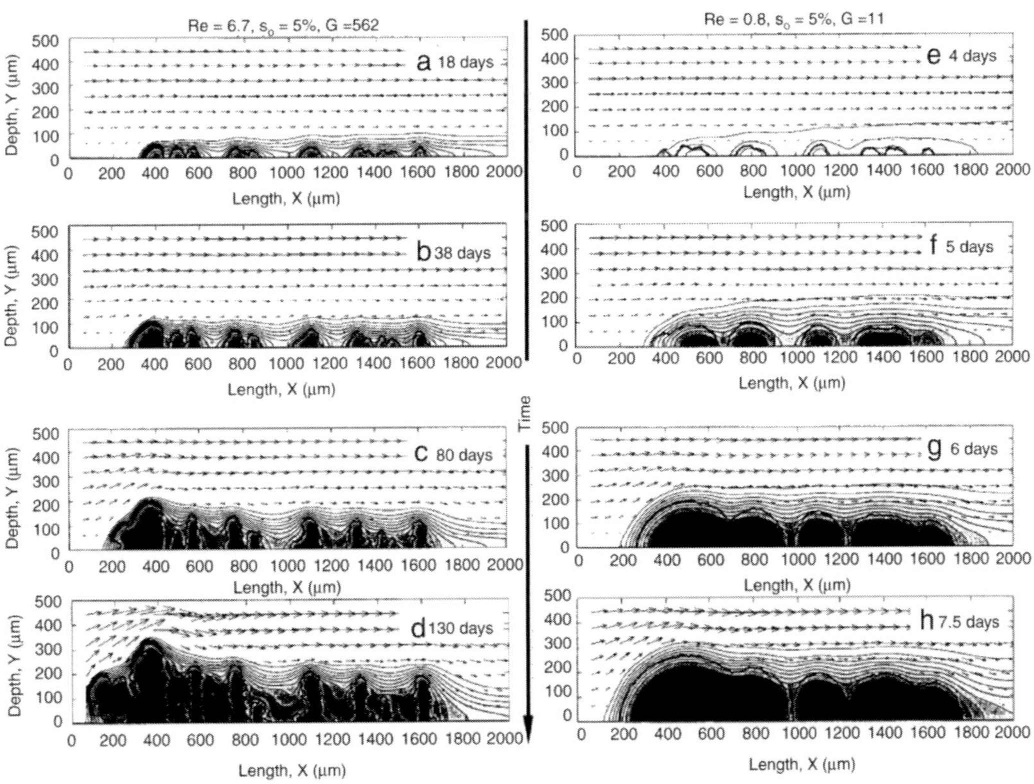

Figure 7. Simulated biofilm evolution in time for *(a-d)* Re = 6.7, $S_0 = 5\%$, $G = 562$ and *(e-h)* Re = 0.8, $S_0 = 5\%$, $G = 11$. The arrows represent the vector velocity at intervals of 16 grid nodes. The thick lines indicate the biofilm's surface. Iso-concentration lines show the decrease of the substrate's concentration from the maximum value in the bulk liquid (white patches) to zero in the biofilm (dark gray patches), with a variation of 10% between the lines.

In [48], a two dimensional model was developed for the detachment based on the internal stress created by a moving fluid past the biofilm. This model used the strategy listed in this section with step 5 implemented, and it can model two biofilm detachment mechanisms, erosion (small-particle loss) and sloughing (large-biomass-particle removal) in a unified way.

The model predicts that erosion makes the biofilm surface smooth and sloughing leads to an increased biofilm-surface roughness. Simulations also indicated an avalanche effect in the biomass's loss, and fast growing biofilms have a faster detachment rate than slow-growing biofilms, and thus suffer an instability in the biofilm's accumulation and an abrupt biomass loss.

The hybrid discrete-continuum biofilm model was a fully quantitative model incorporating many physical, chemical and biological processes affecting the biofilm's development. The quantitative nature of the model not only produced simulation results in good agreement with experimental data, but also predicted the properties of biofilms and shed light on experimental studies. The model was widely accepted and successfully applied to several situations since its development.

Another completely different discrete-continuum biofilm model was proposed by [53,54] based on the Immersed Boundary Method. In this microscale model, the microbes in the biofilm are treated as discrete entities (cells) which can swim through the action of flagellae, the bulk fluid is treated as a continuum, and the formation of the biofilm is achieved by a cell-cell aggregation and a cell-substratum adhesion which are modeled by generating the appropriate binding forces between discrete representations of the organisms. The model includes the hydrodynamic interaction between the biofilm and the fluid, the substrate reaction, diffusion and convection, and the chemotactic responses of swimming microbes. Numerical simulations in 2D predicted that the higher local cohesion forces and cell motility can greatly facilitate the biofilm's formation, cells may adhere preferentially to a saturated surface with chemotaxis, and changes of the channel's geometry also have effects on the surface irregularities in the biofilm's formation.

Multidimensional Continuum Models

The hybrid discrete-continuum models of multiple steps described previously (CA models) can produce results in good agreement with experimental expectations. However, there are some drawbacks associated with the element of randomness in the step of the biomass redistribution. Namely, they are lattice dependent and not invariant to changes of the coordinate system; multiple simulations with the same initial condition generally produce different outputs; the biomass, though a continuous variable according to the growth kinetics, suffers discrete changes during splitting. This randomness can be viewed as an artifact from the detail of the algorithm rather than the growth process modeled. Even though the random feature of these models determines the biofilm's structure to some extent, sometimes a deterministic model is more desirable. Several deterministic continuum biofilm models were developed to accommodate this need.

Eberl et al., [55] developed a spatio-temporal continuum model under which the biomass spreading is described by a nonlinear density-dependent diffusion mechanism. The whole region is divided into a bulk liquid region Ω_1 without biomass and a solid biofilm region Ω_2 where all the biomass is contained. The governing system of equations is given by:

$$\nabla \cdot u = 0, \frac{\partial u}{\partial t} + u \cdot \nabla u = -\frac{1}{\rho} \nabla p + \nabla^2 u, \text{in}\, \Omega_1 = \{x \in \Omega \,|\, m(t,x) = 0\}, u \equiv 0,$$

in

$$\Omega_2 = \{x \in \Omega \mid m(t,x) > 0\}, \ \frac{\partial c}{\partial t} + u \cdot \nabla c = \nabla \cdot (d_1(m)\nabla c) - f(c,m),$$

$$\frac{\partial m}{\partial t} = \nabla \cdot (d_2(m)\nabla m) + g(c,m), \ f(c,m) = \frac{k_1 cm}{k_2 + c}, \ g(c,m) = k_3(f(c,m) - k_4 m).$$

Here $u(t,x)$ is the fluid velocity, $p(t,x)$ is the fluid pressure, $m(t,x)$ is the biomass density, $c(t,x)$ is the nutrient concentration, $d_{1,2}$ are diffusion coefficients for c and m, $f(c,m)$ is the nutrient consumption rate, $g(c,m)$ is the biomass production rate, and $k_1,...k_4$ are constants. In this paper, $d_1(m)$ is considered a constant and the density-dependent diffusion coefficient $d_2(m)$ is given by $d_2(m) = (\frac{\varepsilon}{m_{\max} - m})^a \cdot m^b$.Thus the diffusion operator in the equation for $m(t,x)$ degenerates for a small biomass and is singular at the biomass density bound $m_{\max}$. The parameters a, b, ε are chosen to guarantee the following properties: 1. there exists a "sharp interface" between the biomass and the bulk fluid, 2. biomass spreading is significant only if a certain maximum density is approached, 3. biomass density cannot exceed the maximum bound. Simulations under a hydrostatic condition ($u \equiv 0$) were conducted in 1D and 3D, which yielded the following observations. Biofilms develop a spatial heterogeneous structure under a heterogeneous environment (either nutrient concentration or inoculation density). With limited nutrients, the biofilm tends to develop isolated colonies with channels in between, thus a rough structure. With a sufficient amount of nutrient, the biofilm tends to develop a more compact structure with colonies connected to each other. In either case, mushroom-shaped colonies dominating over their smaller neighbors were observed, which is in good agreement with experiments (e.g. [1]).

Dockery [56] developed a continuum model of a single substrate limited biofilm growing into a static aqueous environment. The region under consideration in R^3 is divided into a biofilm region $z \le h(x,y,t)$ and an aqueous region $z > h(x,y,t)$, with the interface $z = h(x,y,t)$. The biofilm is modeled as a homogeneous, viscous, incompressible fluid with a velocity given by Darcy's law $u = -\lambda \nabla p$ where p is the pressure and λ is a constant. Despite the incompressibility, the growth (or decay) of the biofilm yields a field of sources (or sinks) in the biofilm fluid. Thus $g = \nabla \cdot u$ for some prescribed growth function g. Combining these two equations yield $-\lambda \nabla^2 p = g$. The normal velocity of the biofilm's interface ($x, y, h(x,y,t)$) is $n \cdot u|_{z=h^-} = -n \cdot \lambda \nabla p|_{z=h^-}$, where n is the unit upward normal and h^- means that z should approach h from below, the biofilm side. Since p is proportional is λ^{-1}, the interface velocity is independent of λ. This is true since the velocity u is generated by the biofilm's growth, and thus depends only on g but not on λ. The aqueous region $z > h$ is assumed to be static. Hence the boundary conditions for p are given by $p|_{z=h}=0$ and $p|_{z=-\infty}=0$, where $z=-\infty$ corresponds to the bottom of the domain. The concentration of a single limiting substrate is denoted by $S(x,y,z,t)$. Substrate diffuses through the aqueous region into the biofilm, where it is consumed and diffuses with a different diffusion coefficient. It is assumed that there is a function $H(x,y,z,t)$ with $H\geq h$ such that $S(x,y,z,t)=S_\infty$ for $z\geq H$. The substrate's equations are $S_t-D_1\nabla^2S=0, h<z<H$, $S_t-D_2\nabla^2S=-u(S), z<h$, with boundary conditions $S|_{z=H}=S_\infty, S|_{z=-\infty}=0$, and matching conditions $S|_{z=h^+}=S|_{z=h^-}$, $D_1\nabla S|_{z=h^+}=D_2\nabla S|_{z=h^-}$. The substrate uptake rate $u(S)$ is defined by the Monod function $u(S)=u_{\max}\dfrac{S}{S+K_S}$ and the growth function $g(S)$ is taken as a function of $u(S)$, namely, $g=g(u(S))$. The equations are non-dimensionalized by introducing a characteristic time scale T and length scale w. Since the substrate diffusion occurs much faster than the biofilm's growth, it is assumed to be in a steady state and S_t is set to zero in the equation. The dimensionless equations of the model are given by $\nabla^2p=-g(u(S)), z<h$, $-\nabla^2S=0, h<z<H$ $-\nabla^2S=-Gu(S), z<h$, with boundary conditions described above. Here $G=w^2u(S_\infty)/D_2S_\infty$ is the non-dimensional growth number. One dimensional moving front solutions are analyzed by writing the equation for S as a first order system of ODEs, and analytical solutions are presented for the particular choices of u and g. It shows that the penetration depth of the substrate into the biofilm is $O(G^{-1/2})$. Linear stability analysis shows that if the substrate concentration S is sufficient low, the planar front solutions suffer a fingering instability with preferred wavelength of the order of $O(G^{-1/2})$. Nonlinear evolution of the finger instability in 2D is studied through numerical simulations, where the level set method is used to track the interface $z=h$. This continuum model provides scaling predictions of the biofilm's growth process and can be compared to experimental data, and it is among the early work of the rigorous mathematical treatment of biofilm models.

Alpkvist [57] extended the work in [56] and proposed a multidimensional continuum model for the heterogeneous growth of biofilm systems with multiple species and multiple substrates while retaining Darcy's law for the velocity. The model was tested for two systems. System 1 is the biofilm consisting of an active and inert biomass with oxygen as a substrate; system 2 is the biofilm consisting of heterotrophs, autotrophs, and inerts with limiting substrates of oxygen, acetate, and ammonium. Numerical simulations were performed in 2D and 3D. It predicted that for system 1, the inactivation rate has an effect on the stability of the biofilm's interface, namely, a higher inactivation rate increases the depth of the active layer by locally decreasing the amount of active biomass. For system 2, the 2D simulation produced results in agreement with [9]; but the 3D simulation yielded the total averaged mass fractions of the biofilm components not adding to 1 due to the heterogeneity of the biofilm surface, a multidimensional phenomenon cannot be described by the 1D model in [9].

Cogan [58] proposed a continuum model in which the biofilm is treated as a biological gel composed of an extra-cellular polymeric substance (EPS) and water. The bacteria are enmeshed in the network and are the producers of the polymer. Gels absorb or expel solvent causing a swelling or contraction due to the osmotic pressure gradients, where the physical and chemical environmental conditions affecting the biofilm morphology are built into the osmotic pressure. The model assumes that the primary forces which induce the gel's motion are applied to the fluid solvent and the EPS, and provide a mechanism of biomass redistribution through swelling of the EPS and the constitutive relationships. Here both the network and the solvent are treated as Newtonian fluids with a much larger viscosity for the network. This fundamental formulation of the model follows the two-fluid model [59]. However, inertia is neglected so that the governing equation for the force balance is the Stokes equation for multi-fluids.

The volumes of the substrate and the bacteria are assumed to be negligible and the volume fractions of the polymer network and solvent are denoted by θ_n and θ_s respectively. Four forces acting on the network are assumed, including the viscous stress $\nabla\cdot(\theta_n\sigma_n)$, the frictional drag between the network and the solvent, the gradient of an osmotic pressure $\nabla\Psi(\theta_n)$ representing the force induced by the colligative properties of the gel, and the hydrostatic pressure. The forces acting on the solvent include all the forces described above except the osmotic pressure. Here σ_n is the viscous stress tensor given by $\sigma_n = \frac{\eta_n}{2}(\nabla\overrightarrow{U_n} + \overrightarrow{U_n^T})$, $\overrightarrow{U_n}$ is the network's velocity and η_n is the network's viscosity. The osmotic pressure $\Psi(\theta_n)$ is obtained by a Taylor's expansion of the Flory–Huggins formula up to θ_n^3 around $\theta_n = 0$, and the expression is $\Psi(\theta_n) = \frac{k_B T}{3\upsilon_1}\theta_n^2(\theta_n - 3(\chi_1 - \frac{1}{2}))$. Here k_B is the Boltzmann's constant, T is the temperature, υ_1 is the volume of one monomer of the network constituent, and χ_1 is the Flory–Huggins interaction parameter. The analysis is done for a simplified model obtained by ignoring the inertial term, setting the frictional force to zero and assuming a zero solvent velocity, which in turn yields a zero hydrostatic pressure gradient. Thus the simplified network's momentum equation is

$$\nabla\cdot(\theta_n\sigma_n) - \nabla\Psi(\theta_n) = \eta_n\nabla\cdot(\frac{\theta_n}{2}(\nabla\overrightarrow{U_n} + \overrightarrow{U_n^T})) - \nabla\Psi(\theta_n) = 0 \quad (2.8)$$

The network's redistribution is governed by the principle of conservation of mass, and the governing equation is $\frac{\partial\theta_n}{\partial t} + \nabla\cdot(\theta_n\overrightarrow{U_n}) = g_n$ (2.9)

The substrate concentration c is governed by the following convection-diffusion-reaction equation: $\frac{\partial(\theta_s c)}{\partial t} + \nabla\cdot(c\overrightarrow{U_s}\theta_s - D\theta_s\nabla c) = -g_c$. Here $\overrightarrow{U_s}$ is the solvent velocity and the scaling factor θ_s reflects the fact that the substrate is dissolved in the solvent. Since $\theta_s = 1 - \theta_n$, $\overrightarrow{U_s} = 0$, the substrate is assumed to be in quasi-steady-state, and the consumption rate is taken as $g_c = A\theta_n c$, the substrate equation becomes

$D\nabla\cdot((1-\theta_n)\nabla c)=A\theta_n c$ (2.10). The growth of the EPS network is modeled by Monod kinetics, i.e., $g_n=\varepsilon\mu\theta_n\frac{c}{K_c+c}$, where K_c is the half-saturation constant and μ is the maximum production rate. In the substrate limited regime (c), g_n can be approximated by $g_n=\varepsilon A\theta_n c$, where. The system of the governing equations of the model are given by (2.8)-(2.10), and the variables are θ_n, $\overrightarrow{U_n}$, c. The interface between the biofilm region ($\theta_n>0$) and the solvent region ($\theta_n=0$) is not explicitly tracked, but is characterized by a rapid change in θ_n. Linear stability analysis of the flat interface shows that there is a maximally unstable mode and the higher frequency perturbations are damped out, which is in qualitative agreement with the result of [56]. Numerical simulations were performed in 2D. This model lays out the framework of studying the effect of the material and the chemical properties (network viscosity, osmotic pressure) on the biomass's distribution, and produces results in qualitative agreement with previously developed models. The idea of using the network volume fraction θ_n to represent the biofilm provides the possibility of modeling the whole biofilm system as a multi-phase fluid.

Experimental results in [60] indicate that the detachment of multicellular clumps is more significant than the erosion of single cells, possibly because clumps retain certain advantages of the biofilm. Klapper et al.,[61] proposed a more general multi-fluid model to explain this phenomenon by the cohesion energy. Let ϕ_b and ϕ_s be the volume fractions of the biomaterial and the solvent with $\phi_b+\phi_s=1$, u_b and u_s being the velocities of the biomaterial and solvent, and the whole system is incompressible so the volume average velocity $u=\phi_b u_b+\phi_s u_s$ is divergence free. A cohesive energy $E(\phi_b)$ of the form $E(\phi_b)=\int(f(\phi_b)+\frac{\kappa}{2}|\nabla\phi_b|^2)dV$ is introduced. Here $f(\phi_b)$ is the mixing energy density (Figure 7a) with a minimum at $\phi_{b,0}$ and an inflection point at $\phi_{b,\inf l}$. The term $\frac{\kappa}{2}|\nabla\phi_b|^2$ penalizes the rough change of ϕ_b at small-scales. Using $(d/dt)\phi_b+\nabla\cdot(\phi_b u_b)=0$ (the mass conservation of ϕ_b with growth term neglected), the time derivative of E is calculated as $\frac{dE}{dt}-\int(\phi_b\nabla\cdot\Pi)\cdot u_b dV$ (2.11) where $\Pi=-[f'(\phi_b)-\kappa\nabla^2\phi_b]I$ is a cohesive stress tensor. Eq. (2.11) is in the form of a work integral $\dot{E}=\int f\cdot u_b dV$, hence the force term $\phi_b\nabla\cdot\Pi$ must be included in the momentum balance equation for the biomaterial. Neglecting the inertial and viscous and viscoelastic stresses terms, the momentum equations for ϕ_b and ϕ_s are $0=\phi_b\nabla\cdot\Pi-\zeta(u_b-u_s)+\phi_b\nabla p$, $0=-\zeta(u_b-u_s)+\phi_s\nabla p$ where p is the hydrostatic pressure and $\zeta(u_b - u_s)$ is the frictional coupling between biomaterial and solvent with $\zeta = \zeta_0\phi_b\phi_s$. Some algebraic manipulations yield an equation for ϕ_b as

$$\frac{\partial\phi_b}{\partial t} + \nabla\cdot(\phi_b u) = \nabla\cdot[a(\phi_b)f''(\phi_b)\nabla\phi_b] - \kappa\nabla\cdot[a(\phi_b)\nabla\nabla^2\phi_b]$$

where $a(\phi_b) = \zeta_0^{-1}\phi_b(1-\phi_b)$. Analysis and a numerical simulation were performed in the 1D case, where the boundary conditions and the incompressibility yield $u = 0$ and the equation for ϕ_b becomes a modified Cahn–Hilliard equation. The model predicted that a minimization of the cohesive energy results in a spontaneous formation of free surface films. Furthermore, in the presence of the cohesion energy for a sufficiently low density biomaterial, the one-dimensional system appears to spontaneously separate into microcolonies due to a spinodal decomposition instability (Figure 8). This model suggests that the cohesion energy is fundamental to many aspects of the biofilm's mechanics. We remark that this mode and the model proposed by Keener and Cogan are essentially the same. They only differ in minor details, both of which belong to the class of binary or two fluid models.

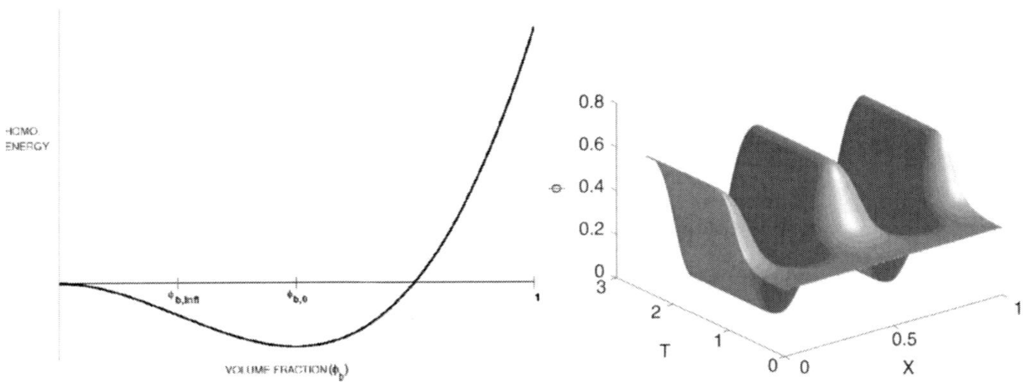

(a) Mixing energy density $f(\varphi_b)$ **(b) Solution starting with a random perturbation of the constant solution**

Figure 8. Cohesion energy model.

Experiments [43,62] also indicate that the biofilm responds elastically to a mechanical stress on short time scales and behaves as a viscous fluid on long time scales. Early work of modeling biofilm as a viscoelastic fluid includes [63]. Researchers are now starting to build biofilm models using more sophisticated constitutive equations to capture the viscoelastic nature of the biofilm.

Most of the models described above either track the biofilm–fluid interface explicitly [46-48, 56], or assume that the solvent has zero velocity [55, 58]. Motivated by the work from [56, 58, 61] and other researchers, but in particular frustrated by the lack of mathematical machinery to handle the inlet and outlet boundary conditions in the multi fluid models, [64, 65] proposed a phase-field biofilm model. The model is based on the one-fluid two-component formulation in which the combination of the extracellular polymeric substances

(EPS) and the bacteria is effectively modeled as one fluid component, while the collective ensemble of the substrate and the solvent are modeled as the other. The biofilm is assumed to be an incompressible continuum, in which the relative motion of the polymer network and the solvent relative to the average velocity is accounted for by mixing kinetics. This model can be shown to be a closure model of a more general kinetic theory for biofilm mixtures [66].

Let v be the average velocity, p the pressure, ϕ_n and ϕ_s the volume fraction of the polymer network and the solvent respectively, and c the substrate concentration, the phase field model for biofilms (treated as viscous fluids) consists of the following four equations in dimensionless form: $\nabla \cdot v = 0$,

$$\rho \frac{dv}{dt} = \nabla(\phi_n \tau_n + \phi_s \tau_s) - [\nabla p + \Gamma_1 \nabla \cdot (\nabla \phi_n \nabla \phi_n)],$$

$$\frac{\partial}{\partial t}(\phi_s c) + \nabla \cdot (cv\phi_s - D_s \phi_s \nabla c) = -g_c,$$

$$\frac{\partial \phi_n}{\partial t} + \nabla \cdot (\phi_n v) = \nabla \cdot (\Lambda \phi_n \nabla \frac{\delta f}{\delta \phi_n}) + g_n.$$

where

$$\tau_n = \frac{2}{\mathrm{Re}_n} D, \quad \tau_s = \frac{2}{\mathrm{Re}_s} D, \quad D = \frac{\nabla v + \nabla v^T}{2}, \quad g_c = A\phi_n \frac{c}{K_n}, \quad g_n = \varepsilon \mu \phi_n \frac{c}{K_c + c}.$$

The mixing free energy density is given by $f = \frac{\Gamma_1}{2} \| \nabla \phi_n \|^2 + \Gamma_2 \times [\frac{\phi_n}{N} \ln \phi_n + (1-\phi_n) \ln(1-\phi_n) + \chi \phi_n (1-\phi_n)]$. Here we consider that the network and the solvent have the same density ρ, Re_n and Re_s are the Reynolds numbers of the network and the solvent respectively, D_s is the diffusion coefficient for the substrate, and Λ, Γ_1, Γ_2 are dimensionless constants associated with the mobility, strength of the distortional free energy and that of the bulk energy. Here f is an extended Flury–Huggins free energy density including the distortional free energy term $\frac{\Gamma_1}{2} \| \nabla \phi_n \|^2$, which accounts for the surface tension effect at the biofilm–solvent interface defined by $\phi_n = 0$, N is the generalized polymerization index, and χ is the Flory–Huggins mixing parameter. Steady state solutions are identified, and the linear stability analysis shows that the flat biofilm–solvent interface is unstable for a finite interval of perturbation modes, with a single maximally unstable mode, which is in agreement with [56,58]. Numerical simulations in 2D captured various phenomena observed in experiments including a mushroom shaped growth (Figure 9), detachment due to shear stress of the flow (Figure 10), rippling and erosion, etc. The advantage of modeling biofilms as a multicomponent material includes the robust treatment of the physics and the interacting dynamics among the components. Meanwhile, deriving a model consisting of a single fluid eliminates several difficulties associated with the coupled biofilm-bulk fluid flow such as velocity, boundary conditions, etc. In particular, the interface conditions are dramatically simplified, since the interface is not separated from the rest of the system. In addition, influent and effluent boundary conditions are natural in the single fluid case. It also provides a framework in which various constitutive relations for each constituent can be investigated in conjunction with the motion of the bulk fluid. The model presented above is the viscous limit of a more general viscoelastic model which we will discuss later.

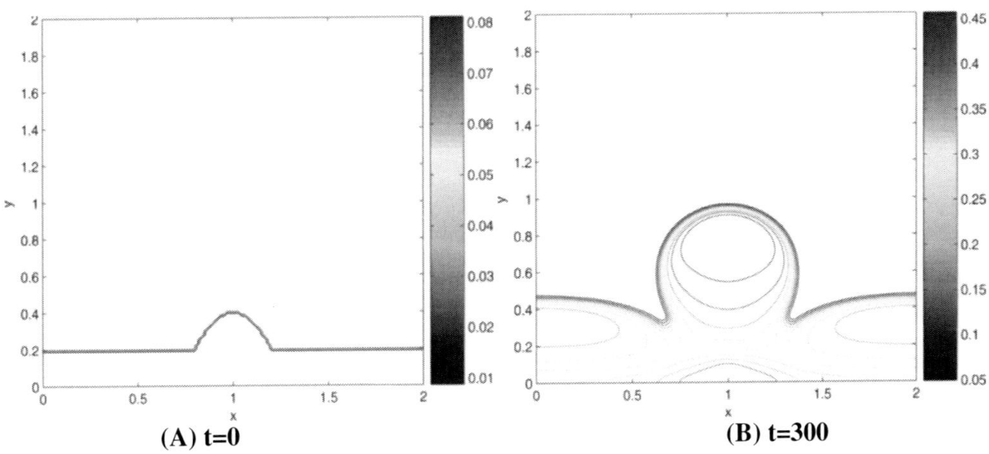

Figure 9. Growth of mushroom shaped biofilm.

Individual-Based Modeling (IbM)

Kreft [67,68] proposed the Individual-based Modeling (IbM) of biofilms. The model consists of two parts: one deals with the growth and behavior of individual bacteria as autonomous agents; the other deals with the substrate and product diffusion and reaction (Figure 10a). Bacterial cells are represented as hard spheres, with each cell having, besides a variable volume and mass, a set of variable growth parameters. Each cell grows by consuming the substrate and divides when a certain volume is reached. The pressure buildup due to the growth of biomass is released by maintenance of a minimum distance between the neighboring cells. For each cell, the vector sum of all positive overlap radii with the neighboring cells is calculated and then the position of the cell is shifted in the direction opposite to this vector. The substrate concentration is governed by a reaction-diffusion equation, thus IbM is a hybrid discrete-continuous model. The biomass spreading mechanism in IbM allows cell movement with a continuous distance and direction, which makes it a deterministic model and overcomes the drawbacks associated with the randomness in the CA based models. Figure 10b shows the computation domain of the model. Numerical simulations in 2D showed that the IbM model produced a more confluent and rounded biofilm structure than the CA based models, due to its deterministic and directionally unconstrained spreading of the biomass. [69] extended the IbM model and proposed a Particle-Based Multidimensional Multispecies Biofilm Model.

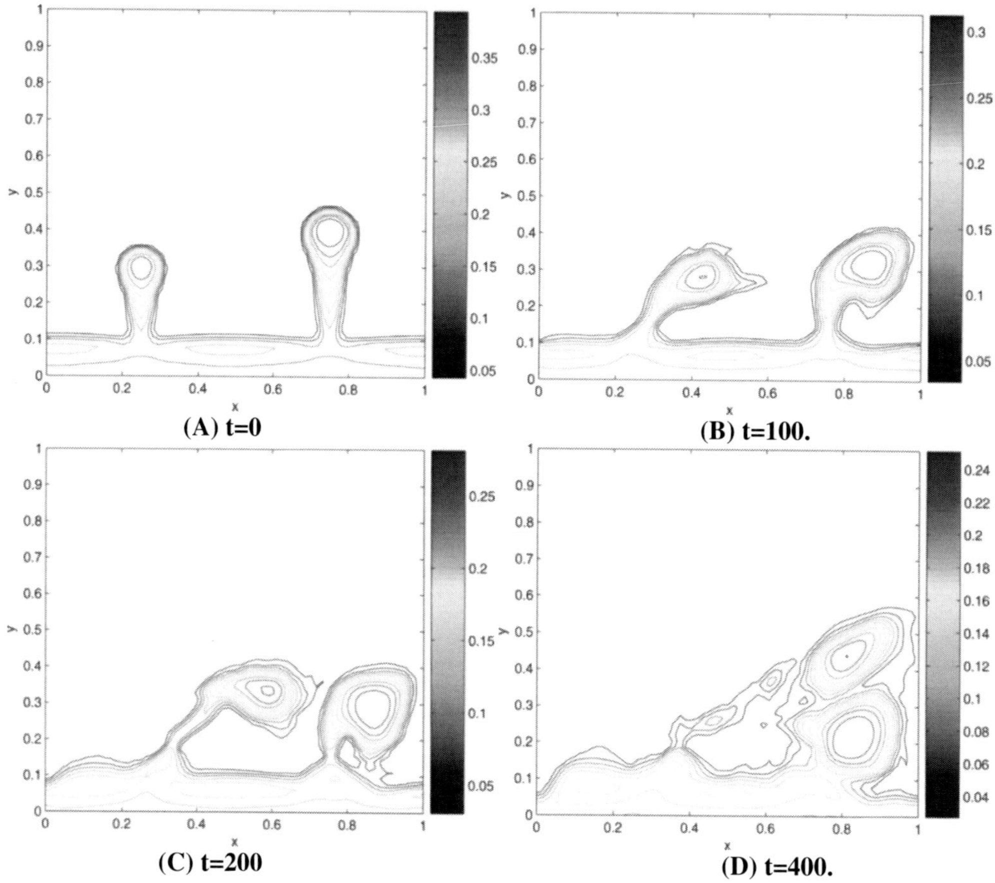

Figure 10. Detachment induced by shear flow over two mushroom shaped biofilm.

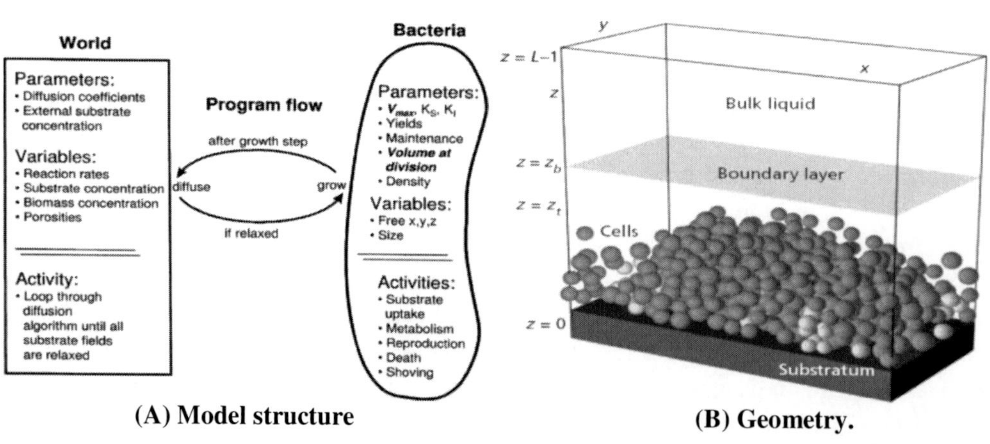

Figure 11. IbM model structure and schematic system geometry.

This model allows variable size biomass particles in order to model the systems with a large-scale heterogeneity. Numerical simulations predicted that if only the average flux of nutrients needs to be known, the 2D and 1D models are very similar. However, the behavior of intermediates, which are produced and consumed in different locations within the biofilm, is better described in the 2D and 3D models because of the multidirectional concentration gradients. The predictions of the 2D or 3D models are also different from those of 1D models for slowly growing or minority species in the biofilm (see Figure 11).

Other Models Related to the Biofilm Properties

Most of the models described above concentrate on the biofilm's structure resulted from biomass growth and spreading, substrate transport and consumption. Experimental results also indicate that the biofilm exhibits complicated physical, chemical and biological properties during the interactions with the surrounding environment such as the external fluid flow and the antimicrobial agents. There are extensive studies on these biofilm properties and many models associated with that, and we will describe a few of them in this section.

Antimicrobial Penetration Barrier

Bacterial cells inside the biofilm generally have a higher tolerance to chemical attack from antimicrobial agents compared to planktonic cells [70-72], and a widely accepted explanation is that the biofilm matrix acts as a protective barrier against attack. [73-75] proposed a theoretical framework to explain this phenomenon based on diffusion processes in biofilms. In this model, the biofilm is assumed to be a uniformly thick planar slab of thickness L_f with one side adjoining an impermeable boundary and the other side exposed to the bulk fluid. Fick's law of diffusion is used and the mass transfer resistance between the biofilm-fluid interface and the bulk fluid is neglected. The biofilm contains cells (volume fraction ε_c), extracellular polysaccharide (volume fraction ε_p), and water (volume fraction $\varepsilon_w = 1 - \varepsilon_c - \varepsilon_p$). The antimicrobial is modeled as a solute in the biofilm with a concentration C, and the spatial variable is z. Five cases were considered based on different properties of the solute in which the governing equations are given as follows:

1. Diffusion of a non-interacting solute.

$$\varepsilon_w \frac{\partial C}{\partial t} = D_e \frac{\partial^2 C}{\partial z^2}.$$

2. Diffusion of a reversibly sorbing, non-reacting solute.

$$[\varepsilon_w + (1 - \varepsilon_w)\kappa] \frac{\partial C}{\partial t} = D_e \frac{\partial^2 C}{\partial z^2}.$$

Here κ is a partition coefficient quantifying the equilibrium distribution of solute between the aqueous and biofilm phases.

3. Diffusion of an irreversibly sorbing, non-reacting solute
$\varepsilon_w \frac{\partial C}{\partial t} = D_e \frac{\partial^2 C}{\partial z^2} - k_s C X_s$, $\frac{\partial X_s}{\partial t} = -\frac{k_s C X_s}{Y_{cx}}$, $X_s(0,z) = X_s^0 \quad 0 \le z \le L_f$. Here X_s is the concentration of free binding sites, k_s is the rate coefficient for the sorption process, Y_{cx} is a yield coefficient describing the mass of solute sorbed per mass of biofilm in which binding sites have been saturated.
4. Diffusion of stoichiometrically reacting solute. The governing equation is identical to the one in the previous case of irreversible sorption. Here Y_{cx} is the mass of solute neutralized per mass of biofilm reacted, and the sorption rate coefficient k_s is replaced with the reaction rate coefficient k_r.
5. Diffusion of a catalytically reacting solute

$$\varepsilon_w \frac{\partial C}{\partial t} = D_e \frac{\partial^2 C}{\partial z^2} - k_r C.$$

In all the cases described above, C satisfied the same initial and boundary conditions given by $C(0,z) = 0$, $0 \le z \le L_f$, $\frac{\partial C(t,z)}{\partial z}|_{z=0} = 0$, $C(t, L_f) = C_0$.

The parameter D_e is the effective diffusion coefficient in the biofilm, and it is expected to be smaller than the diffusion coefficient of water D_{aq}. Here D_e is determined by the following empirical formula $\frac{D_e}{D_{aq}} = (\frac{1-\varepsilon_c}{1+\varepsilon_c/2})(\frac{(1-\varepsilon_p^*)^3}{(1+\varepsilon_p^*)^2})$ where is the polymer volume fraction in the intercellular space. Numerical simulations of the model predicted the following properties. A non-interacting solute was predicted to penetrate the biofilms of up to 1 mm relatively quickly, within seconds or minutes. In this case the solute does not sorb or react in the biofilm, therefore, the diffusion barrier is not large enough to account for the reduced susceptibility of biofilms to antibiotics. Reversible and irreversible sorption retards the antibiotic penetration. A catalytic reaction, provided it is sufficiently rapid, can lead to severe antibiotic penetration failure. This model provides guidelines for the design and analysis of experiments to test the mechanisms of reduced biofilm susceptibility to antibiotics.

Persisters

Besides the penetration barrier described above, another important hypotheses concerning the resistance mechanism to antimicrobial for the bacterial biofilms is the existence of "persister" cells which are extremely tolerant of the antimicrobials. Experimental results [76, 77] demonstrate the existence of persisters and indicate that these are not genetic variants nor cells in a spore-like state nor cells caught coincidentally in a quiescent phase of cell division.

A biofilm with persisters typically exhibits the bi-phasic disinfection curve, namely, the bacterial population versus disinfection time is of the form $p(t)=C_1e^{-k_1t}+C_2e^{-k_2t}$ with $k_1>k_2$, and $C_1>>C_2$. Two different types of persister model will be described here.

Cogan [78] proposed a phenotype-switching model. The bacterial phenotypes are denoted by B_S and B_P for the susceptible and the persister cell density, and there is one growth-limiting substrate, denoted by S, and one antimicrobial, denoted by A. The population of the susceptible cells changes due to growth, death due to antibiotic action, loss due to transition to persister cells and gain as the persistent cells revert back to the susceptible cells. The governing equation of B_S is given by $\frac{dB_S}{dt}=\frac{\mu_{\max}}{Y}\frac{S}{K_S+S}B_S-k_d(A,t)\mu_{\max}\frac{S+\alpha}{K_S+S}B_S-k_l(A,t)\mu_{\max}\frac{S}{K_S+S}B_S+k_g(A,t)B_P$.

Here the parameters k_d, α, k_l, k_g are assumed to be linearly proportional to A and the proportional constants are chosen to fit data from [77]. Persister cells are not killed by the antibiotic, and the population changes are only due to cells converting to and from the susceptible cells. The equation for B_P is $\frac{dB_P}{dt}=k_l(A,t)\mu_{\max}\frac{S}{K_S+S}B_S-k_g(A,t)B_P$.

The substrate is consumed only by the susceptible cells and the equation for S is $\frac{dS}{dt}=-\mu_{\max}\frac{S}{K_S+S}B_S$. Numerical simulation and analysis of a simplified model indicates that the relative dose/withdrawal times are important in determining the effectiveness of a periodic dosing treatment, and there is an optimal withdrawal length specific to the bacterial kinetics.

Klapper [79] proposed a model which treats the persister cells as senescent cells based on the idea that the senescent cells are slow-growing, and slow-growing cells are more tolerant to antimicrobials. Let $b(a,t)$ denote the bacterial population density at time t of the senescence level a. The governing equation of b is $\frac{\partial b}{\partial t}(a,t)+\frac{\partial b}{\partial a}(a,t)=-(k_D+r_K(a,d))b$, $a>0,t>0$ (2.12) with the killing rate function r_K defined by $r_K(a,d)=\begin{cases}k_K, & a\le\delta d\\ 0, & a>\delta d\end{cases}$ where k_K is the killing coefficient. Here k_D is the natural death rate, d is the given antimicrobial concentration and δ is an adjustable tolerance coefficient. For a fixed d value, a small δ means that the cells become tolerant at a relatively young age, and a large δ means that the cells become tolerant at a relatively old age. Eq. (2.12) is a linear wave equation for $a>0,t>0$, with characteristics emanating from positive a- and t-axes. It is supplied with the initial condition $b(a,0)=b_0(a)$ and a birth boundary condition at $a=0$ given by $b(0,t)=\int_0^\infty r_b(a,C)b(a,t)da$ where r_b is a birth rate function depending on the cell's age a and the substrate's concentration C. The evolution of C is governed by the ODE $\frac{dC}{dt} = -\int_0^\infty \int_0^\infty r_C(a,C)b(a,t)da$ where r_C is a substrate usage function. A typical numerical simulation [79] is shown in Figure 12. It predicted the bi-phasic disinfection curve and the relative scarcity of persisters during the growth phase as opposed to during the stationary phase. Thus this model based on senescence provides a natural explanation for persistence-related phenomena.

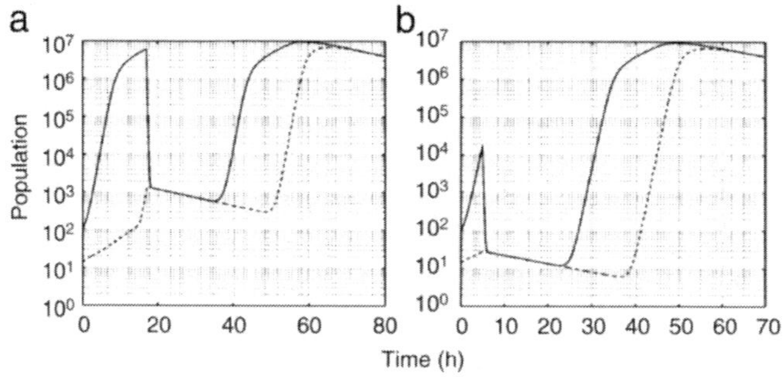

Figure 12. Exposure of a batch culture to an antimicrobial. (a) Antimicrobial applied during the stationary phase at t=17 h and removed at t=27 h, at which time the surviving cells are recultured at the t=0 h nutrient level. (b) Antimicrobial applied during the exponential phase at t=5 h and removed at t=15 h, at which time the surviving cells are recultured at the t=0 h nutrient level. Solid lines, all cells; dashed lines, persisters.

Viscoelastic Model

Towler et al., [80] developed a finite element model of the biofilm response to flows, in which a linear viscoelastic model (Burger material law) is coupled to finite element fluid solver sequentially to study the mechanical response of a single semi-spherical cluster to incompressible and turbulent flows. They concluded that the deformation paths of the viscoelastic biofilm are largely insensitive to the specific material coefficients in the model. Burger's material law is given by a linear integral constitutive equation:

$$\varepsilon_{ij}(t) = \delta_{ij}[(\frac{1}{R_1} - \frac{1}{G_1})\sigma_{kk}(t) + \int_0^t [-\frac{1}{G_2}(1-e^{-G_2(t-\tau)/\eta_2}) - \frac{t-\tau}{\eta_1}]\frac{\partial \sigma_{kk}}{\partial \tau}] + \frac{1}{G_1}\sigma_{ij}(t) + \int_0^t [\frac{1}{G_2}(1-e^{-G_2(t-\tau)/\eta_2}) - \frac{t-\tau}{\eta_1}]\frac{\partial \sigma_{kk}}{\partial \tau}]\frac{\partial \sigma ij(\tau)}{\partial \tau}d\tau$$

(2.13). where ε is the train tensor, δ is the Kroneker delta, σ is the stress tensor, t is time, R_1 is the bulk elastic modulus, G_1 and η_1 are the shear modulus and viscous coefficient, respectively, for the Maxwell component of the Burger model, and G_2 and η_2 are the shear modulus and viscous coefficient, respectively for the Kelvin component. Essentially, the biofilm cluster is treated as a piece of the viscoelastic solid material in this model. Nonlinear viscoelastic models are developed in the context of one fluid multicomponent formulations, in which various viscoelastic constitutive equations can be incorporated depending on the microstructure of the EPS network [64, 66].

Modeling Biofilms Using Other Computational Cellular Tools

Poplawski [81] used the Glazier-Graner-Hogeweg (GGH) model coupled with a discrete-continuum model (PLH) developed by [52] to study the biofilm's dynamics provided by the CompuCell3D modeling environment, a cell-oriented framework designed to simulate the growth and the pattern formation due to the biological cells' behaviors. They simulated the growth of a single-species bacterial biofilm, and studied the roles of the cell-cell and the cell-flow-field interactions in determining the biofilm's morphology. In their simulations, which generalized the PLH model by treating cells as spatially extended deformable bodies, differential adhesion between cells and their competition for a substrate (nutrient), surface to produce a fingering instability that generates the finger shapes of biofilms.

Quorum Sensing (QS)

Biofilms are complex systems that can be viewed simultaneously as microbial ecological communities and as mechanical objects. The formation and growth of a biofilm is the result of a complex dynamic process that involves the transport of organic and inorganic molecules and microbial cells to the surface, a subsequent adsorption to the surface and finally an irreversible attachment aided by the production of EPS [83]. Due to its complexity, the formation of biofilms is regulated at different stages via diverse mechanisms [84, 85]. The most studied regulatory mechanism that has been found to control the production of EPS, biofilm formation and differentiation is quorum sensing (QS) regulation [82, 84-88]. QS is generally interpreted as a cell-cell communication mechanism used by several bacterial taxa to coordinate gene expression and behaviour in groups,based on population densities. Initially, bacteria cells produce and release low amounts of signalling molecules, called autoinducers (e.g., acyl-homoserine lactones(AHL) in Gram-negative bacteria). Concurrently, the cells measure the environmental concentration of the autoinducer. When a critical concentration is reached, changes in gene expressions are induced. QS allows bacteria to maintain cell-cell communication and also regulate the expression of specific genes in response to changes in cell population density [87,89]. In general, the QS process involves the production, release and detection of chemical signaling molecules, thus allowing microbial cells to regulate gene expression in a cell-density-dependent manner [90]. At a given population density, the genes involved in biofilm differentiation and maturation are activated [82, 85].

Quorum sensing(QS) is an important cellular mechanism in biofilm formation and growth dynamics. There have been some local and nonlocal models for QS, passively or actively coupled with the biofilm dynamics [82, 84-88, 98]. More are listed in the review article [98, 99] and the references therein. The models of [99-102] detail biochemical pathways in quorum sensing systems, also describing anti-quorum sensing treatments for applications in the medical field. However, none of them have the full hydrodynamic coupling including the momentum conservation and with the ambient fluid. Two QS processes, the autoinducer-1 and autoinducer-2, have been adopted to describe bacteria biofilms[85, 90]. The autoinducer-1 (AI-1) type is mainly involved in intra-species communication and the autoinducer-2 (AI-2) type is associated with inter-species interaction [92]. Gram-negative bacteria produce and release AI molecules, which are generally N-actyl homoserine lactone (AHL) molecules that serve as a function of controlling the cell-population density. Bacteria detect the accumulation of AHL signals. Above a certain threshold concentration, these signals are present in sufficient quantity to enable similar transcriptional effectors to activate silent genes. This alters their cell-density dependent gene expression and therefore their behavior [84, 86]. In Gram-positive bacteria, communication is carried out with modified oligopeptides generating the signals and membrane-bound sensor histidine kinases acting as receptors. Signalling is mediated by many phosphorylation steps, which control the activity of a response regulator. However, peptide signals are not diffusible across the membrane and therefore the signal release is mediated by oligopeptide exporters. Normally, signal release occurs concurrently with signal processing and modification [85, 91].

It has been estimated that 6% to 10% of all genes in *P. aeruginosa* are controlled by AHL quorum sensing signaling systems[103]. Probably one of the best studied QS processes is the AHL-mediated QS system first described for the bioluminescent marine bacterium *Aliivibrio fischeri* (formerly *Vibrio fischeri* [93]). This system is considered to be the ideal model for the QS paradigm in most gram-negative bacteria. There are two proteins, LuxI and LuxR, which control the expression of the luciferase operon required for light production in *A. fischeri.* LuxI is the autoinducer synthase that produces AHL inducers via S-adenosylmethionine (SAM) and LuxR is the cytoplasmic inducer receptor/DNA-binding transcriptional activator that requires AHL coinducers to initiate expression of the luciferase-coded function [85, 86, 91]. Once produced, AHL molecules diffuse in and out of the cell membrane and increase in concentration when the cell-population density increases. When the critical threshold concentration is reached, the AHL is bound by LuxR. The resulting LuxR-AHL complex activates the transcription of the luciferase operon, and also the expression of LuxI and other genes involved in different behavioural responses, creating a positive feedback loop, which results in the production of light [85, 87, 91].

Quorum-sensing-controlled processes are often crucial for successful bacterial–host relationships, both symbiotic and pathogenic in other ways as well[98]. And QS is regarded as one of the regulatory pathways for EPS production and biofilm formation in bacteria [84, 87, 90, 91, 94,98]. Also, phosphate and polyphosphate metabolism has been associated with biofilm development and the QS regulatory pathway [92, 95]. However, since the QS regulatory system and biofilm formation and maintenance mechanisms are diverse among different bacterial species, the role of QS in biofilm formation cannot be described in general terms [90]. For example, in *Pseudomonas aeruginosa*, QS is essential for adhesion, proper biofilm formation and virulence factors [85, 96]. The QS system of *P. aeruginosa* is unusual because it has two somewhat redundant regulatory QS systems: LasI/LasR and RhlI/RhlR, which work the same way as the LuxI/LuxR system in *A. fischeri* [91]. The first QS system was shown to regulate expression of the elastase LasB and was therefore named the *las* system. The two enzymes, LasB elastase and LasA elastase, are responsible for elastolytic activity which destroys elastin-containing human lung tissue and causes pulmonary hemorrhages associated with *P. aeruginosa* infections. The second QS system in *P. aeruginosa* is named the *rhl* system because of its ability to control the production of rhamnolipid. Rhamnolipid has a detergent-like structure and is responsible for the degradation of lung surfactant and inhibits the mucociliary transport and ciliary function of human respiratory epithelium. Mutant *P. aeruginosa* cells that did not produce any QS signals were found to be more densely populated with a thinner biofilm than the wild type. In addition, mutation of the LasI gene resulted in an abnormal and undifferentiated biofilm formation process [88]. In *E. coli*, cellular functions are controlled by the QS LsrR/LsrK system. Biofilm formation and architecture were found to be significantly altered in lsrR and lsrK mutants. There were differences observed in the cell fimbriae and matrix structure and in the thickness of the mutants compared to the wild type [97]. Lastly, a QS system, AfeI/AfeR, has recently been identified in *Acidithiobacillus ferrooxidans*, which is similar to the LuxI/LuxR proteins [84, 87]. It was found that the amount of lipopolysaccharides present was increased in phosphate starved A. *ferrooxidans*. There was also an increase in the transcription of the *afeI* gene, and therefore AHL levels, when cells were cultured in a low-phosphate medium compared to that of a high-phosphate medium. Since the AHL communication system is present in *A. ferrooxidans*, this suggests that QS could regulate the formation of EPS and biofilms for attachment to solid substrates [84, 92,98].

Conclusion

We have revisited a few families of the biofilm models chronologically. The models described in this review article have already been used to explain many complicated phenomena in the biofilm's dynamics. However, there are still many unanswered questions in the mathematical biofilm modeling, especially in the effort of incorporating all the physical, chemical, biological and ecological processes occurring at various time- and length-scales into a comprehensive analytical or computational model. We also introduced bacterial quorum sensing and its regulatory mechanism. No matter how complicated, quorum sensing (QS) is indeed a very fantastic phenomenon, and it has also attracted the attention of many scientists. But now the research of quorum sensing is only at the preliminary stage.

With the improvement on one's understanding of the biofilm's dynamics, QS and fast advancing computing technology and experimental techniques, we anticipate another surge in research activities in this scientifically and technologically important and economically even more important topic. Note that, here we do not model the phenomenon of bacterial quorum sensing, and this is also currently a biofilm research hotspot.

References

[1] J.W. Costerton, Z. Lewandowski, D.E. Caldwell, D.R. Korber, H.M. Lappin-Scott(1995). Microbial biofilms. *Annu. Review Microbiol, 49,* 711–745.

[2] M.E. Davey, G.A. O'Toole(2000). Microbial biofilms: from ecology to molecular genetics. Microbiol. Mol. Biol. Rev, 64, 847–867.

[3] G.A. O'Toole, H.B. Kaplan, R. Kolter(2000). Biofilm formation as microbial development. *Annu.Rev.Microbiol, 54,* 49–79.

[4] C. Picioreanu, J.B. Xavier, M.C.M. van Loosdrecht(2004). Advances in mathematical modeling of biofilm structure, *Biofilm,1,*337–349.

[5] B. E. Rittmann and P. L. McCarty(1980). Evaluation of steady-state biofilm kinetics. *Biotech. Bioeng, 22,* 2359–2373.

[6] B. E. Rittmann and P. L. McCarty(1980). Model of steady-state-biofilm kinetics. *Biotech. Bioeng, 22,* 2243–2357.

[7] B. E. Rittmann(1982). The effect of shear stress on biofilm loss rate. *Biotech. Bioeng. 24,* 501–506.

[8] B. E. Rittmann and P. L. McCarty(1980). Model of steady-state-biofilm kinetics. *Biotech. Bioeng. 22,* 2243–2357.

[9] O. Wanner and W. Gujer(1986). A multispecies biofilm model. *Wat. Sci. Tech, 28,* 314–328.

[10] J. C. Kissel, P. L. McCarty, and R. L. Street(1984). Numerical simulation of mixed-culture biofilm. *J. Environ. Eng, 110,* 393–411.

[11] O. Wanner, P. Riechert, Mathematical modeling of mixed-culture biofilms. *Biotech. Bioeng, 49 (1996),* 172–184.

[12] T.A. Witten, L.M. Sander(1981). Diffusion limited aggregation, A kinecal critical phonemenon. *Phys Rev. Lett, 47, 1400–1403.*

[13] H. Fujikawa, M. Matsushita (1989). Fractal growth of *Bacillus subtilis* on agar plates. *J. Phys.Soc.Japan, 58,* 3875–3878.

[14] M. Matsushita, H. Fujikawa(1990). Diffusion-limited growth in bacterial colony formation. *Physica, A168,* 498–506.

[15] H. Fujikawa(1994). Diversity of the growth patterns of *Bacillus subtilis* colonies on agar plates. *FEMS Microbiol. Ecol, 13,* 159–168.

[16] S. Tolman, P. Meakin, M. Matsushita(1989).Cluster-size distribution in the incremental growth of DLA clusters. *J. Phy. Soc. Jpn, 58,* 2721–2726.

[17] E.R. Berlekamp, J.H. Conway, R.K. Guy(1982). Winning Ways for Your Mathematical Plays. *Academic Press, New York,* NY.

[18] S. Wolfram(1984). Cellular automata as models of complexity. *Nature (London), 311,* 419–424.

[19] D.G. Green(1990). Cellular automata models in biology. *Math. Comp. Mod, 13,* 69–74.

[20] G.B. Ermentrout(1993). Cellular automata approaches to biological modeling. L. Edelstein-Keshet. *J. Theoret. Biol, 160,* 97–133.

[21] R. L. Colasanti(1992). Cellular automata models of microbial colonies. *Binary Comput Microbiol, 4,* 191.

[22] G. C. Barker and M. J. Grimson(1993). A cellular automaton model of microbial growth. *Binary Comput.Microbiol, 5,* 132–137.

[23] E. Ben-Jacob, O. Schochet, A. Tenenbaum, I. Cohen, A. Czirok, and V. Tamas(1994). Generic modelling of cooperative growth patterns in bacterial colonies. *Nature(London), 368,* 46–49.

[24] J. W. T. Wimpenny and R. Colasanti(1997). A unifying hypothesis for the structure of microbial biofilms based on cellular automaton models. *FEMS Micro. Ecol, 22,* 1–16.

[25] G. Pizarro, D. Griffeath, D.R. Noguera (2001). Quantitative cellular automaton model for biofilms. *J. Environ Eng, 127,*782–789.

[26] G. Pizarro, R. Moreno C. Garcia, and M. E. Sepulveda(2004). Two-dimensional cellular automaton model for mixed-culture biofilm. *Wat. Sci. Tech, 49,* 193–198.

[27] C. Picioreanu, M. C. M. Loosdrecht, and J. J. Heijnen(1998). A new combined differential discrete cellular automaton approach for biofilm modeling: application for growth in gel beads. *Biotech.Bioeng, 57,* 718–731.

[28] R.H. Wijffels, C.D. de Gooijer, S. Kortekass, J. Tramper(1991). Growth and substrate consumption of *Nitrobacter agilis* cells immobilized in carrageenan. *Biotech. Bioeng, 38,* 224–231.

[29] R.H. Wijffels, C.D. de Gooijer, S. Kortekass, J. Tramper(1991). Growth and substrate consumption of *Nitrobacter agilis* cells immobilized in carrageenan. *Biotech. Bioeng, 38,* 232–240.

[30] R.H. Wijffels(1994). Ph.D. Thesis. Nitrification by immobilized cells. *Wageningen Agricult- ural University, Wageningen, The Netherlands.*

[31] D.R. Noguera, G. Pizarro, D.A. Stahl, B.E. Rittman, (1999). Simulation of multispecies biofilm development in three dimensions. *Wat. Sci. Tech, 39,* 123–130.

[32] S. W. Hermanowicz(2001). A simple 2d biofilm model yields a variety of morphological features. *Math. Biosci, 169,* 1–14.

[33] D. DeBeer, P. Stoodely, Z. Lewandowski(1994). Liquid flow in heterogenous biofilms. *Biotech. Bioeng, 44,* 636–641.

[34] D. DeBeer, P. Stoodely, Z. Lewandowski(1996). Liquid flow and mass transport in heterogeneous biofilms.*Water Res, 30,* 2761–2765.

[35] D. DeBeer, P. Stoodely, F. Roe, Z. Lewandowski(1994). Effects of biofilm structure on oxygen distribution and mass transport. *Biotech. Bioeng, 43,*1131–1138.

[36] S.M. Hunt, M.A. Hamilton, J.T. Sears, G. Harkin, J. Reno(2003). A computer investigation of chemically mediated detachment in bacterial biofilms. *Microbiology,149,*1155–1163.

[37] S.M. Hunt, E.M. Werner, B. Huang, M.A. Hamilton, P.S. Stewart(2004). Hypothesis for the role of nutrient starvation in biofilm detachment. *Appl. Environ. Microbiol, 72,*7418–7425.

[38] J.D. Chambless, P.S. Stewart(2007). A 3D computer model analysis of three hypothetical biofilm detachment mechanisms. *Biotech. Bioeng, 97,* 1573–1584.

[39] J.D. Chambless, S.M. Hunt, P.S. Stewart(2006). A three-dimensional computer model of four hypothetical biofilm detachment mechanisms. *Appl. Environ. Microbiol, 72,* 2005–2013.

[40] A. Gjaltema, P.A.M. Arts, M.C.M. van Loosdrecht, J.G. Kuenen, J.J. Heijnen(1994). Heterogeneity of biofilms in rotating annular reactors: Occurrence, structure, and consequences. *Biotech. Bioeng, 44,* 194–204.

[41] T.C. Zhang, P.L. Bishop(1994). Density, porosity, and pore structure of biofilms. *Water Res, 28,* 2267–2277.

[42] T.C. Zhang, P.L. Bishop(1994). Evaluation of tortuosity factors and effective diffusivities in biofilms. *Water Res, 28,* 2279–2287.

[43] P. Stoodley, Z. Lewandowski, J. D. Boyle, and H. M. Lappin-Scott(1999). Structural deformation of bacterial biofilms caused by short term fluctuations in flow velocity: an in-situ demonstration of biofilm viscoelasticity. *Biotech. Bioeng, 65,* 83–92.

[44] P. Stoodley, Z. Lewandowski, J.D. Boyle, H.M. Lappin-Scott(1999). The Formation of Migratory Ripples in a Mixed Species Bacterial Biofilm Growing in Turbulent Flow. *Environ. Microbiol, 1,* 447–457.

[45] P. Stoodley, F. Jorgensen, P. Williams, H.M. Lappin-Scott(1999), in: R. Bayston, et al., (Eds.), Biofilms: The Good, The Bad, and The Ugly. *Bioline Press, Cardiff, United Kingdom, pp,* 323–330.

[46] C. Picioreanu, M. C. Loosdrecht, and J. J. Heijnen(1999). Discrete-differential modelling of biofilm structure. *Wat. Sci. Tech. 39,* 115–122.

[47] C. Picioreanu, M. C. Loosdrecht, and J. J. Heijnen(2000). Effect of diffusive and convective substrate transport on biofilm structure formation: a two-dimensional modeling study. *Biotech. Bioeng, 69,* 504–515.

[48] C. Picioreanu, M. C. Loosdrecht, and J. J. Heijnen(2001). Two-dimensional model of biofilm detachment caused by internal stress from liquid flow. *Biotech. Bioeng, 72,* 205–218.

[49] S.P. Dawson, S. Chen, G.D. Doolen(1993). Lattice Boltzmann computations for reaction-diffusion equations. *J. Chem. Phys, 98,* 1514–1523.

[50] S. Chen, S.P. Dawson, G.D. Doolen, D.R. Janecky, A. Lawniczak, (1995). Lattice methods and their applications to reacting systems. *Computers Chem Eng, 19,* 617–646.

[51] Y. Chen, H. Ohashi, M. Akiyama(1997).Simulation of laminar flow over a backward-facing step using the lattice BGK method. *JSME Int. J. Series B, 40,* 25–32.

[52] C. Picioreanu, M.C. Loosdrecht, J.J. Heijnen(1998). Mathematical modeling of biofilm structure with a hybrid differential-discrete cellular automaton approach. *Biotech. Bioeng, 58,*101–116.

[53] R. Dillon, L. Fauci, A. Fogelson, D. Gaver(1996). Modeling biofilm processes using the immersed boundary method. *J. Comput. Phys, 129,*57–73.

[54] R. Dillon, L. Fauci(2000). A microscale model of bacterial and biofilm dynamics in porous media.*Biotech. Bioeng, 68,* 536–547.

[55] H. J. Eberl, D. F. Parker, and M. C. M. van Loosdrecht(2001). A new deterministic spatio-temporal continuum model for biofilm development. *J. Theor. Mde, 3,* 161–175.

[56] J. D. Dockery and I. Klapper(2001). Finger formation in biofilm layers. *SIAM. J. Appl. Math, 62,* 853–869.

[57] E. Alpkvist, I. Klapper(2007). A microscale model of bacterial and biofilm dynamics in porous media. *Bull Math. Biol, 69,* 765–789.

[58] N. G. Cogan and J. P. Keener(2004). The role of the biofilm matrix in structural development. *Math. Med. Biol, 21,*147–166.

[59] A. N. Beris and B. Edwards(1994). Thermodynamics of flowing system. *Oxford University Press.*

[60] C.A. Fux, S. Wilson, P. Stoodley*(2004).* Detachment characteristics and oxacillin resistance of *Staphylococcus aureus* biofilm emboli in an in-vitro infection model. *J. Bacteriol, 186,* 4486–4491.

[61] I. Klapper and J. Dockery(2006). Role of cohesion in material description of biofilms. *Phys. Rev. E,74,*031902.

[62] B.W. Towler, C.R. Rupp, A. Cunningham, P. Stoodley(2003). Viscoelastic properties of a mixed culture biofilm from rheometer creep analysis. *Biofouling,19,* 279–285.

[63] I. Klapper, C. J. Rupp, R. Cargo, B. Purevdorj, and P. Stoodley(2002). A viscoelastic fluid description of bacterial biofilm material properties. *Biotech. Bioeng, 80,* 289–296.

[64] T. Zhang, N. C. Cogan, and Q. Wang(2008). Phase field models for biofilms. i. theory and One-dimensional simulations. *SIAM J. Appl. Math, 69,* 641–669.

[65] T. Zhang, N. Cogan, and Q. Wang(2008). Phase-field models for biofilms ii. 2-d numerical simulations of biofilm-flow interactions. *Comm. Comput. Phys, 4,*72 –101.

[66] Q. Wang and T. Zhang(2010). Kinetic theories for biofilms. *Discrete and Continuous Dynamical Systems: Series B.*

[67] J.U. Kreft, G. Booth, J.W.T. Wimpenny(1998). a simulator for individual-based modelling of bacterial colony growth. *Microbiology,144,* 3275–3287.

[68] J.U. Kreft, C. Picioreanu, J.W.T. Wimpenny, M.C.M. van Loosdrecht(2001). Individual-based modelling of biofilms. *Microbiology, 147,* 2897–2912.

[69] C. Picioreanu, J.-U. Kreft, M.C. Loosdrecht(2004). Particle-based multidimensional multispecies biofilm model. *Appl. Environ. Microbiol, 70,* 3024–3040.

[70] C.A. Gordon, N.A. Hodges, C. Marriot(1988). Antibiotic interaction and diffusion through alginate and exopolysaccharide of cystic fibrosis-derived *Pseudomonas aeruginosa. J. Antimicrob. Chemother, 22,* 667–674.

[71] W.W. Nichols, S.M. Dorrington, M.P.E. Slack, H.L. Walmsley(1988). Inhibition of tobra- mycin diffusion by binding to alginate. *Antimicrob. Agents Chemother, 32,* 518–523.

[72] D. DeBeer, R. Srinivasan, P.S. Stewart(1994). Direct measurement of chlorine penetration into biofilms during disinfection. *Appl. Environ. Microbiol, 60,* 4339–4344.

[73] P.S. Stewart, J. Raquepas(1995). Implication of reaction-diffusion theory for the disinfection of microbial biofilms by reactive antimicrobial agents. *Chem. Eng. Sci, 50,* 3099–3104.

[74] P.S. Stewart(1996). Theoretical aspects of antibiotic diffusion into microbial biofilms. *Antimicrob. Agents Chemother, 40,* 2517–2522.

[75] P.S. Stewart(2002).Mechanisms of antibiotic resistance in bacterial biofilms. *Int. J. Med.Microbiol,292,* 107–113.

[76] K. Lewis(2001). Riddle of bioflm resistance. *Antimicrob. Agents. Chemo, 45,* 999–1007.

[77] I. Keren, N. Kaldalu(2004), A. Spoering, Y. Wang, and K. Lewis. Persister cells and tolerance to antimicrobials. *FEMS Microbiol. Lett, 230,* 13–18.

[78] N.G. Cogan (2006). Effects of persister formation on bacterial response to dosing. *J. Theoret. Biol, 238,* 694–703.

[79] I. Klapper, P. Gilbert, B.P. Ayati, J.Dockery, P.S. Stewart(2007). Senescence can explain microbial persistence. *Microbiology 153,* 3623–3630.

[80] B.W. Towler, A.Cunningham, P.Stoodley, L.McKittrick(2007). A model of fluid-biofilm interaction using a Burger material law. *Biotech. Bioeng, 96,* 259–271.

[81] N.J. Poplawski, A. Shirinifard, M. Swat, J.A. Glazier(2008). Simulation of single-species bacterial-biofilm growth using the Glazier-Graner-Hogeweg model and the CompuCell3D modeling environment. *Math. Biosciences Eng, 8,* 355–388.

[82] R.M. Donlan (2002). Biofilms: Microbial life on surfaces. Emerging Infect. *Dis, 8,* 881-890.

[83] I.B. Beech(2004). Corrosion of technical materials in the presence of biofilms—current understanding and state-of-the art methods of study. *Int. Biodeterior. Biodeg, 53,* 177–183.

[84] L.M. Ruiz, S. Valenzuela, M.Castro,A.Gonzalez, M.Frezza, L.Soulère, T.Rohwerder, Y.Queneau, A.Doutheau, W. Sand, C.A. Jerez, N.Guiliani(2008). AHL communication is a widespread phenomenon in biomining bacteria and seems to be involved in mineral-adhesion efficiency. *Hydrometallurgy, 94,* 133–137.

[85] C.M. Waters and B.L Bassler(2005). Quorum sensing: cell-to-cell communication in bacteria. *Annu. Rev.Cell Dev. Biol, 21,* 319–346.

[86] S.B. von Bodman, D.R.Majerczak, D.L.Coplin(1998). A negative regulator mediates quorum sensing control of exopolysaccharide production in *Pantoea stewartii* subsp. *Proc. Natl. Acad. Sci. USA, 95,*7687–7692.

[87] M. Rivas, M. Seeger, D.S. Holmes, E. Jedlicki(2005). A Lux-like quorum sensing system in the extreme acidophile *A cidithiobacillus ferrooxidans. Biol. Res, 38,* 283–297.

[88] D.G. Davies, M.R. Parsek, J.P. Pearson, B.H. Iglewski, J.W. Costerton, E.P. Greenberg, (1998). The involvement of cell-to-cell signals in the development of a bacterial biofilm. *Science, 280,* 295–298.

[89] L. Valenzuela, A. Chi, S. Beard, A.Orell, N. Guiliani, J. Shabanowitz, D.F. Hunt, C.A. Jerez(2006). Genomics, metagenomics and proteomics in biomining microorganisms. *Biotechnol. Adv, 24,* 197–211.

[90] S. Hooshangi, W.E.Bentley(2008). From unicellular properties to multicellular behavior: bacteria quorum sensing circuitry and applications. *Curr. Opin. Biotechnol, 19,* 550–555.

[91] M.B. Miller, B.L. Bassler(2001). Quorum sensing in bacteria. *Annu. Rev. Microbiol, 55,*165-199.

[92] C. Farah, M. Vera, D.Morin, D.Haras, C.A.Jerez, N.Guiliani(2005). Evidence for a functional quorum-sensing type AI-1 system in the extremophilic bacterium *Acidithiobacillus ferr- ooxidans. Appl. Environ. Microbiol, 71,* 7033-7040.

[93] H. Urbanczyk J.C. Ast, M.J.Higgins, J.Carson, P.V.Dunlap(2007). Reclassification of *Vibrio fischeri. Int. J. Syst. Evol. Microbiol, 57,* 2823-2829.

[94] L. Hall-Stoodley, P. Stoodley(2002). Developmental regulation of microbial biofilms. *Curr. Opin. Biotechnol, 13,* 228–233.

[95] M.H. Rashid, K.Rumbaugh, L. Passador,D.G. Davies, A.N.Hamood, B.H. Iglewski, A. Kornberg(2000). Polyphosphate kinase is essential for biofilm development, quorum sensing, and virulence of Pseudomonas aeruginosa. *Proc. Natl. Acad. Sci. USA, 97,* 9636–9641.

[96] S.Nakamura, Y. Higashiyama, K.Izumikawa, M. Seki, H. Kakeya,Y.Yamamoto, K.Yanagihara, Y.Miyazaki, Y.Mizuta, S. Kohno(2008). The roles of quorum-sensing system in the release of extracellular DNA, lipopolysaccharide, and membrane vesicles from Pseudomonas aeruginosa. *Jpn. J. Infect. Dis, 61,* 375–378.

[97] J. Li, C.Attila, L.Wang, T.K.Wood, J.J.Valdes, W.E.Bentley(2007). Quorum sensing in Escherichia coli is signaled by AI-2/LsrR: effects on small RNA and biofilm architecture. *J. Bacteriol, 189,* 6011–6020.

[98] I. Klapper and J. Dockery(2010). Mathematical description of microbial biofilms. *Siam Review, 52,* 1–45.

[99] J.P. Ward, J.R. King and A.J. Koerber(2001). Mathematical modelling of quorum sensing in bacteria, *IMA J. Mathematics Applied in Medicine and Biology, 18*, 263-292.

[100] K. Anguige, J.R.King, J.P.Ward, P.Williams(2004). Mathematical modelling of therapies targeted at bacterial quorum sensing. *Mathematical Biosciences, 192*,39-83.

[101] K. Anguige, J.R.King, J.P.Ward(2005). Modelling antibiotic and anti-quorum sensing treatment of a spatially structured Pseudomonas aeruginosa population. *Journal of Mathematical Biology, 51*,557-594.

[102] K. Anguige, J.R.King, J.P.Ward(2006). A multi-phase mathematical model of quorum sensing in a maturing Pseudomonas aeruginosa biofilm. *Mathematical Biosciences, 203(2),* 240-276.

[103] V.E. Wagner, D. Bushnell, L. Passador, A.I. Brooks, and B.H. Iglewski(2003). Microarray analysis of Pseudomonas aeruginosa quorum-sensing regulons:Eff ects of growth phase and environment, *J. Bacteriol., 185*, 2080–2095.

In: Biofilms in Bioengineering
Editors: Manuel Simões and Filipe Mergulhão

ISBN: 978-1-62948-161-6

Chapter VI

Ecology of Environmental Biofilms

Gavin Collins[1,2*], Florence Abram[1], Therese Mahony, Vincent O'Flaherty[1], William T. Sloan[2] and Christopher Quince[2]

[1]Microbiology, School of Natural Sciences and Ryan Institute, National University of Ireland, Galway, University Road, Galway, Ireland

[2]School of Engineering, University of Glasgow, Rankine Building, Glasgow, UK

Abstract

Biofilms are ubiquitous and underpin many biogeochemical cycles, as well as a wide range of biotechnological applications. In fact, no serious environmental microbiology study can consider microbial cells in isolation out-with the biofilm scenario, as the planktonic mode is unlikely in almost every environment. Instead, microbial individuals are part of communities, or systems, which are usually diverse and metabolically complex relying on successive or interdependent reactions and interactions, mutualism and syntrophies. To understand, exploit and extend the application of, these biofilm systems the 'ecophysiological' roles must be determined. The vast majority of microbes from natural, and even, engineered biofilms, cannot be cultured in isolation in the laboratory. Thus, modern microbial ecology is largely based on targeting informative, 'marker' molecules, such as DNA, RNA and proteins. A range of powerful tools is now available for detailed and elegant studies that identify, and quantify, microbial groups and species, and link taxonomy to activity, even at the level of single cells. In this Chapter, the range of tools available to study biofilm ecology is outlined. Complex biofilms, such as those from anaerobic digestion and microbial fuel cell technologies, are considered. A unification of metabolomics and other 'omics approaches – particularly metagenomics – is described to provide a high-throughput solution to link taxonomy with function in biofilm communities. This approach can also enable the construction and validation of 'ecosystems biology' models operating at the level of the whole biofilm community.

Keywords: Biofilm ecology; Ecophysiology; Ecosystems biology; Metagenomics

* E-mail: gavin.collins@nuigalway.ie.

Introduction

Microbial communities are central to health, sustainable cities, agriculture and most of Earth's biogeochemical cycles. Ecologists and microbiologists recognise this and there are reams of scientific publications that support it, but there is little evidence that such broad statements impact on industry and policy makers. Where microorganisms are recognized as a hazard, then their importance is recognised. Thus, every child in most countries knows to wash their hands after a visit to the toilet; so in public health the fundamental role of bacteria in perturbing the human ecosystem is known and we have changed policies over the years to mitigate against their worst effects. Even now, there is a huge effort in mapping and preventing the spread of emerging pathogens, particulary those with antibiotic resistances. This public health-centric view has skewed perspectives on the importance of microbes so, for example, in large-scale, multi-disciplinary research projects on aquatic environments, there is often only a commitment to study species that index pathogens from faeces. The fact that the vast majority of bacteria are not harmful to humans – and, indeed, that we rely on them to sustain our existence – is lost to almost everyone outside academic circles. Facts such as there being 10 times as many bacterial cells on, and in, a human than there are human cells; or there being at least 10^{30} bacteria on Earth [39]; or there being a billion times more bacteria in the world than stars in the sky [40]; or the shear diversity of microbial species and metabolic functions should be awe-inspiring but are often met with only mild curiosity. This is, in part, because we cannot see microorganisms and the information is perceived to be abstract and irrelevant to our everyday existence, in the same way that the quarks and neutrinos of particle physics are perceived to have little bearing on our exploitation of the material world. Yet we believe that we can – either accidentally or by design – effect change in microbial communities and their functions. Indeed, microbial communities in bioengineered systems are 'exploited' to provide services, such as waste treatment, in 'black box' technologies and usually without information on the species present, their metabolic capacity or their likely response to different process operating parameters. In the case of bioengineered systems of mixed microbial communities, this approach results in over-dimensioning of bioreactor volumes based on empirical data only, and often leads to sub-optimal performance or system failure. Our exploration of the microbial world is, however, in its very early stages and, therefore, our ability to manipulate communities and predict change now is currently limited.

There is a diverse and vast array of microbial diversity in nature where the predominant mode of bacterial growth consists of sessile, matrix-enclosed biofilm communities. Microbial biofilms may be defined as populations of micro-organisms that are concentrated at an interface (usually solid or liquid) and typically surrounded by an extracellular polymeric slime (the extracellular polysaccharides, or EPS) matrix [147]. Biofilms represent a distinct lifestyle for bacteria providing protection from deleterious environmental conditions [77] and greater resistance to biocides and antimicrobials. The formation and structure of these ubiquitous micro-communities provides an area of intensive microbiological and environmental research, where advances in analytical and molecular technology allow observation and analysis of the dynamics of biofilm formation processes and development.

Efficient – and extended – application of microbial communities for environmental management, therefore, relies on knowledge of the microbial ecosystem. This will likely include the Bacteria, Archaea, eukaryotes and viruses. However, in some cases, bioengineers interested in microbial ecology are faced with the task of persuading stakeholders that microbial communities are central to the success of their activities, and demonstrating that gaining a fundamental understanding of the way environmental biofilms form and respond to environmental change will allow a more certain and quantitative design of a range of technologies, environments and activities, from wastewater treatment plants (WWTP) to river catchment management strategies. Ultimately, only demonstrative outputs, such as tools that rapidly provide useful information on bioprocesses, give stakeholders hands-on experience of working with bacteria to make the communities more tangible and less abstract. The challenge is to demonstrate that information on the ecology of biofilms in bioengineering can support decision-making and avert process problems.

Environments and instances where ecological information on the prevailing biofilms can be used to enhance human-biofilm interaction include natural environments, such as the human mouth and gingivo-dental juncture [56]; the human gut [12]; fluvial biofilms in river catchments and their effect on sediment stability [180]; as well as engineered environments, such as slow sand filters for water purification [61]; aerobic biological wastewater treatment in trickling filters [98,162] and activated sludge systems [165], including those with aerobic granules [177]; anaerobic digesters [32,66,122] and microbial fuel cells [17,22].

Bioremediation, and waste and wastewater treatment processes comprise the biggest and most important engineered application of microorganisms worldwide. Effective waste treatment is critical to maintaining ecosystem and human health. Furthermore, concerns about mitigating greenhouse gas emissions, in the context of increasing energy costs, mean that efficient treatment of wastes and wastewaters is one of the most important engineering challenges facing society. As the world's population grows, and as cities expand, the need for sustainable, robust approaches for the management of domestic and industrial wastewaters increases. Water and energy management are important and interrelated issues. The energy required to treat sewage is high; for example, the water industry is the fourth-most energy-intensive sector in the United Kingdom, accounting for 3-5% of national electricity use [179]. Further tightening of water quality standards suggest energy costs will increase. Thus, new, low-energy technologies for sustainable waste conversion with net bioenergy production are now a key focus. This Chapter considers strategies to determine the microbial ecology of two related technologies: anaerobic digestion (AD) and microbial fuel cells (MFCs).

AD is the biological mineralisation of organic material to biogas, in the absence of oxygen, by the sequential activity of several microbial groups (Figure 1). AD is an established waste treatment technology for residues from various sources, including industrial processes and agriculture. It has advantages over aerobic biological treatment, including lower operating costs and, primarily, biogas production [92]. Biogas is similar to natural gas and consists of a mixture of 50-85% methane and 15-50% carbon dioxide with trace amounts of other gases. It is a renewable energy source and is used for generation of heat and electricity, or as vehicle fuel. The anaerobic reactors used for full-scale wastewater treatment and biogas production are high-rate systems based on self-immobilised, or granular, biofilms (Figure 1), e.g. the upflow anaerobic sludge blanket (UASB) reactor [92,93]. Biofilms are a potentially multi-species community of microbes – more usually living on a surface – embedded in a self-generated matrix of EPS and, in the case of the AD applications, each self-aggregating granule theoretically contains all of the trophic groups required for the waste conversion (Figure 1). Increased application of AD could provide benefits, both economically – by the reduction of energy and operational costs – and environmentally – by preserving fossil fuels and reducing polluting emissions. Recently AD research has moved towards: (i) energy-efficient, low-temperature (psychrophilic) waste treatment applications [108]; (ii) biorefinery applications for the conversion of a range of non-food feedstocks – such as grass and solid wastes – to valuable products, including organic acids with industrial value, bioplastics and alcohols (iii) anaerobic fermentation processes for biohydrogen production [130]; and electricity generation in microbial fuel cells [28].

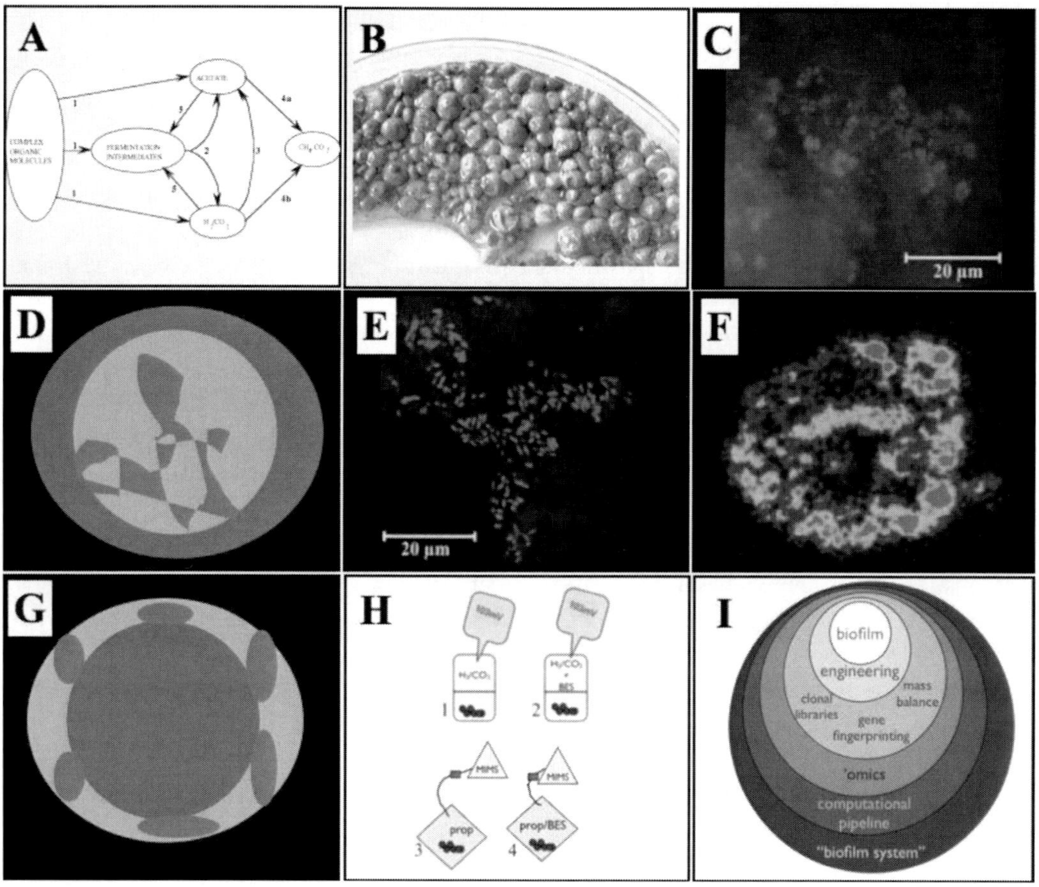

Figure 1. A: Carbon flow in anaerobic sludge ganules: 1 = Hydrolytic/fermentative bacteria; 2 = obligate hydrogen-producing acetogenic (OHPA) bacteria (syntrophs); 3 = homoacetogenic bacteria; 4a = acetoclastic methanogens (Archaea); 4b = hydrogenophilic methanogens (Archaea); 5 = fatty acid-synthesising bacteria. B: Granular anaerobic sludge; each granule: Ø, ca. 2 mm; C: FISH microimage illustrating the juxtaposed arrangement of cocoid (red) and filamentous methanogens (green) on the cross-section of a typical anaerobic sludge granule; **D:** Model illustrating the conventionally-described, basic microbial architecture of anaerobic granules, including the position of microscopically observed Bacteria (red); methanogens (green) and Crenarchaeota (blue); E: fluorescence microimage illustrating a cluster of Crenarchaeota at the surface (edge of cross-section) of an anerobic granule; F: granule cross-section illustrating uptake of ^{14}C-labelled acetate: uptake 'hotspots' are indicated in red and coincide with zones of intermingled Crenarchaeota and methanogens; G: the alternative anaerobic granule structure, including a bacterial core (red) and an archaeal envelope (methanogens and clusters of Crenarchaeota on the surface), based on FISH experiments and radiolabel uptake data; H: homoacetogenic activity tests, with pressure measurements in vials and MIMS from gas-tight bags; I: the idealized approach to describe environmental biofilm ecosystems at the community level.

MFCs are bioreactors, divided into two compartments separated by means of a membrane that is permeable to cations but is impermeable to oxygen; electrodes are inserted into the compartments and connected through an electrical circuit [113]. Typically, anaerobic microbial consortia oxidize an organic substrate (e.g. glucose or secondary wastewater) in the anodic compartment and the electrons generated in the reaction are transferred to the anode and delivered to the cathodic compartment, in which they reduce an oxidized substrate (e.g. oxygen or oxidized metal). Thus, the MFC converts chemical energy directly into electrical energy. The energy potentially available in an MFC is significant but the efficiency of MFCs is limited by several factors: the presence of electron acceptors in the medium, the incomplete oxidation of the substrate, and the kinetic limitations in electron transfer from microorganisms to the anode. This was thought to occur strictly through redox mediators that shuttle the electrons but, more recently, microbes capable of direct electron transfer have been discovered. Although not currently capable of large-scale electricity generation, MFCs present an attractive technology for self-sufficient wastewater treatment with modest electricity generation.

Complex microbial communities underpin AD and MFC applications but the metabolic roles of microbial individuals are still largely undetermined. Moreover, the species are strongly connected through syntrophies where the waste products of one provide resources for another. Consequently, to understand, exploit and extend the application of, these biofilm systems the 'ecophysiological' roles must be determined. Questions, such as how the resulting syntrophic interactions structure the system-level behaviour, can then be addressed to support full-scale engineering and operation.

In this Chapter, the range of tools available now to study biofilm ecology is outlined. A unification of metabolomics and other 'omics approaches – particularly metagenomics – is proposed to provide a high-throughput solution to link taxonomy with function in biofilm communities. This approach could also enable the construction and validation of 'ecosystems biology' models operating at the level of the whole community.

Microbial Ecology

Microbial ecology is a long-standing scientific discipline involving the study of relationships between microorganisms and their environment, which is currently undergoing remarkable, if not revolutionary changes [153]. The most significant recent advances in microbial ecology involve interdisciplinary approaches that link the physiology of microorganisms in situ with the complex biotic and abiotic factors that shape microbial niches. Thus, microbial ecophysiology as a burgeoning division of microbial ecology, attempts to rectify the pervading knowledge gaps which exist regarding the ecological functional roles and niches occupied by the vast micro flora which exist throughout Earth. One of the greatest challenges in contemporary microbial ecology is the exploration of the identity and function of the estimated 90-99% of previously unidentified and uncultured microorganisms extant in the biosphere. Although great strides have been made recently in identifying and characterizing the staggering diversity of microorganisms facilitating primary and secondary production, nutrient transformation, biogeochemical cycling and mineralization processes [140] – all of which are applied in bioengineered systems underpinned by biofilms – our current understanding of the extent of microbial life on Earth still remains quite rudimentary.

Indeed, the vast majority of organisms in AD biofilms and in MFCs have not yet been cultured. To study these organisms requires the application of techniques from the 'molecular toolbox' to characterize the complex, mixed microbial consortia present. Those techniques and approaches should be reliable and reproducible, and the data generated should directly support the operation of, and decision-making in, bioengineering processes.

Historical Challenges of Biofilm Microbial Ecology

Whilst microbial ecology arguably represents one of the most important areas of biological research, for many years our understanding of microorganisms in the environment has been based primarily on the analysis of those microorganisms that we could culture purely on media [43,136] or with a complex series of morphological, biochemical and physiological tests. Indeed, it is by these simple and classical techniques that some of the numerous species involved in AD were initially identified and subsequently described on the basis of physiological and biochemical characteristics. However, it is a common misconception that microorganisms isolated in pure culture from environmental biofilms represent the numerically dominant or functionally significant species [70]. It is estimated that in the region of 99% of microorganisms typically observable in nature (i.e. by non-discriminatory microscopy) have failed to be cultivated in the laboratory [4,71] and remain refractory to culturing by standard methods. Even if an organism could be cultivated, properties determined in the laboratory may not necessarily reflect the activities and physiology of their counterparts in the environment [54] and studies would still be required on their in situ activity and distribution. For example, in many biofilm studies, different genes are expressed in planktonic and biofilm growth modes [1,166].

Understanding the presence and functional role of a particular organism in its environment requires techniques for rapid in situ phylogenetic (taxonomic) identification and an analysis of its metabolic activities. In the early part of the 20th century, Sergei Winogradsky and Marinus Beijerinck paved the way for more appropriate microbial ecology research, whereby the inherent metabolic interactions and functioning between microbial communities was the investigative focus rather than the analysis of single isolated species to elucidate ecological relationships. Increasingly it became more evident that culture-independent tactics were required if precise and accurate depictions of natural and bioengineered microbial communities were to be obtained. A major breakthrough occurred following the proposal by Zuckerlandl and Pauling [196] that macromolecules, such as gene sequences, act as documents of evolutionary history and could be used to relate organisms.

The 16S rRNA Gene and PCR

A wide body of research has focused on extracting DNA from biofilms in bioengineered environments and amplifying 16S rRNA genes by polymerase chain reaction (PCR). This gene is present in all prokaryotes and can be used as a marker for the taxa present [29].

In the 1970s, Carl Woese and colleagues were the first to propose that variation in the sequences of DNA encoding the small subunit ribosomal RNA (SSU rRNA), present in all organisms, would provide valuable information regarding evolutionary relatedness and the true phylogenetic analysis of bacteria, and indeed of all other forms of life [190]. This incisive formulation allowed the establishment of a molecular sequence-based phylogenetic tree, which could be used to relate all organisms and reconstruct the history of all life [70,138,190]. The highly conserved nature of rRNA allows for the synthesis of universal primers that can anneal to sequences conserved in the 16S rRNA genes from all three phylogenetic domains [138].

After comparing the small variations between the 16S rRNA genes from numerous organisms, Woese demonstrated that the Archaea constitute an exclusive domain of life [190,191]. The domain Archaea is divided into the four kingdoms; Euryarchaeota, Crenarchaeota, Korarchaeota and Nanoarchaeota [50,68]. Despite being considered exclusively as life's extremists, it is now apparent that Archaea are astounding in their ubiquity [26,158] and although they are relatively inconspicuous they remarkably succeed as inhabitants of a breathtaking array of ecological niches [44,97,100,146]. Crenarchaeal phylogeny has been fundamentally changed by the isolation of novel SSU rRNA gene clones from thermo-, meso- and psychro-philic habitats, including anaerobic digester biofilms [32]. Within the *Crenarchaeota*, most cultivated representatives to date originate almost exclusively from geothermally heated habitats, such as submarine vents and terrestrial hot springs [171]. However, Preston et al., [144] successfully cultivated a psychrophilic, symbiotic crenarchaeon *Cenarchaeum symbiosum* closely related to other nonthermophilic crenarchaeotes that inhabit diverse marine and terrestrial environments. Only a hypothetical framework for the role of *Crenarchaeota* in AD biofilms has yet been formulated, as will be discussed later in this chapter.

In the late 1980s, the invention of the polymerase chain reaction (PCR) facilitated the unlimited progress of protein- and nucleic acid-based techniques for molecular analysis, thereby paving the way for successful molecular phylogenetic studies of microbes in particular. PCR is one of the most powerful techniques to have been developed, allowing the amplification of even minute DNA traces, thus permitting the highly-refined detection and identification of diverse microbial communities present in environmental biofilms.

Now, PCR-amplification of 16S rRNA genes comprises the mainstay of biofilm microbial ecology. For most cases, this approach continues to be the most suitable, convenient and accessible, not least because researchers have access to a vast repository of 16S rRNA gene sequences for comparative analyses and to support phylogenetic reconstructions. The molecule is also likely to underpin the majority of future technological innovations, particularly those developed as diagnostic tools to support bioengineering processes.

16S rRNA Alternatives

Functional genes are an obvious alternative target for ecological surveys. Using this approach, diversity across a metabolically-related set of organisms can be detected. Another option is to work with alternative targets, such as the rpoB gene, which has been used as a biomarker for PCR-based microbial analyses of mixed communities. Peixoto et al., [141] compared fingerprinting analyses of rpoB and 16S rRNA genes, and found fewer bands using rpoB, which were easier to delineate and therefore to analyse. Comparison of the community profiles revealed that the methods were complementary. The gene for the beta subunit of the RNA polymerase, rpoB, is a single copy gene unlike 16S rRNA. Multiple copies of 16S rRNA genes in bacterial genomes complicate diversity assessments made from fingerprinting profiles, meaning that rpoB may be useful for diversity studies from biofilms. Other studies have similarly targeted this gene (e.g. [78]), and the potential of alternative housekeeping genes to study community structure and to map population dynamics, with enhanced resolution for fined-scale analyses, was tested and discussed in detail by Case et al., [25].

Cloning and Sequencing

One of the great advantages of using PCR in molecular microbial ecology studies is that different groups within a microbial community may be targeted and studied independently of the other groups present, simply by choosing specific primers for a specific population or species. Following amplification of specific DNA sequences, or 16S rRNA genes, by PCR, the amplified genes must be separated prior to subsequent analysis as they constitute a heterogeneous mixture of sequences from across the community. Cloning in Escherichia coli cells is one of the most widely used methods to separate PCR-amplified DNA that is identical in length but different in sequence [189] from environmental samples. Once cloned, the 16S rRNA gene library can be screened by a variety of methods [63], allowing the grouping of similar sequences into operational taxonomic units (OTUs; [114]), which can then be analysed phylogenetically to infer relatedness and identity of the species detected. In general, cloning and gene library construction have been applied in combination with other techniques to study wastewater treatment biofilms [157].

Community Fingerprinting Techniques

Numerous genetic profiling techniques, which are less laborious than cloning, have been developed to permit larger-scale characterisation of multiple environmental samples. A range of PCR-associated procedures amplifying the 16S rRNA genes may be applied to explore the diversity of microbial species present in natural and biotechnological samples. Examples of these include: Amplified rDNA Restriction Analysis (ARDRA; [114]), 16S rDNA cloning and sequence analysis [8,19, 47,183], Single-Strand Conformation Polymorphism (SSCP) analysis [89,197], Denaturing Gradient Gel Electrophoresis (DGGE; [94,118]), Length Heterogeneity PCR [173], Temperature Gradient Gel Electrophoresis (TGGE) [129], Terminal Restriction Fragment Length Polymorphism (TRFLP) analysis [16,135] and Ribosomal Intergenic Spacer Analysis (RISA). Genetic fingerprinting approaches have gained major recognition as rapid profiling methods to reveal minor increases in abundance in target organisms, and thus direct the optimization of reactor conditions in a very sensitive manner. In contrast to the more laborious cloning procedures, fingerprinting methods facilitate monitoring of temporal and spatial changes in communities with minimal effort [136], which is important in environments like AD bioreactors and MFCs where the microbial populations change and develop with time.

Due to its sensitivity and high-throughput, TRFLP is an ideal technique for comparative community analyses [102]. Following DNA extraction, 16S rRNA genes are amplified using primers which are end-labeled with a fluorescent dye, following which, resultant PCR products are then restricted with tetrameric enzymes. The subsequent sequence fragments are electrophoresed on an automated sequencer but only fluorescent terminal fragments (TRFs) are detected allowing an investigative focus on specific microbial groups, if required [103,135]. TRFLP therefore allows the analysis and detection of spatial and temporal changes and developments within microbial communities [80], as each TRF should theoretically represent a different group or member of the microbial community, such as those observed in complex anaerobic granular biofilms.

TRFLP was first reported as a means of strain comparison [24] using randomly primed PCR reactions and has since been applied for various sample types and specific microorganisms including; Mycobacteria [6], marine sediments [16], rice fields, soil [11], and also in anaerobic digestion [36,107,139]. High achievable resolution through the use of automated sequencer technology, immediate digital output, the presence of an internal standard in all sample runs as well as the ability to make direct reference to the already extensive database makes TRFLP a useful and applicable fingerprinting technique in microbial ecology studies. The possibility to employ degenerate primers also increases the potential range of markers targeted by a single primer set via TRFLP and given that the technique is not dependent on cultivation, it views PCR-amplifiable targets from all populations, even those that have yet to be cultivated [103]. This approach therefore provides a useful tool in assessing diversity within mixed communities as well as comparative distribution across microbial populations.

DGGE involves the separation of PCR products on polyacrylamide gels with an increasing linear gradient of a denaturant; either in the form of chemicals such as urea and formamide (DGGE) or simply temperature in the case of Temperature-GGE (TGGE). DGGE allows fragments of DNA of the same length but different sequence to be resolved electrophoretically [63,118,187], thus generating band patterns that directly reflect the genetic biodiversity of the sample [157]. Separation in DGGE is based on the electrophoretic mobility of a partially melted DNA molecule in a polyacrylamide gel, which is decreased, compared with that of the completely helical form of the DNA molecule [117,118]. Each different microorganism present will have a distinct denaturation point, which is visible as a single band. DGGE has been widely used to investigate the genetic diversity of microbial communities, population dynamics within communities, gene expression in mixed populations and to monitor the enrichment of particular species during bacterial isolation. Muyzer et al., [118] initially employed DGGE to separate amplified 16S rRNA genes from microbial mats and hydrothermal vents and since then the technique has been extensively applied to provide information on the genetic diversity of groundwater communities [76,184], marine habitats [9], soils [119,120], as well as anaerobic and aerobic wastewater treatment systems [27,155]. This technique can be utilized to monitor community changes over time as some bands may disappear or decrease in intensity, while others may emerge depending on bioreactor conditions.

RNA-based Ecology

RNA targets propose data based more accurately on the activity of biofilm constituents, since it is assumed that only active cells will transcribe DNA to nuclear and messenger RNA, which will thus be indicative of current activity. Comparisons of community profiles based on DNA- and RNA-targeted (cDNA) PCRs should, therefore, identify the active fractions of biofilm communities [58]. However, experiments of this kind are not without difficulties – this work can be technically challenging due to the instability of mRNA. By the same token, however, this usually provides instructive data when the target is mRNA i.e. a functional gene. It is less straightforward, however, in the case of rRNA, which may not degrade as quickly and might not, therefore, provide an insight to the active fraction [178].

Limitations of PCR-based Methods

Despite its universal application, PCR and related techniques do present some negative aspects and limitations in their inherent applications. There are several potential biases introduced during initial DNA retrieval and subsequent amplification. Firstly the environmental matrices of sludge, wastewater, sediment and soil are 'dirty' which can create a significant problem for extracting cellular material and avoiding interferences [153]. Several techniques have been described for direct detection and extraction of DNA and RNA from environmental samples [176] and the effective extraction of nucleic acids is an essential starting point in the molecular analysis of microbial communities in the environment [145].

Methods currently employed for DNA extraction vary widely, from direct methods of in situ lysis to indirect methods of initial cell extraction prior to lysis [58]. Many workers have attempted to increase DNA yields from biofilm samples by using severe physical treatments, such as mechanical bead-beating and sonication to lyse microbial cells. However, such treatments can shear DNA to sizes of 5 to 10 kb or less, thereby rendering the DNA unsuitable for community analysis based on PCR-fingerprinting techniques, due to the risk of creating chimeric products [195]. Insufficient or preferential lysis of cells will also bias the view of the composition of microbial diversity as DNA or RNA which is not released from the cells will not contribute to the final analysis of diversity [189]. This is particularly relevant to members of the kingdom Archaea, some of which possess a particularly thick and rigid outer cell layer, namely mithochondroitin [85], which is more resistant to cellular lysis.

Choice and design of primers is an important aspect of any PCR-based technique applied in microbial ecology and must be taken into consideration, particularly when interpreting phylogenetic data from biofilms with diverse microbial communities. Baker et al., [7] assessed and reviewed the specificity of commonly-used 16S rRNA gene primers and concluded that no universal primer sets are truly 'universal' and no single set of primers can be recommended that are guaranteed to amplify all microorganisms. As a consequence, many, if not all, 16S rRNA gene libraries will not be totally representative of microbial communities, especially on a quantitative level [45,151]. However, until more knowledge of uncultured and unidentified environmental microorganisms is retrieved though ongoing microbial ecology surveys, the use of universal 16S rRNA primers will still remain the cornerstone of PCR-mediated analysis of complex microbiota.

With respect to the PCR procedure, several problems may arise, especially when the method is applied to environmental samples: i) inhibition of PCR amplification by co-extracted contaminants, ii) differential amplification and iii) formation of artefactual products [189] are frequent inconveniences which may be encountered. Furthermore, 16S rRNA sequence variations due to rrn operon heterogeneity would unavoidably lead to a biased reflection of microbial diversity while the universal primers employed for the amplification of rRNA genes often contain degeneracies which may influence the formation of primer-template hybrids [174]. In spite of these often varying discrepancies, PCR-mediated analysis of 16S rRNA genes is a powerful tool for the determination of microbial diversity in environmental biofilms.

Real-time (Quantitative) qPCR

Real-time or quantitative PCR is the reliable detection and measurement of products generated during each cycle of PCR, which are directly proportionate to the amount of template prior to the start of the PCR process [53]. Compared with a conventional PCR method employing a forward and reverse primer, an additional fluorescent probe is required in qPCR assays. The fluorescent probe hybridizes with a specific site within the targeted region and each sequence of the primer and probe is designed to be specific for a target organism or microbial group, thereby allowing for increased sensitivity and specificity as three oligonucleotides complementary to the target DNA are simultaneously employed [193]. The amount of emitted fluorescence is proportional to the amount of PCR product and enables monitoring of the PCR in real-time. The qPCR curve is used to define the exponential phase of the reaction, which is a prerequisite for accurate calculation of the initial copy number at the beginning of the reaction [81].

The qPCR assay is sensitive down to sub-attomole levels [21] and is fast becoming the method of choice for the analysis of microorganisms in medical and environmental microbiology, in particular. Due to its high sensitivity, qPCR can be applied for the detection and quantification of microbial populations without prior cultivation in anaerobic processes and environmental samples [193]. For example, the mcrA gene, which encodes methyl coenzyme M reductase I, a key enzyme in methane production and exclusive to methanogenic organisms, has been successfully utilised as a functional marker in real-time PCR assays in anaerobic systems [167,170].

Yu et al., [193] and Lee et al., [90] successfully designed and established Taqman primer and probe sets, which target and quantify each of the four orders of methanogens associated with anaerobic bioreactors: Methanococcales (Methanococcus spp.), Methanobacteriales (Methanobacterium spp., Methanobrevibacter spp., and Methanosphaera spp.), Methanomicrobiales (Methanogenium spp., Methanoculleus spp., Methanocorpusculum spp., and Methanospirillum spp.) and Methanosarcinales (Methanosarcina spp. and Methanosaeta spp.). The assays were shown to be applicable to the determination of the concentrations of acetoclastic methanogens, and allowed for the correlation of microbial community dynamics with operating conditions and performance of various anaerobic digesters [10,193].

The quantification of functional genes could be used as a proxy for potential activity. For example, the potential nitrogen (N) loss in nitrifying-denitrifying biofilms could be assessed by quantifying ammonia monooxygenase and the range of reductases implicated in (N) removal, and those data could then be correlated with off-gas flux measurements. The same concept could be applied for a range of processes in any biofilm sample.

A major advantage of qPCR, of course, is that assays can be used to obtain absolute abundances of even rare organisms. However, this is still only true for a predetermined target, and the choice of primers is still important.

PCR-Independent Techniques

Several other molecular techniques are available, which, although based on rRNA molecules, do not require nucleic acid extraction, are non-invasive to a large extent and have been employed with great success to analyse anaerobic biofilms. PCR and genetic fingerprinting methods cannot – easily or quickly – be used to study the three-dimensional (3-D) structure of biofilm aggregates. However, whole-cell in situ hybridization provides a powerful tool for quantitative analyses, as single cells can be specifically detected and counted under the microscope [189] while still maintaining the structural integrity of the biofilm [150].

Prokaryotic cells can be identified without prior cultivation by applying fluorescence in situ hybridization (FISH) using rRNA targeted oligonucleotide probes [181] while simultaneously revealing the spatial relationships of microorganisms within the environment [3,23]. The main advantage of FISH is that it does not involve DNA or RNA amplification but it allows for microscopic inspection of intact cells [187]. The 3-D structure or architecture of a community can be visualised by means of confocal laser scanning microscopy (CLSM) in combination with fluorescent molecular probes, which allow extracellular polysaccharide composition and heterogeneity with respect to certain parameters within biofilm microenvironments to be investigated. The principle of CLSM involves excitation of a fluorophore by a focused laser beam [187], which provides greater focal depth and the high resolution required to produce in situ imaging of microbial forms. FISH has been extensively applied to anaerobic biofilm studies [32, 33, 73, 94, 150, 154, 161, 164] with much success to establish which species are present and their micro-location in the biofilms.

A more recent development of FISH, namely catalyzed reporter deposition-FISH (CARD-FISH), introduced to microbial ecology by Pernthaler et al., [142], has become a potent alternative to conventional FISH as it includes an enzymatic amplification step of the fluorescent signal [157]. Probes used for CARD–FISH are labeled with the enzyme horseradish peroxidase (HRP), which catalyzes the deposition of fluorochrome-labeled tyramides within the target cells [74]. With this additional catalytic step a cell containing only a single ribosome can theoretically be detected as the fluorescence signal is amplified many times by the massive deposition of labeled tyramide [157]. Therefore CARD-FISH is ideal for samples, which yield low signal or high background fluorescence, as can often be the case with environmental samples.

FISH and AD Sludge Granules

Anaerobic granules, which are found in anaerobic bioreactors, are particulate biofilms, formed spontaneously by the auto-immobilization of anaerobic bacteria in the absence of a support material. The stable and efficient operation of anaerobic bioreactors is primarily dependent on the growth and maintenance of sludge granules containing all the microbial trophic groups necessary for the complete transformation of the organic constituents of wastewaters and waste in general [59]. These dense particles consisting of an intertwined mixture of symbiotic anaerobic microorganisms (Figure 1), which work together in methane formation, are the secret to the successful operation of AD technology [105]. Granule formation allows for loading rates in anaerobic bioreactors, far beyond the typical loading rates applied in conventional aerobic activated sludge processes [72].

Methanosaeta spp. are regarded as being significant in the onset of granulation and the maintenance of stable granules during system perturbations [57,110] with M. concilii being highly regarded as a key organism in anaerobic sludge granulation [72]. FISH experiments have helped construct a picture of the highly-organised biofilms involved: generally the ultra-structure of granules is concentrically layered, where methanogenic Archaea are surrounded by fermentative bacterial species [46,86,99,185]. The unique architectural formation of granular sludge is considered to be a result of substrate profiles formed within the granules [163] and the specific spatial orientation of microbial consortia within the anaerobic granule is essential since the distance between microorganisms, such as acetogens and methanogens, must be sufficiently small to ensure and maintain a low hydrogen partial pressure [62] and allow for the transfer of metabolic products.

Linking Identity with Function

The fundamental question in microbiology and biofilm research has been the same for at least a decade now: 'which organisms do what'? This represents the single biggest knowledge gap in biofilm ecology and is the single greatest barrier to the extended application of yet-to-be-cultured species in bioengineered systems. Indeed, addressing this is the raison d'être of many research groups. The next sections consider approaches to this grand challenge.

Stable Isotope Probing (SIP)

FISH is well suited to microbial community analyses; nevertheless, this and other methods are inadequate to determine reliably the general physiological activity and ecophysiological traits of the detected cells [123,181]. However, techniques involving the use of substrates labeled with stable- and radio-isotopes can achieve the goal of linking identity, activity and function simultaneously under conditions which approach those occurring in situ.

SIP is a term used to describe the use of ^{13}C-labeled growth substrates to link microbial function with identity via the selective recovery of 'heavy' ^{13}C-labeled DNA [121,149,182]. SIP exploits the intrinsic physical properties of the atoms that constitute all cellular components, in particular the isotopes of carbon. Radajewski et al., [148] demonstrated that ^{13}C DNA synthesized during the growth of metabolically distinct microbial groups on a ^{13}C enriched carbon source could be resolved from the lighter ^{12}C DNA by equilibrium density gradient centrifugation. The DNA recovered may then be analysed by more conventional molecular methods, along with the ^{12}C DNA, to identify the metabolically active microbial groups, which incorporated this ^{13}C DNA into their genome. The valuable aspect of DNA-SIP, or RNA-SIP [186], is that incorporation of a labeled substrate into DNA, or RNA, implies the microbial species represented were metabolically active under the conditions tested [149]. Stable isotopes have been used extensively as tracers for rate measurements in microbial ecology. Labeling experiments with ^{13}C-bicarbonate have been used to determine the autotrophic organisms involved in CO_2 uptake [14] as well as biogeochemical methane oxidation [13,15].

There are some limitations and problems with SIP. The application of substrate tests can theoretically lead to bias, tending to emphasise the presence of organisms, which may not play the key metabolic roles in situ. These cross-feeding effects might then be inadvertently detected as a successional change occurring during a DNA-SIP time-course experiment [149].

Microautoradiography-Fluorescent *In Situ* Hybridization (MAR-FISH)

Cultivating yet-to-be characterized microbes from envionmental biofilms and deciphering their in situ functions is an important challenge but a difficult task [163]. However, some of the limitations of recently applied techniques to investigate metabolic function and microbial interaction may be circumvented by combining microautoradiography with FISH (MAR-FISH; [182]). MAR-FISH is an elegant approach, which is exploited to evaluate active uptake of radiolabeled substrate by different phylogenetic groups in complex microbial communities [131]. Microbiologists and others have been using MAR for several decades to visualize and count cells within microbial communities that can metabolize radioactively-labeled substrates. Measurement of in situ nutrient uptake by heterotrophic bacteria via microautoradiography was first suggested in the mid 1960's by Brock and Brock [18] to determine the nutritional requirements of microorganisms and to estimate cell growth-rate and reproduction [131]. However, in most cases, the microorganisms being examined could not be identified [182] by MAR alone. Since then, however, MAR has been revived and applied in combination with FISH for the effective identification of active microorganisms [91,137].

The combined MAR-FISH technique is an efficient method to obtain reliable quantifiable information regarding the ecophysiology of microorganisms at the single-cell level in mixed communities [124,126,127,185]. With the tools available today the resolution of MAR to individuals (ca. 1 μm) can be obtained [65], depending on the sample type. The archetypal MAR-FISH procedure is composed of five key steps:

1) incubation with radiolabeled substrates.
2) fixation and handling.
3) FISH probe application.
4) autoradiographic development procedure (MAR)
5) microscopic observation.

For incubation of anaerobic samples it is critical to completely remove trace amounts of O_2, which may cause a false substrate uptake within the initial few minutes [82,127,131]. The radiotracers most commonly used in MAR-FISH are typically the soft beta emitters ^{3}H, ^{14}C, ^{33}P and in a few cases the stronger emitter ^{32}P. Common to all radiotracers is their ability to form silver grains on top and near the radiolabeled bacteria that can be clearly visualized by bright-field or phase-contrast microscopy [126]. In general, a higher resolution is obtained with ^{3}H and other low energy emitters due to the lower energy of the β-particles present [5]. Resolution levels of less than 0.5 μm have been attained with ^{3}H in contrast to ^{14}C and ^{33}P which have resolutions of 1-3 μm [126]. The use of pasteurized samples as negative controls should also be incorporated. MAR is usually considered a quantitative or semi-quantitative procedure, especially when ^{14}C, which is a relatively strong beta emitter, is applied [131].

Using MAR-FISH, substantial ecophysiological data have been obtained regarding a plethora of diverse microorganisms: Achromatium in freshwater sediment biofilms [55], alpha-Proteobacteria [83], anammox bacteria [60,160], filamentous microbes in activated sludge [128], iron-reducing bacteria [125], the filamentous methane oxidizer Crenothrix [172], halophilies from saltern ponds [156], pelagic and marine bacteria [2,137], phosphate-accumulating organisms (PAOs; [82,91]), glucose accumulating organisms (GAOs; [20]), marine proteobacteria [37], methylotrophic bacteria [52], sulfate-reducing bacteria [75], uncultured Nitrospira-like bacteria [41] and marine planktonic Archaea [137].

Extensive MAR-FISH studies have been done with particular focus on flocs (loose biofilms) in activated sludge (i.e. aerobic) samples and the ecophysiological traits of filamentous microbes, particularly the filamentous strains *Microthrix parvicella* and *Thiotrix* spp., associated with the notorious phenomenon of filamentous bulking [5]. However, despite the influx of data corresponding to activated sludge-associated microbial populations, there is a lack of analogous information regarding anaerobic sludge granules.

The MAR-FISH technique does demonstrate some limitations, primarily concerning its practical application. As with SIP experiments, MAR-FISH is susceptible to substrate cross-feeding effects [131] or partial digestion [126], predominantly when long incubation times are involved. However, this may be dealt with by the addition of effective inhibitors. Transfer of the radio-label from primary consumers within a niche to other microorganisms can also occur [182]. Furthermore, in marked contrast to fluorescently-stained cells, the 3-D distribution of silver grains in the emulsion layer cannot be accurately analysed, even by CLSM [131]. Not all substrates of interest are easily available in a radioisotope-labeled form [182], thereby possibly limiting attempts to accurately reflect in situ conditions when analyzing certain microbial environments. It must also be considered that incorporation rates of radiolabeled carbon into biomass might vary significantly depending on the type of microorganism present and the activity of the species [131].

Regardless of some of its inherent limitations MAR-FISH is an encompassing approach allowing the simultaneous in situ identification of substrate uptake patterns of individual microbial cells [91], whilst permitting an in-depth investigation of the ecophysiology of uncultivated bacteria under their natural conditions. Armed with this PCR-independent technique, radioactively labeled substrates that are incorporated into single uncultured microbial cells can be quantified and the microbial populations involved may be identified by application of specific 16S rRNA-targeted FISH probes. Therefore, MAR presents – alone or in combination with other in situ techniques – enormous potential for application in comprehensive biofilm studies [126].

nanoSIMS: A New Window on Activity

The requirement to use radioactively labeled substrates for MAR-FISH limits the application of that technique to suitable radioisotopes. The use of MAR to study nitrogen cycling, for example, is not convenient. Secondary ion mass spectrometry (SIMS) coupled with FISH is, however, a revolutionary breakthrough [131]. In this technique, an ion beam is used to sputter biomass and generate secondary ions to be analysed by mass spectrometry. Isotopically labeled RNA probes can be followed, as well as labeled substrates. More recently, the approach has been refined with the development of nanoSIMS technology, which makes it possible to determine the chemical, radioisotopic or stable isotopic composition of biomass at the submicron level [84,87,115]. The high sensitivity and spatial resolution of nano-SIMS opens a novel possibility for coupling phylogenetic identity and metabolic function in studies of mixed microbial communities from the environment [84]. The technique has already been applied to demonstrate $^{15}N_2$ fixation by bacteria inhabiting the gills of the shipworm Lyrodus pedicellatus [88] and the uptake of fixed nitrogen by the eukaryotic host. It is almost certain that this powerful technique will be widely used for biofilm studies in the future. Indeed, despite the high cost of acquiring nano-SIMS infrastructure, the application of nano-SIMS to pulse-chased experiments with labeled substrates and isotopically-labeled RNA probes opens wide a new window to the ecophysiology of complex biofilm systems.

A Case Study in Microbial Ecophysiology

Using a careful combination of molecular and analytical techniques it is possible to link taxa to function in complex systems. We illustrate this with a case study: the detection of Crenarchaeota in sludge granules from AD bioreactors. These organisms were first observed in the bioreactors using 16S rRNA gene cloning. FISH was used to determine the spatial location of crenarchaeal cells, which were observed in close association with acetate-consuming methanogens (Figure 1; [34]). Based on this, it was hypothesized that the Crenarchaeota in AD bioreactors are H_2-oxidising autotrophs (homoacetogens) that compete with methanogens for H_2 uptake and then generate acetate for other syntrophic, methanogenic partners. In the case of complex, methanogenic biofilms, this cascade of metabolic activity starts with H_2 production by acetogenic microbes feeding on organic acids, which supports the homoacteogen-methanogen partnership. A realistic approach to assess the activity of the consortium would be to measure H_2 depletion by the biofilm with, and without, selective methanogenic inhibitors (Figure 1). Indeed, an elegant approach would then be to determine the pool of H_2 available by incubating biofilms with organic acids, again with and without methanogenic inhibitors, and to follow the dynamics of hydrogen production and consumption using membrane-inlet mass spectrometry (Figure 1). A range of options is then, of course, open to combine these strategies with microscopy and labeling studies.

Incubations of granules with radiolabeled acetate (^{14}C-sodium acetate), followed by biofilm cross-sectioning and beta-microimaging to detect radiolabel uptake, revealed that the acetate was confined to zones dominated by the co-located Crenarchaeota and acetoclastic methanogens (Figure 1; [34]). Subsequently, stable isotope probing (DNA-SIP) experiments using ^{13}C-bicarbonate and sludge granules, with separate PCR-amplification of 16S rRNA genes from 'heavy' and 'light' DNA fractions, indicated the autotrophic – or at least, mixotrophic – potential of crenarchaeal clones. This is a fascinating example of how culture-independent techniques, and disparate datasets, can together reveal the metabolic function of target groups, but it is not a high-throughput and generically applicable approach.

Minorities Often Do the Majority of the Work in Biofilms

The 'uncultivated majority' is oft-cited in the microbial ecology literature (e.g. [84]), but it is worth outlining the important role of minority populations in microbial communities. Minor taxa have been shown as the predominant carbon and nutrient cyclers in several habitats (e.g. [116]), which underscores the importance of not overlooking those populations in genomic datasets.

Next-Generation Sequencing and Metagenomics

To date, most 16S rRNA studies have used either gel-based, fingerprinting techniques, such as DGGE, which provide a 'barcode' of community diversity but without direct sequence information. Equally, many studies have relied on clone libraries, which do provide direct sequence reads but, through cost and time constraints, are limited in size [143]. Both these methods will fail to resolve the rare taxa. Until recently there was no economical method capable of resolving the complete community structure. Next-generation sequencing technologies have transformed this situation by providing orders-of-magnitude more sequence data at the same cost [101]. This is driving the 'omics revolution allowing the sequencing of complete genomes from isolated organisms and environmental DNA. Application to MFCs and AD systems is still in its infancy. The genomes of potentially important species have been sequenced; for example, *Geobacter sulfurreducens* [111] and *Shewanella oneidensis* [64], which appears capable of direct mediator-less electron transfer in MFCs. These two genomes have been extensively studied and their genes annotated to functions, which have been confirmed by experiments and regulatory networks inferred from transcription profiles. This information has been integrated into metabolic pathway models. However, even for these two highly studied organisms, a complete systems-level understanding is still some way off [48]. Indeed, to extend this highly reductionist approach to all organisms that could be present in a system that is open to the environment – although potentially desirable – is impossible.

Metagenomics can be considered the extension of genome sequencing to a community. In shotgun metagenomics, short reads are taken indiscriminately from all organisms and then assembled into longer individual genome fragments or contigs. When only a few genomes are present, it is possible to obtain long contigs and even complete genomes, from which the same inferences regarding gene function and putative metabolic pathways as for a single genome can be made. This technique can prove very effective for simple communities such as those associated with the breakdown of a particular compound in anaerobic bioreactors [96]. However, gene annotation is only the suggestion of potential function, the methods are far from perfect, and that gene may not be actively expressed. For more complex communities, metagenomics will fail to recover substantially sized contigs unless the sequencing depth is very high. This motivates the amplification of focal genes, most often marker genes such as the 16S rRNA. These amplified gene fragments can then be sequenced to very high depth without dissipating effort through the whole metagenome and sequences recovered from a substantial portion of the community. Unfortunately, the diversity of information associated with the genome fragments of metagenomics is lost and the connection of taxonomy to function is even more tenuous.

The 'OMICS Revolution

Recently, the development and application of numerous 'omics-based approaches – for example, metagenomics, proteomics and transcriptomics – has become widespread. Advances in bioinformatics and the development of more powerful sequencers, and better cloning vectors, have facilitated the expansion of metagenomic studies [175,188], which now comprise a versatile approach for the exploration of biofilms for the structure, gene content and function of the respective microorganisms present. For example, microarray technology, which is based on the ability of complementary sequences of nucleic acids to hybridize or bind to one another [95], has become a powerful functional genomics tool that allows researchers to view the physiology of living cells [51,152,194]. Microarrays can be classified as one of two main types: microarrays for genotyping and detecting single nucleotide polymorphisms, and microarrays for analysis of genome diversity and gene expression profiling. The recent introduction of 454 pyro-sequencing has revolutionised the study of microorganisms in numerous ecological niches [69,159,169]. Pyro-sequencing involves the use of a tag sequencing strategy that combines the use of amplicons of the V6 hypervariable region of small-subunit (SSU) rRNA as proxies for the presence of individual phylotypes [operational taxonomic units (OTUs)] with massively parallel sequencing. In addition, techniques such as, FISH-SIMS (FISH-secondary ion mass spectrometry; [133]), which retains the spatial information of individuals within a multispecies association while enabling geochemical or isotopic analysis of identified microorganisms, show real promise in advancing our understanding of syntrophic associations and microbial interactions in the environment [132].

Relating Population Dynamics with Process Parameters and Performance

Linking ecological datasets with process performance in bioengineered environments underscores the utility of microbial ecology for bioengineers. Collins et al., [31] observed temporal shifts in the structure of the archaeal community in anaerobic digesters in response to varying operational conditions. In general, TRFLP data were in good agreement with reactor efficiency data, thus underlining the inextricable link between reactor performance and the microbial population dynamics underpinning the AD process. The practical potential of molecular diagnostic biomonitoring was illustrated by, for example, shifts in the relative abundance of key methanogenic organisms (e.g. *Methanosaeta* sp.) significantly prior (approx. 100 days) to process disturbance (granule disintegration).

The vision, of course, comprises rapid, inexpensive and reliable tools, or kits, which can be used on-site to detect keystone signals – either key taxa or key activities. Tools based on qPCR are now most likely, but a wide-ranging, holistic, systems-based approach, as will be outlined in the final sections of this Chapter, is also desirable.

Reproducibility of Ecological Data

The exploitation of multivariate statistical analyses is increasingly evident in microbial ecology as the magnitude of biological and environmental information generated is exceptional. Multivariate statistical approaches to microbial community analysis allows for the relationship between samples to be demonstrated, by plotting the samples of similar community structures close together, on a low dimensional space; while accordingly, dissimilar samples are depicted further apart [106]. This is useful for temporal and spatial analyses, and can provide direct information for bioengineers looking to link biofilm ecology with process performance. In this case, and for purpose of using the data to make predictions for either the ecology or the process performance under similar future conditions, the reproducibility of the ecological techniques, and of the diversity and abundance data generated, is absolutely important. It is, therefore, obviously important to include as many replicates as possible.

Descriptive statistical tools, like hierarchical clustering (dendrograms) and ordination methods, namely, principle component analysis (PCA) or non-metric multidimensional scaling (NMS), may be applied to classify or compare community fingerprints generated by molecular methodologies [175], and are commonly employed in wastewater treatment studies. Diversity indices can be calculated from fingerprinting data to investigate changes in the global diversity of a given community or in species composition. The species diversity is measured not only by taking into consideration the number of different species, but also by including the inequality in relative abundance as a component of diversity [106]. Detailed descriptions and reviews of the statistical methods and diversity indices frequently applied in microbial ecology studies of environmental and engineered biofilms have been published (e.g. [38,49,104,106,175]).

Microsensors and In Situ Chemical Profiling

Direct observations on chemical micro-environments in biofilms are possible using micro-electrodes and biosensors. As with mass balance analyses from bioreactor-level systems, localized chemical profiling at the biofilm-level complements ecological information on community structure. Many studies employing microsensors have focused on nitrifying biofilms (e.g. early studies, such as de Beer and Schramm [42]), but anaerobic granules, for example, present ideal environments from which to construct detail profiles along the biofilm, vertically and horizontally. The combination of microsensor and molecular techniques is highly useful for studies on the microbial ecology of biofilms.

An Integrated Approach for 'Ecosystems Biology'

An integration of 'omics techniques could provide the high-throughput approach required to link taxonomy with function, and to deconstruct the complex biofilms found in AD bioreactors and MFCs, and in other biofilm systems. Next-generation sequencing of marker genes will provide a high-resolution picture of community structure generating relative frequencies of even low-abundance taxa. It is important to sample the minor community constituents as it has been hypothesized that this so-called 'rare biosphere' may be important in ecosystem functioning [168]. This can be coupled to techniques capable of resolving the absolute abundance of higher-level taxonomic groups, for example microfluidic cell-counting coupled to quantitative PCR or fluorescence labeling of marker genes. This approach can determine 'who' is there – and in what numbers; metagenomics then allows determination of the functional genes they possess. Metagenomics is a high-throughput technique but the connection to function is not direct. Annotating those genes can provide a framework of potential metabolic pathways on which further information can be hung. Linking the same marker genes that have been amplified through co-occurrence on contigs to the functional genes could allow the compartimentalisation of metabolic functions to particular species and, consequently, enable a connection to taxa abundance.

The next step is to resolve from these potential pathways the true metabolic processes occurring in the community. Central to this is metabolomics. Using spectroscopic methods such as GC/MS, it is possible to determine the concentrations of small metabolite intermediates along the active pathways. Even more powerful is to couple this with experiments pulsing stable isotope-labeled substrates through the community to – through the application of flux balance analysis – calculate the rates of reaction along targeted pathways. At the same time, mass balance measurements based on resolving the concentrations of input substrates and ions in MFCs and bioreactors, and reaction endpoints, provide a useful corollary.

Further 'omics information may be incorporated to resolve ambiguities, or to provide verification of the predictions. For example, transcriptomics can be used to determine which functional genes are expressed and metaproteomics to search for enzymes. However, the study of all of the proteins collectively expressed by a microbial consortium can provide critical insights into microbial functionality. The success of metaproteomics is obvioulsy closely linked to metagenomics, and with the exponential increase in the availability of metagenome sequences, an overwhelming amount of metaproteomic data are being generated, which require systematic analyses.

In summary, metagenomics is married to metabolomics to directly elucidate the functional role of whole communities. This is an experimental strategy for obtaining the information necessary to resolve the pathways in individual samples and to link the components of those pathways to species. Its success, however, will be dependent on coupling it to a computational pipeline capable of extracting the information and an experimental strategy that maximizes the statistical power of the complete data set. The individual components of such a pipeline exist: it is possible to identify microbial taxa from marker genes using resources such as the Ribosomal Database Project [30]; metagenomics reads can be annotated to functional genes using MG-RAST [112]; and potential metabolic pathways can be constructed from the KEGG database [79]. Similarly, tools exist for converting GC/MS peaks into predictions of molecule identity [67]. The key will be integrating the information in a single pipeline within a single statistical framework that allows for errors, mislabeling and potentially contradictory information. Deriving the optimal or most likely set of possible pathways and their divisions amongst species will be a highly complex optimization problem. A Bayesian probabilistic approach may provide the best method by which to solve this. This will have the advantage of predicting not a single solution but a complete distribution of possible results. Finally, any statistical technique is limited by the quality of the data set. It is vital to link these tools to experiments that maximise the information available. This should be achieved by running replicate systems to determine the variability of the responses and conversely by using carefully planned perturbations to expand the range of system states explored. These perturbations could be through changing substrate concentrations or operating conditions, such as temperature. These requirements underpin the importance of the high-throughput nature of 'omics' techniques to obtain, at reasonable cost, multiple data sets each one containing huge amounts of information.

The final step of an ecosystems biology approach is integrating this information into mathematical models capable of predicting both community composition and bioreactor function. These will take the metabolic-taxonomic framework elucidated above and add dynamics. The actual mathematical structures to do this are well established: in a well-mixed bioreactor, a system of ordinary differential equations would be appropriate, possibly with the addition of environmental noise; more complex models will be necessary if spatial localization of groups of organisms in anaerobic granules or on the anodes of MFCs is deemed important. The greater challenge than the model structure will be parameterisation. Once again a Bayesian approach will be vital to propagate that uncertainty into predictions that formally account for it.

Conclusion

Linking the taxonomy and activity of environmental microbes continues to be the cornerstone of modern microbial ecology. Powerful new techniques, including nano-SIMS now offer the chance to determine the role of microbial individuals from within complex, heterogenous and dynamic biofilms. Indeed, the merits of applying a reductionist approach to the microbial world and considering individual cells in isolation are questionable, irrespective of the fascinating information that can be found. The seemingly chaotic nature of environmental biofilms instead calls for systems-wide studies. The holistic, polyphasic approach outlined in this Chapter can be viewed as systems biology but at the community level (Figure 1). It is not desirable to reconstruct, and model, every reaction and process in every microbial cell, but it is desirable to identify those pathways, syntrophies and ecological interactions that are critical for controlling the community – and by extension, process – behaviour. The integration of 'omics approaches coupled to process monitoring, isotope studies and advanced microscopy, can generate comprehensive, integrated datasets at microorganism, biofilm and bioreactor level. This will enable the link between processes occurring at microorganism level (scale c. 1 μm-1 mm) and the processes occurring within bioreactors (scale >1 m). This will, in turn, lead to improved bioreactor monitoring capability, reactor design, performance and operational stability.

Acknowledgments

The authors thank Prof. Manuel José Vieira Simões for the invitation to contribute this chapter. Support from the Engineering and Physical Sciences Research Council (UK), Science Foundation Ireland and the European Research Council is acknowledged.

References

[1] Adnan, M., Morton, G. & Hadi, S. (2011). Analysis of rpoS and bolA gene expression under various stress-induced environments in planktonic and biofilm phase using 2(ΔΔ-CT) method. *Molecular Cell Biochemistry* 357(1-2), 275-82.

[2] Alonso, C. & Pernthaler, J. (2006). Roseobacter and SAR11 dominate microbial glucose uptake in coastal North Sea waters. *Environmental Microbiology* 8 (11), 2022–2030.

[3] Amann, R.I., Glöckner, F.O. & Neef, A. (1997). Modern methods in subsurface microbiology: in situ identification of microorganisms with nucleic acid probes. *FEMS Microbiology Reviews* 20, 191-200.

[4] Amann, R.I., Ludwig, W. & Schleifer, K.H. (1995). Phylogenetic identification and in situ detection of individual microbial cells without cultivation. *Microbiological Reviews* 59 (1), 143–169.

[5] Andreasen, K. & Nielsen, H.P. (1997). Application of microautoradiography to the study of substrate uptake by filamentous microorganisms in activated sludge. *Applied Environmental Microbiology* 63 (9), 3662-3668.

[6] Avaniss-Aghajani, E., Jones, K., Chapman, D. & Brunk, C. (1994). A molecular technique for identification of bacteria using small subunit ribosomal RNA sequences. *Biotechniques* 17, 144–149.

[7] Baker, G.C., Smith, J.J. & Cowan, D.A. (2003). Review and re-analysis of domain-specific 16S primers. J*ournal of Microbiological Methods* 55, 541– 555.

[8] Barns, S.M., Takala, S.L. & Kuske, C.R. (1999). Wide Distribution and Diversity of Members of the Bacterial Kingdom Acidobacterium in the Environment. *Applied Environmental Microbiology* 65 (4), 1731–1737.

[9] Bernard, L., Schafer, H., Joux, F., Courties, C., Muyzer, G. & Lebaron, P. (2000). Genetic diversity of totals, active and culturable marine bacteria in coastal seawater. *Aquatic Microbial Ecology* 23, 1-11.

[10] Bialek, K., Kim, J., Lee, C., Collins, G., Mahony, T. and O'Flaherty, V. (2011). Quantitative and qualitative analyses of methanogenic community development in high-rate anaerobic bioreactors. *Water Research* 45(3), 1298-1308.

[11] Blackwood, C.B., Marsh, T., Kim, S-H. & Paul, E.A. (2003). Terminal restriction fragment length polymorphism data analysis for quantitative comparison of microbial communities. *Applied Environmental Microbiology* 69, 926-932.

[12] Bollinger, R.R., Barbas, A.S., Bush, E.L., Lin, S.S. & Parker, W. (2007). Biofilms in the normal human large bowel: fact rather than fiction. *Gut*. 56, 1481-1482.

[13] Borodina, E., Cox, M.J., McDonald, I.R. & Murrell, C. (2005). Use of DNA-stable isotope probing and functional gene probes to investigate the diversity of methyl chloride-utilizing bacteria in soil. *Environmental Microbiology* 7 (9), 1318-1328.

[14] Boschker, H.T. & Middelburg, J.J. (2002). Stable isotopes and biomarkers in microbial ecology. *FEMS Microbiology Ecology* 40, 85-95.

[15] Boschker, H.T., Nold, S.C., Wellsbury, P., Bos, D., de Graaf, W., Pel, R., Parkes, R.J. & Cappenberg, T.E. (1998). Direct linking of microbial populations to specific biogeochemical processes by ^{13}C-labelling of biomarkers. *Nature* 392, 801-805.

[16] Braker, G., Ayala-del-Río, H.L., Devol, A.H., Fesefeldt, A. & Tiedje, J.M. (2001). Community Structure of Denitrifiers, Bacteria, and Archaea along Redox Gradients in Pacific Northwest Marine Sediments by Terminal Restriction Fragment Length Polymorphism Analysis of Amplified Nitrite Reductase (nirS) and 16S rRNA Genes. *Applied Environmental Microbiology* 67 (4), 1893-1901.

[17] Bretschger, O., Osterstock, J.B., Pinchak, W.E., Ishii, S. & Nelson, K.E. (2010). Microbial fuel cells and microbial ecology: applications in ruminant health and production research. *Microbial Ecology* 59(3), 415-27.

[18] Brock, T.D. & Brock, M.L. (1968). Autoradiography as a tool in microbial ecology. *Nature* 209, 734-736.

[19] Bruggraf, S., Huber, H. & Stetter, K.O. (1997). Reclassification of the Crenarchaeal Orders and Families in Accordance with 16S rRNA Sequence Data. *International Journal Systemic Bacteriology* 47 (3), 657–660.

[20] Burow, L.C., Kong, Y., Nielsen, J.P., Blackall, L.L. & Nielsen, P.H. (2007). Abundance and ecophysiology of Defluviicoccus spp., glycogen-accumulating organisms in full-scale wastewater treatment processes. *Microbiology* 153, 178–185.

[21] Bustin, S.A. (2002). Quantification of mRNA using real-time reverse transcription PCR (RT-PCR): trends and problems. Journal *of Molecular Endocrinology* 29, 23–39

[22] Butler, C.S. & Nerenberg, R. (2010). Performance and microbial ecology of air-cathode microbial fuel cells with layered electrode assemblies. *Applied Microbiology and Biotechnology* 86(5), 1399-1408.

[23] Caldwell, D.E., Korber, D.R. & Lawerence, J.R. (1992). Confocal laser scanning microscopy and computer analysis in microbial ecology. *Advances in Microbial Ecology* 12, 1-68.

[24] Cancilla, M.R., Powell, I.B., Hillier, A.J. & Davidson, B.E. (1992). Rapid genomic fingerprinting of Lactococcus lactis strains by arbitrarily primed polymerase chain reaction with ^{32}P and fluorescent labels. *Applied Environmental Microbiology* 58, 1772–1775.

[25] Case, R.J., Boucher, Y., Dahllof, I, Holmstrom, C., Doolittle, W.F. and Kjelleberg, S. (2007). Use of 16S rRNA and rpoB genes as molecular markers for microbial ecology studies. *Applied Environmental Microbiology* 73(1): 278-88.

[26] Cavicchioli, R. (2006). Cold-adapted Archaea. Nature Reviews Microbiology 4, 331-343.

[27] Chan, O-C., Liu, W-T. & Fang, H.H.P. (2001). Study of microbial community of brewery granular sludge by denaturing gradient gel electrophoresis of 16S rRNA gene. *Water Science* Technology 43 (1), 77-82.

[28] Chaudhuri, S.K. & Lovley, D.R. (2003). Electricity generation by direct oxidation of glucose in mediatorless microbial fuel cells. *Nature Biotechnol.* 21, 1229-1232.

[29] Clarridge, J. (2004). Impact of 16S rRNA gene sequence analysis for identification of bacteria on clinical microbiology and infectious diseases. *Clinical Microbiology Reviews* 17, 840-862.

[30] Cole, J.R., Wang, Q., Cardenas, E., Fish, J., Chai, B., Farris, R.J., Kulam-Syed-Mohideen, A.S., McGarrell, D.M., Marsh, T., Garrity, G.M. & Tiedje, J.M. (2009). The Ribosomal Database Project: improved alignments and new tools for rRNA analysis. *Nucleic Acids Research* 37, D141-D145.

[31] Collins, G., Connaughton, S., Enright, A.-M., Scully, C., Mahony, T. & O'Flaherty, V. (2006). Application and biomolecular monitoring of psychrophilic anaerobic digestion. *Water Science and Technology* 54(2), 41-47.

[32] Collins, G., Foy, C., McHugh, S., Mahony, T. & O'Flaherty, V. (2005b). Anaerobic biological treatment of phenolic wastewater at 15-18ºC. *Water Research*, 39 (8), 1614-1620.

[33] Collins, G., Kavanagh, S., McHugh, S., Connaughton, S., Kearney, A., Rice, O., Carrigg, C., Scully, C., Bhreathnach, N., Mahony, T., Madden, P., Enright, A-M. & O'Flaherty, V. (2006). Accessing the Black Box of Microbial Diversity and Ecophysiology: Recent Advances through Polyphasic Experiments. *J. Environ Sci. Health Part A* 41, 897-922.

[34] Collins, G., O'Connor, L., Mahony, T., Gieseke, A., de Beer, D. & O'Flaherty, V. (2005). Distribution, localization and phylogeny of abundant populations of Crenarchaeota in anaerobic granular sludge. *Applied Environmental Microbiology* 71(11), 7523-7527.

[35] Collins, G., O'Connor, L., Mahony, T, Gieseke, A., deBeer, D. & O'Flaherty, V. (2005a). Distribution, Localization, and Phylogeny of Abundant Populations of Crenarchaeota in Anaerobic Granular Sludge. *Applied Environmental Microbiology* 71(11), 7523-7527.

[36] Collins, G., Woods, A., McHugh, S., Carton, M. & O'Flaherty, V. (2003). Microbial community structure and methanogenic activity during start-up of psychrophilic anaerobic digesters treating synthetic industrial wastewaters. *FEMS Microbiology Ecology* 46, 159-170.

[37] Cottrell, M.T. & Kircham, D.L. (2000). Natural Assemblages of Marine Proteobacteria and Members of the Cytophaga-Flavobacter Cluster Consuming Low- and High-Molecular-Weight Dissolved Organic Matter. *Applied Environmental Microbiology* 66 (4), 1692–1697.

[38] Culman, S.W., Gauch, H.G., Blackwood, C.B. & Thies, J.E. (2008). Analysis of T-RFLP data using analysis of variance and ordination methods: A comparative study. *Journal Microbiological Methods* 75, 55–63.

[39] Curtis, T. (2006). Microbial ecologists: it's time to 'go large'. *Nature Reviews in Microbiology* 4, 488

[40] Curtis, T. (2007). Crystal ball – 2007: Theory and the microbial world. *Environmental Microbiology* 9(1), 1–11.

[41] Daims, H., Nielsen, J.L., Nielsen, P.H., Schleifer, K.-H. & Wagner, M. (2001). In situ characterization of Nitrospira-like nitrite-oxidizing bacteria active in wastewater treatment plants. *Applied Environmental Microbiology* 67, 5273–5284.

[42] de Beer, D. & Schramm, A. (1999). Micro-environments and mass transfer phenomena in biofilms studied with microsensors. *Wat. Sci. Technol.* 39(7), 173-178

[43] DeLong, E. (2002). Microbial population genomics and ecology. *Current Opinion in Microbiology* 5, 520-524.

[44] Fang, J., Shizuka, A., Kato, C & Schouten, S. (2006). Microbial diversity of cold-seep sediments in Sagami Bay, Japan, as determined by 16S RNA gene and lipid analysis. *FEMS Microbiology Ecology* 57, 429-441.

[45] Farelly, V., Rainey, F.A. & Stackebrandt, E. (1995). Effect of genome size and rrn gene copy number on PCR amplification of 16S rRNA genes from a mixture of bacterial species. *Applied and Environmental Science* 61, 2798– 2801.

[46] Fermoso, F.G., Collins, G., Bartacek, J., O'Flaherty, V. and Lens, P. (2008). Acidification of methanol-fed anaerobic granular sludge bioreactors by cobalt deprivation: Induction and microbial community dynamics. *Biotechnology & Bioengineering* 99 (1), 49-58

[47] Fernández, A., Huang, S., Seston, S., Xing, J., Hickey, R., Criddle, C. & Tiedje, J. (1999). How Stable is Stable? Function versus Community Composition. *Applied and Environmental Science* 65, 3697-3704.

[48] Fredrickson, J.K. et al., (2008). Towards environmental systems biology of *Shewanella*. *Nature Reviews in Microbiology* 6, 592-603.

[49] Fromin, N., Hamelin, J., Tarnawski, S., Roesti, D., Jourdain-Miserez, K., Forestier, N., Teyssier-Cuvelle, S., Gillet, F., Aragano, M. & Rossi, P. (2002). Statistical analysis of denaturing gradient gel electrophoresis (DGE) fingerprinting studies. *Environmental Microbiology* 4 (11), 634-643.

[50] Garcia, J.L., Patel, B.P. and Ollivier, B. (2000). Taxonomic, Phylogenetic, and Ecological Diversity of Methanogenic Archaea. *Anaerobe* 6, 205-226

[51] Gilbride, K.A., Lee, D.Y. and Beaudette, L.A. (2006). Molecular techniques in wastewater: Understanding microbial communities, detecting pathogens, and real-time process control. *Journal of Microbiological Methods* 66, 1–20.

[52] Ginige, M.P., Hugenholtz, P., Daims, H., Wagner, W., Keller, J. & Blackall, L.L. (2004). Use of stable isotope probing, full-cycle rRNA analysis, and fluorescence in situ hybridization-microautoradiography to study a methanol-fed denitrifying microbial community. *Applied and Environmental Microbiology* 70, 588–596.

[53] Ginzinger, D.G. (2002). Gene quantification using real-time quantitative PCR: An emerging technology hits the mainstream. *Experimental Hematology* 30, 503–512.

[54] Gray, N.D & Head, I.M. (2001). Linking genetic identity and function in communities of uncultured bacteria. *Environ. Microbiol.* 3(8), 481-92.

[55] Gray, N.D., Howarth, R., Pickup, R.W., Jones, J.G. & Head, I.M. (2000). Use of combined microautoradiography and fluorescence in situ hybridization to determine carbon metabolism in mixed natural communities of uncultured bacteria from the genus Achromatium. *Appl. Environ. Microbiol.* 66, 4518-4522.

[56] Greenberg, E.P. (2003). Bacterial communication: tiny teamwork. Nature. 424, 134.

[57] Griffen, M.E., McMahon, K.D., Mackie, R.I. & Raskin, L. (1998). Methanogenic population dynamics during start-up anaerobic digestors treating municipal solid waste and biosolids. *Biotechnol. Bioengin.* 57, 342-355.

[58] Griffiths, R.I., Whitely, A.S., O'Donnell, A.G. & Bailey, M.J. (2000). Rapid Method for Coextraction of DNA and RNA from Natural Environments for Analysis of Ribosomal DNAand rRNA-Based Microbial Community Composition. *Appl. Environ. Microbiol.* 66(12), 5488–5491.

[59] Grotenhius, J.T.C., Smit, M., Plugge, C.M., Yuansheng, X., vaN Lammen, A.A.M., Stams, A. and Zehdner, AJ.B. (1991). Bacteriological composition and structure of granular sludge adapted to different substrate. *Appl. Environ. Microbiol.* 57, 1942-1949.

[60] Güven, D., Dapena, A., Kartal., B., Schmid, M.C., Maas, B., van de Pas-Schoonen, K., Sozen, S., Mendez, R., Op Den Camp, H.J., Jetten, M. S., Strous, M. & Schmidt, I. (2005). Propionate Oxidation by and Methanol Inhibition of Anaerobic Ammonium-Oxidizing Bacteria. *Appl. Environ. Microbiol.* 71(2), 1066-1071.

[61] Haig, S.J., Dorea, C.C., Davie[s,] R.L., Collins, G. & Quince, Q. (2011). Biological Aspects of Slow Sand Filtration: Past, Present and Future. International Water Association Young Water Professionals Meeting 2011. Edinburgh, UK.

[62] Harmsen, H.J.M., Akkermans, A.D.L., Stams, A.T. & De Vos, W.M. (1996). Population dynamics of propionate-oxidizing bacteria under methanogenic and sulfidogenic conditions. *Appl. Environ. Microbiol.* 62(6), 2163–2168.

[63] Head, I., Saunders, J.R. & Pickup, R.W. (1998). Microbial Evolution, Diversity, and Ecology: A Decade of Ribosomal RNA Analysis of Uncultivated Microorganisms. *Microbial Ecol.* 35, 1-21.

[64] Heidelberg, J.F. et al., (2002). Genome sequence of the dissimilatory metal ion-reducing bacterium Shewanella oneidensis. *Nature Biotechnol.* 20, 1118-1123.

[65] Hesselsoe, M., Nielsen, J.L., Roslev, P. & Nielsen, P.H. (2005). Isotope labeling and microautoradiography of active heterotrophic bacteria on the basis of assimilation of $^{14}CO_2$. *Appl. Environ. Microbiol.* 71(2), 646-655.

[66] Hofman-Bang, J., Zheng, D., Westermann, P., Ahring, B. & Raskin, L. (2003). Molecular Ecology of Anaerobic Reactor Systems, B. K. Ahring (Ed.), Biomethanation, Advances in Biochemical Engineering/Biotechnology, Springer-Verlag, Inc., 81, 151-203.

[67] Horai, H. et al., (2010). MassBank: a public repository for sharing mass spectral data for life sciences. *J. Mass Spectrom.* 45, 703-714.

[68] Huber, H., Hohn, M.J., Stetter, K.O. and Rachel, R (2003). The phylum Nanoarchaeota: Present and future perspectives of a unique form of life. *Research Microbiology* 154, 165-171.

[69] Huber, J.A., Welch, D.B., Morrison, H.G., Huse, S.M., Neal, P.R., Butterfield, D.A. and Sogin, M.I. (2007). Microbial Population Structures in the Deep Marine Biosphere. *Science* 318 (5), 97-100.

[70] Hugenholtz, P. (2002). Exploring prokaryotic diversity in the genomic era. *Genome Biology* 3 (2), 1-8.

[71] Hugenholtz, P., Goebel, B.M. and Pace, N. (1998). Impact of culture-independent studies on the emerging phylogenetic view of bacterial diversity. *J. Bacteriol.* 180 (18), 4765–4774.

[72] Hulshoff Pol, L.W., de Castro Lopes, S.I., Lettinga, G. & Lens, P.N.L. (2004). Anaerobic sludge granulation. *Wat. Res.* 38, 1376–1389.

[73] Imachi, H., Sekiguchi, Y., Kamagata, Y., Ohashi, A. & Harada, H. (2000). Cultivation and in situ detection of a thermophilic bacterium capable of oxidizing propionate in syntrophic association with hydrogenotrophic methanogens in a thermophilic methanogenic granular sludge. *Appl. Environ. Microbiol.* 66, 3608–3615.

[74] Ishii, K., Mußmann, M., MacGregor, B.J. & Amann, R. (2004). An improved fluorescence in situ hybridization protocol for the identification of bacteria and archaea in marine sediments. *FEMS Microbiol. Ecol.* 50, 203–212.

[75] Ito, T., Nielsen, J.L., Okabe, S., Watanabe, Y. & Nielsen, P.H. (2002). Phylogenetic identification and substrate uptake patterns of sulfate-reducing bacteria inhabiting an aerobic-anaerobic sewer biofilm. *Appl. Environ. Microbiol.* 68, 356–364.

[76] Iwamoto, T., Tani, K., Nakamura, K., Suzuki, Y., Kitagawa, M., Eguchi, M. & Nasu, M. (2000). Monitoring impact of in situ biostimulation treatment on groundwater bacterial community by DGGE. *FEMS Microbiol. Ecol.* 32, 129-141.

[77] Jackson, D. W., Suzuki, K., Oakford, L., Simecka, J., Hart, M.E. & Romeo, T. (2002). Biofilm Formation and Dispersal under the Influence of the Global Regulator CsrA of Escherichia coli. *J. Bacteriol.* 184, 290-301.

[78] Kalliopi, R., Comi, G. and Cocolin, L. (2004). The rpoB gene as a target for PCR-DGGE analysis to follow lactic acid bacterial population dynamics during food fermentations. *Food Microbiol.* 21(4), 481-7.

[79] Kanehisa, M., Goto, S., Kawashima, S., Okuno, Y. & Hattori, M. (2004). The KEGG resource for deciphering the genome. *Nucleic Acids Res.* 32, D277-D280.

[80] Kitts, C.L. (2001). Terminal restriction fragment patterns: a tool for comparing microbial communities and assessing community dynamics. *Current Issue Intestinal Microbiology* 2 (1), 17-25.

[81] Klein, D. (2002). Quantification using real-time PCR technology: applications and limitations. *Trends Molec. Medicine* 8(6), 257-260.

[82] Kong, Y., Nielsen, J.L. & Nielsen, P.H. (2004). Microautoradiographic study of Rhodocyclus-related poly-P accumulating bacteria in full-scale enhanced biological phosphorus removal plants. *Appl. Environ. Microbiol.* 70, 5383–5390.

[83] Kragelund, C., Kong, Y., van der Waarde, J., Thelen, K., Eikelboom, D., Tandol, V., Thomsen, T.R. & Nielsen, P.H. (2006). Ecophysiology of different filamentous Alphaproteobacteria in industrial wastewater treatment plants. *Microbiol.* 152, 3003-3012.

[84] Kuypers, M.M.M. & Jørgensen, B.B. (2007). Crystal ball- 2007: The future of single-cell environmental microbiology. *Environ. Microbiol.* 9(1), 1–11.

[85] Lange, M., and Ahring, B.K. (2001). A comprehensive study into the molecular methodology and molecular biology of methanogenic Archaea. *FEMS Microbiol. Rev.* 25, 553–555.

[86] Kongjan, P., O-Thong, S. and Angelidaki, I. (2011). Biohydrogen production from desugared molasses using thermophilic mixed cultures immobilized on heat treated anaerobic sludge granules. *Int. J. Hydrogen Energy*, 36(21), 14261-14269.

[87] Lechene, C., Hillion, F., McMahon, G., Benson, D., Kleinfeld, A.M., Kampf, J.P., et al., (2006). High-resolution quantitative imaging of mammalian and bacterial cells using stable isotope mass spectrometry. *J. Biol.* 5, 20.

[88] Lechene, C.P., Luyten, Y., McMahon, G. & Distel, D.L. (2007). Science. 317, 1563.

[89] Leclerc, M., Delbes, C., Moletta, R. and Godon, J.J. (2001). Single strand conformation polymorphism monitoring of 16S rDNA Archaea during start-up of an anaerobic digester. *FEMS Microbiology Ecology* 34, 213-220.

[90] Lee, C., Kim, J., Shin, S.G. & Hwang, S. (2008). Monitoring bacterial and archaeal community shifts inamesophilic anaerobic batch reactor treating a high-strength organic wastewater. *FEMS Microbiol.* Ecol. 65, 544–554.

[91] Lee, N., Nielsen, P-H., Andreasen, K.H., Juretschko, S., Nielsen, J.l., Schleifer, K-H. & Wagner, M. (1999). Combination of Fluorescent In Situ Hybridization and Microautoradiography – a New Tool for Structure-Function Analyses in Microbial Ecology. *Appl. Environ. Microbiol.* 65 (3), 1289–1297.

[92] Lettinga, G. (1995). Anaerobic digestion and wastewater treatment systems. *Antonie Leeuwenhoek* 67, 3-28.

[93] Lettinga, G., van Velsen, A.F.M., Hobma, S.W., De Zeeuw, W. & Klapwijk, A. (1980). Use of the Upflow Sludge Blanket (USB) reactor concept for biological wastewater treatment. *Biotechnol. Bioeng.* 22, 699-734.

[94] Liu, W.T., Chan, O.C. & Fang, H.P. (2002). Characterization of microbial community in granular sludge treating brewery wastewater. *Wat. Res.* 36, 1767-1775.

[95] Lucchini, S., Thompson, A. & Hinton, J.C. (2001). Microarrays for microbiologists. *Microbiol.* 147, 1403-1414.

[96] Lykidis A, Chen, C.-L., Tringe, S.G., McHardy, A.C., Copeland, A., Kyrpides, N.C., Hugenholtz, P., Macarie, H., Olmos, A., Monroy, O. & Liu, W.-T (2011). Multiple syntrophic interactions in a terephthalate-degrading methanogenic consortium. *ISME Journal.* 5, 122-130.

[97] MacGregor, B.J., Moser, D.P., Alm, E., Nealson, K.H. & Stahl, D.A. (1997). Crenarchaeota in Lake Michigan Sediment. *Appl. Environ. Sci.* 63 (3), 1178–1181.

[98] Mack, N., Mack, J.P. & A. O. Ackerson. (1975). Microbial film development in a trickling filter. *Microbial Ecol.* 2(3), 215-226.

[99] MacLeod, F.A., Guiot, S.R. & Costerton, J.W. (1990). Layered structure of bacterial aggregates produced in an upflow anaerobic sludge bed and filter reactor. *Appl. Environ. Microbiol.* 56 (6), 1598–607.

[100] Marchesi, J., Weightman, A., Cragg, B.A., Parkes, R.J. & Fry, J.C. (2001). Methanogen and bacterial diversity and distribution in deep gas hydrate sediments from the Cascadia Margin as revealed by 16S rRNA molecular analysis. *FEMS Microbiol. Ecol.* 34, 221-228.

[101] Margulies, M. et al., (2005). Genome sequencing in microfabricated high-density picolitre reactors. *Nature.* 437, 376-380.

[102] Marsh, T. (1999). Terminal restriction fragment length polymorphism (T-RFLP): an emerging method for characterising diversity among homologous populations of amplification products. *Curr. Opinion Microbiol.* 2, 323-327.

[103] Marsh, T. (2005). Culture-independent microbial community analysis with terminal restriction fragment length polymorphism. In: Methods in Enzymology, Leadbetter, J.R. (ed.), Academic Press, San Diego.

[104] Marzoroti, M., Wittebolle, L., Boon, N., Daffonchio, D. & Verstraete, W. (2002). How to get more out of molecular fingerprints: practical tools for microbial ecology. *Environ. Microbiol.* 10(6), 1571–1581.

[105] McCarty, P.L. (2001). The development of anaerobic treatment and its future. *Wat. Sci.* Technol. 44(8), 149–156.

[106] McCune, B. & Grace, J.B. (2002). Analysis of Ecological Communities. MJM Software Design, Gleneden Beach, USA.

[107] McHugh, S., Carton, M., Collins, G. & O'Flaherty, V. (2004). Reactor performance and microbial community dynamics during anaerobic biological treatment of wastewaters at 16-37°C. *FEMS Microbiol. Ecol.* 48, 369-378.

[108] McHugh, S., O'Reilly, C., Mahony, T., Colleran, E. & O'Flaherty, V. (2003). Anaerobic granular sludge bioreactor technology. *Rev. Environ. Sci. Bio/Technol.* 2, 225-245.

[109] McMahon, G., Glassner, B.J. & Lechene, C.P. (2006). Quantitative imaging of cells with multi-isotope imaging mass spectrometry (MIMS) – nanoautography with stable isotope tracers. *Appl. Surf. Sci.* 252, 6895–6906.

[110] Merkel, W., Manz, W., Szewzyk, U. & Krauth, K. (1999). Population dynamics in anaerobic wastewater reactors: modelling and in situ characterisation. *Wat. Res.* 33, 2392-2402.

[111] Methe, B.A. et al., (2003). Genome of Geobacter sulfurreducens: Metal reduction in subsurface environments. *Science*. 302, 1967-1969.

[112] Meyer, F. et al., (2008). The metagenomics RAST server - a public resource for the automatic phylogenetic and functional analysis of metagenomes. *Bmc Bioinformatics* 9.

[113] Min, B., Cheng, S. & Logan, B.E. (2005). Electricity generation using membrane and salt bridge microbial fuel cells. *Wat. Res.* 39(9), 1675–86.

[114] Moyer, C.L., Dobbs, F.C. & Karl, D.M. (1994). Estimation of diversity and community structure through restriction fragment length polymorphism distribution analysis of bacterial 16S rRNA genes from a microbial mat in an active hydrothermal vent system, Loihi Seamount, Hawaii. *Applied Environmental Science* 60, 871-879.

[115] Musat, N., Foster, R., Vagner, T., Adam, B. and Kuypers, M.M. (2012). Detecting metabolic activities in single cells, with emphasis on nanoSIMS. *FEMS Microbiol Rev.* 36(2), 486-511

[116] Musat, N., Halm, H., Winterholler, B., Hoppe, P., Peduzzi, S., Hillion, F., Horreard, F., Amann, R., Jørgensesen, B.B. and Kuypers, M.M. (2008). A single-cell view on the ecophysiology of anaerobic phototrophic bacteria. *Proc Natl Acad Sci U S A*. 105(46), 17861-6.

[117] Muyzer, G. & Smalla, K. (1998). Application of denaturing gradient gel electrophoresis (DGGE) and temperature gradient gel electrophoresis (TGGE) in microbial ecology. *Antonie van Leeuwenhoek 73*, 127-141.

[118] Muyzer, G., Waal, E.C. & Uitterlinden, A.G. (1993). Profiling of complex microbial populations by denaturing gradient gel electrophoresis analysis of polymerase chain reaction amplified genes coding for 16S rRNA. *Appl. Environ. Microbiol.* 59, 695–700.

[119] Nakatsu, C.H. (2007). Soil microbial community analysis using denaturing gradient gel electrophoresis. *Soil Science Soc. America* J. 71, 562-571.

[120] Nakatsu, C.H., Torsvi, V. & Øvereås, L. (2000). Soil community analysis using DGGE of 16S rRNA polymerase chain reaction products. Soil Science Soc. America J. 64, 1382-1388.

[121] Neufeld, J.D., Vohra, J., Dumont, M.G., Lueders, T., Manefeld, M., Friedrich, M.W. and Murrell, J.C. (2007). DNA stable-isotope probing. *Nature Protoc.* 2(4), 860-6.

[122] Niclas Krakat, A. Westphal, A., Schmidt, S. & Scherer, P. (2010). Anaerobic Digestion of Renewable Biomass: Thermophilic Temperature Governs Methanogen Population Dynamics. *Appl. Environ. Microbiol.* 76(6), 1842-1850.

[123] Nielsen, J.L., Christensen, D., Kloppenborg, M. & Nielsen, P.H. (2003). Quantification of cell-specific substrate uptake by probe-defined bacteria under in situ conditions by

microautoradiography and fluorescence in situ hybridization. *Environ. Microbiol.* 5(3), 202-211.

[124] Nielsen, J.L. & Nielsen, P.H. (2005a). Advances in microscopy: Microautoradiography of single cells. In: Methods in Enzymology, Leadbetter, J.R. (ed.), Academic Press, San Diego.

[125] Nielsen, J.L., Juretschko, S., Wagner, M. & Nielsen, P.H. (2002). Abundance and phylogenetic affiliation of iron reducers in activated sludge as assessed by fluorescence in situ hybridization and microautoradiography. Appl. Environ. Microbiol. 68, 4629–4636.

[126] Nielsen, P.H. & Nielsen, J.L. (2005b). Microautoradiography: recent advances within the studies of the ecophysiology of bacteria in biofilms. *Wat. Sci. Technol.* 52, 187-194.

[127] Nielsen, P.H., Andreasen, K., Lee, N. & Wagner, M. (1999). Use of microautoradiography and fluorescent in situ hybridization for characterization of microbial activity in activated sludge. *Wat. Sci. Technol.* 39 (1), 1–9.

[128] Nielsen, P.H., Aquino de Muro, M. & Nielsen, J.L. (2000). Studies on the in situ physiology of Thiothrix spp. in activated sludge. *Environ. Microbiol.* 2, 389–398.

[129] Nubel, U., Engelen, B., Felske, A., Snaidr, J., Wieshuber, A., Aman, R.I., Ludwig, W. & Backhaus, H. (1996). Sequence Heterogeneities of Genes Encoding 16S rRNAs in Paenibacillus polymyxa Detected by Temperature Gradient Gel Electrophoresis. *J. Bacteriol.* 178(19), 5636–5643.

[130] Oh, S.E. & Logan, B.E. (2005). Hydrogen and electricity production from a food processing wastewater using fermentation and microbial fuel cell technologies. *Wat Res.* 39, 4673-4682.

[131] Okabe, S., Kindaichi, T. & Ito, T. (2004). MAR-FISH – An ecophysiological approach to link phylogenetic affiliation and in situ metabolic activity of microorganisms at a single cell level. *Microbes Environment* 19, 83-98.

[132] Orphan, V.J. (2009). Methods for unveiling cryptic microbial partnerships in nature. *Curr. Opinion Microbiol.* 12, 231–237.

[133] Orphan, V.J. House, C.H., Hinrichs, K.U., McKeegan, K.D. & DeLong, E.F. (2001). Methane-consuming Archaea revealed by directly coupled isotopic and phylogenetic analysis. *Science* 293, 484-487.

[134] Orphan, V.J., House, C.H., Hinrichs, K.-U., McKeegan, K.D., & DeLong, E.F. (2001) Methane-consuming Archaea revealed by directly coupled isotopic and phylogenetic analysis. *Science*. 293, 484-487.

[135] Osborn, A.M., Moore, E.R. & Timmis, K.N. (2000). An evaluation of terminal-restriction fragment length polymorphism (T-RFLP) analysis for the study of microbial community structure and dynamics. *Environ. Microbiol.* 2 (1), 39-50.

[136] Osborn, M. A. & Smith, C. J. (2005). Molecular Microbiol. Ecol. Taylor and Francis Group, Abington, UK.

[137] Ouverney, C.C. & Fuhrman, J.A. (1999). Combined microautoradiography-16S rRNA probe technique for determination of radioisotope uptake by specific microbial cell types in situ. *Appl. Environ. Microbiol.* 65, 1746-1752.

[138] Pace, N.R. (1997). A Molecular View of Microbial Diversity and the Biosphere. *Science* 276, 734-740.

[139] Padmasiri, S.I., Zhang, J., Fitch, M., Norddahl, B., Morgenroth, E. & Raskin, L. (2007). Methanogenic population dynamics and performance of an anaerobic membrane

bioreactor (AnMBR) treating swine manure under high shear conditions. *Wat. Res.* 41(1), 134–144.

[140] Paerl, H.W. & Steppe, T.F. (2003). Scaling up: the next challenge in environmental microbiology. *Environ. Microbiol.* 5(11), 1025-1038.

[141] Peixoto, R.S., da Costa Coutinho, H.L., Rumjanek, N.G., Macrae, A. & Rosado, A.S. (2002) Use of rpoB and 16S rRNA genes to analyse bacterial diversity of a tropical soil using PCR and DGGE. *Lett. Appl. Microbiol.* 35(4), 316-20.

[142] Pernthaler, A., Pernthaler, J. & Amann, R. (2002). Fluorescence in situ hybridization and catalysed reporter deposition for the identification of marine bacteria. *Appl. Environ. Microbiol.* 68, 3094–101.

[143] Phung, N.T., Lee, J., Kang, K.H., Chang, I.S., Gadd, G.M. & Kim, B.H. (2004). Analysis of microbial diversity in oligotrophic microbial fuel cells using 16S rDNA sequences. *FEMS Microbiol. Lett.* 233, 77-82.

[144] Preston, C.M., Wu, K.Y., Molinskit, T. & DeLong, E.F. (1996). A psychrophilic crenarchaeon inhabits a marine sponge: Cenarchaeum symbiosum gen. nov., sp. nov. *PNAS.* 93, 6241-6246.

[145] Purdy, K.J (2005). Nucleic Acid Recovery from Complex Environmental Samples. In: Methods in Enzymology, Leadbetter, J.R. (ed.), Academic Press, San Diego.

[146] Purdy, K.J., Embley, T.M. & Nedwell, D.B. (2002). The distribution and activity of sulfate reducing bacteria in estuarine and coastal marine sediments. *Antonie van Leewenhoek.* 81, 181-187.

[147] Purevdorj, L. & Stoodley, P. (2003). The importance of interspecies signalling in biofilm development. In: biofilms in Industry, Medicine and Environmental Biotechnology Vol. 1- analysis, monitoring and control. Lens, P., Moran, A., Mahony, T., Stoodley, P. & O' Flaherty, V. (eds). IWA Press, London. pp. 385-418.

[148] Radajewski, S., Ineson, P., Parekh N.R. & Murrell. J.C. (2000). Stable-isotope probing as a tool in microbial ecology. *Nature* 403, 646-649.

[149] Radajewski, S., McDonald, I.R. & Murrell, J.C. (2003). Stable-isotope probing of nucleic acids: a window to the function of uncultured microorganisms. *Curr. Opinion Biotechnol.* 14, 296–302.

[150] Raskin, L., Stromley, J.M., Rittmann, B.R., Stahl, D.A. (1994). Group-specific 16SrRNA hybridization probes to describe natural communities of methanogens. *Appl. Environ. Microbiol.* 60, 1232-1240.

[151] Reysenbach, A.L., Giver, L.J., Wickham, G.S. & Pace, N.R. (1992). Differential amplification of rRNA genes by polymerase chain reaction. *Appl. Environ. Microbiol.* 58, 3417– 3418.

[152] Riesenfeld, C.S., Schloss, P.D. & Handelsman, J. (2004). METAGENOMICS: Genomic Analysis of Microbial Communities. Annual Reviews Genetics 38, 525–52

[153] Rittmann, B., Hausner, M., Löffler, F., Love, N.G., Muyzer, G., Okabe, S., Oerther, D. B., Peccia, J., Raskin, L. & Wagner, M. (2006). A vista for Microbial Ecology and Environmental Biotechnology. *Environ. Sci. Technol.* 15, 1096-1103.

[154] Rocheleau, S., Greer, C.W., Lawrence, J.R., Cantin, C., Larame, L. & Guiot, S. (1999). Differentiation of Methanosaeta concilii and Methanosarcina barkeri in Anaerobic Mesophilic Granular Sludge by Fluorescent in Situ Hybridization and Confocal Scanning Laser Microscopy. *Appl. Environ. Microbiol.* 65 (5), 2222–2229.

[155] Roest, K., Heilig, H., Smidt, H., de Vos, W., Stams, A.J. & Akkermans A.D. (2005). Community analysis of a full-scale anaerobic bioreactor treating paper mill wastewater. Systematic *Appl. Microbiol.* 28, 175–185.

[156] Rosselló-Mora, R., Lee, N., Antó, J. & Wagner, M. (2003). Substrate uptake in extremely halophilic microbial communities revealed by microautoradiography and fluorescence in situ hybridization. *Extremophiles* 7, 409–413.

[157] Sanz, J.L. & Kochling, T. (2007). Molecular biology techniques used in wastewater treatment: An overview. *Proc. Biochem.* 42, 119-133.

[158] Schleper, C., Jurgens, G. & Jonuscheit, M. (2005). Genomic studies of uncultivated Archaea. *Nature Reviews* 3, 479-489.

[159] Schlüter, A., Bekel, T., Diaz, N.N., Dondrup, M., Eichenlaub, R., Gartemann, K.H., Krahn, I., Krause, L., Krömeke, H., Kruse, O., Mussgnug, J.H., Neuweger, H., Niehaus, K., Pühler, A., Runte, K.J., Szczepanowskil, R., Tauch, A., Tilker, A., Viehöver, P. & Goesmann, A. (2008). The metagenome of a biogas-producing microbial community of a production-scale biogas plant fermenter analysed by the 454-pyrosequencing technology. *J. Biotechnol.* 136, 77–90.

[160] Schmid, M.C., Maas, B., Dapena, A., van de Pas-Schoonen, K., van de Vossenberg, J., Kartal, B., van Niftrik, L., Schmidt, I., Cirpus, I., Kuenen, G., Wagner, M., Damste, J.S., Kuypers, M., Revsbech, P., Mendez, R., Jetten, M. & Strous, M. (2005). Biomarkers for In Situ Detection of Anaerobic Ammonium-Oxidizing (Anammox) Bacteria. *Applied Environmental Microbiology* 71 (4), 1677–1684.

[161] Schramm, A., D. de Beer, M. Wagner & Amann, R. I. (1998). Identification and activities in situ of Nitrosospira and Nitrospira spp. as dominant populations in a nitrifying fluidized bed reactor. *Appl. Environ. Microbiol.* 64, 3480–3485.

[162] Schramm, A., Hauer, L., Revsbech, N.P., Ramsing, N.B., Amann, R. & Scheifer, K.H. (1996). Structure and Function of a Nitrifying Biofilm as Determined by In Situ Hybridization and the Use of Microelectrodes. *Appl. Environ. Microbiol.* 62(12), 4641–4647.

[163] Sekiguchi, Y. (2006). Yet-to-be Cultured Microorganisms Relevant to Methane Fermentation Processes. *Microbes and Environment* 21 (1), 1-15.

[164] Sekiguchi, Y., Kamagata, Y., Nakamura, K., Ohashi A. & Harada, H. (1999). Fluorescent in situ hybridisation using 16S rRNA-targeted oligonucleotide reveals localisation of methanogens and selected uncultured bacteria in mesophilic and thermophilic sludge granules. *Appl. Environ. Microbiol.* 65, 1280-1288.

[165] Seviour, R. & Nielsen, P.H. (2010). Microbial Ecology of Activated Sludge. London, UK: IWA Publishing.

[166] Shemesh, M., Tam, A. and Steinberg, D. (2007) Differential gene expression profiling of Streptococcus mutans cultured under biofilm and planktonic conditions. *Microbiology* 153(5), 1307-17.

[167] Shigematsu, T., Tang, Y., Kawaguchi, H., Ninomiya, K., Kijima, J., Kobayashi, T., Morimura, S. & Kida, K. (2003). Effect of dilution rate on structure of a mesophilic acetate-degrading methanogenic community during continuous cultivation. *J. Biosci. Bioeng.* 96(6), 547–558.

[168] Sogin, M.L., Morrison, H., Huber, J. A., Welch, D.M., Huse, S.M., Neal, P.R., Arrieta, J.M. & Herndl, G. (2006). Microbial diversity in the deep sea and the underexplored "rare biosphere". *PNAS.* 103(32), 12115-12120.

[169] Sogin, M.L., Morrison, H.G., Huber, J.A., Welch, D.M., Huse, S.M., Neal, P.R., Arrieta, J.M. & Herndl, G.J. (2006). Microbial diversity in the deep sea and the underexplored "rare biosphere". *Proc. Nat. Acad. Sci. USA*. 103, 12115-12120.

[170] Steinberg, L. & Regan, J.M. (2009). mcrA-Targeted Real-Time Quantitative PCR Method To Examine Methanogen Communities. Appl. Environ. Microbiol. 75 (94), 4435-4442.

[171] Stetter, K.O. (1996). Hyperthermophilic prokaryotes. *FEMS Microbiol. Rev*. 18, 149-158.

[172] Stoecker, K., Bendinger, B., Schöning, B., Nielsen, P.H., Nielsen, J.L., Baranyi, C., Toenshoff, E.R., Daims, H. & Wagner, M. (2006). Cohn's Crenothrix is a filamentous methane oxidizer with an unusual methane monooxygenase. *PNAS*. 103 (7), 2363-2367.

[173] Suzuki, M., Rappé, M.S. & Giovannoni, S.J. (1998). Kinetic bias in estimates of coastal picoplankton community structure obtained by measurements of small-unit RNA gene PCR amplicon lenght heterogeneity. *Appl. Environ. Microbiol*. 64, 4522-4529.

[174] Suzuki, M.T. & Giovannoni, S.J. (1996). Bias Caused by Template Annealing in the Amplification of Mixtures of 16S rRNA Genes by PCR. *Appl. Environ. Mircrobiol*. 62 (2), 625–630.

[175] Talbot, C., Topp, E., Palin, M.F. & Masse, D.I. (2008). Evaluation of molecular methods used for establishing the interactions and functions of microorganisms in anaerobic bioreactors. *Wat. Res*. 42, 513 – 537.

[176] Tsai, Y-L. & Olson, B.H. (1991). Rapid Method for Direct Extraction of DNA from Soil and Sediments. *Appl. Environ. Microbiol*. 57(94), 1070-1074.

[177] Tsuneda, S., Nagano, T., Hoshino, T., Ejiri, Y., Noda, N. & Hirata, A. (2003). Characterization of nitrifying granules produced in an aerobic upflow fluidized bed reactor. *Wat. Res*. 37(20), 4965-4973

[178] Urich, T., Lanzén, A., Qi, J., Huson, D.H., Schleper C, Schuster, S.C. (2008). Simultaneous Assessment of Soil Microbial Community Structure and Function through Analysis of the Meta-Transcriptome. *PLoS ONE* 3(6), e2527.

[179] UK Parliamentary Office for Science and Technology. (2007). Postnote: Energy and sewage. Number 282, April 2007.

[180] Vignaga, E., Sloan, W., Haynes, H., Phoenix, V. & Beattie, T. (2009). Entrainment of biostabilized non-cohesive sediments: modelling biofilm adhesion as an elastic force. 33rd IAHR Congress Water Engineering for a Sustainable Future, Vancouver, Canada.

[181] Wagner, M., Horn, M. & Daims, H. (2003). Flourescence in situ hybridisation for identification and characterisation of prokaryotes. Curr. Opinion Microbiol. 6, 302-309.

[182] Wagner, M. (2004). Deciphering the function of uncultured microorganisms. *ASM News* 70, 63-70.

[183] Ward, D.M., Weller, R. & Bateson, M.M (1990). 16S rRNA sequences reveal numerous uncultured microorganisms in a natural community. *Nature* 345, 63-65.

[184] Watanabe, K, Kodama, Y. & Harayama, S. (2001). Design and evaluation of PCR primers to amplify bacterial 16S ribosomal DNA fragments used for community fingerprinting. *J. Microbiol. Meth*. 44, 253-262.

[185] Watanabe, K. & Baker, P.W. (2000). Environmentally-relevant microorganisms. *J. Biosci. Bioeng*. 89, 1-11.

[186] Whiteley, A.S., Thomson, B., Lueders, T. and Manefield, M. (2007). RNA stable-isotope probing. *Nature Protoc*. 2(4), 838-44.

[187] Wilderer, P.A., Bungartz, H-J., Lemmer, H., Wagner, M., Keller, J. & Wuertz, S. (2002). Modern scientific methods and their potential in wastewater science and technology. *Wat. Res.* 36, 370–393

[188] Wilmes, P. & Bond, P.L. (2006). Metaproteomics: studying functional gene expression in microbial ecosystems. *Trends Microbiol.* 14, 92-97.

[189] Wintzingerode, F.V., Gobel, U.B. & Stackebrandt, E. (1997). Determination of microbial diversity in environmental samples: pitfalls of PCR-based rRNA analysis. *FEMS Microbiol. Rev.* 21, 213– 229.

[190] Woese, C. and Fox, G.E. (1977). Phylogenetic structure of the prokaryotic domain: the primary kingdoms. *Proc. Natl. Acad. Sci. USA* 74(11), 5088-90.

[191] Woese, C., Kandler, O. and Wheelis, M.L. (1990). Towards a natural system of organisms: Proposal for the Archaea, Bacteria and Eucarya. *Proc. Natl. Acad. Sci. USA* 87, 4576-9.

[192] Yu, Y., Kim, J. & Hwang, S. (2005b). Use of real-time PCR for groupspecific quantification of aceticlastic methanogens in anaerobic processes: population dynamics and community structures. Biotechnol. Bioeng. 93(3), 424-433.

[193] Yu, Y., Lee, C., Kim, J. & Hwang, S. (2005a). Group-Specific Primer and Probe Sets to Detect Methanogenic Communities Using Quantitative Real-Time Polymerase Chain Reaction. *Biotechnol. Bioeng*. 89(6), 670-679.

[194] Zhou, J. & Thompson, D.K. (2002). Challenges in applying microarrays to environmental studies. *Curr. Opinion Biotechnol*. 13, 204–207.

[195] Zhou, J., Bruns, M-A. & Tiedje, J.M. (1996). DNA Recovery from Soils of Diverse Composition. *Appl. Environ. Microbiol.* 62(2), 316–322.

[196] Zuckerlandl, E. & Pauling, L. (1965). Molecules as documents of evolution. *Theoretical Biology* 8, 357–366.

[197] Zumstein, E., Moletta, R. & Godon, J.J. (2000). Examination of two years of community dynamics in an anaerobic bioreactor using fluorescence polymerase chain reaction (PCR) single-strand conformation polymorphism analysis. *Environ. Microbiol.* 2, 69–78.

In: Biofilms in Bioengineering
Editors: Manuel Simões and Filipe Mergulhão

ISBN: 978-1-62948-161-6

Chapter VII

Biofilms in Drinking Water

Lúcia Chaves Simões[1,2*]

[1]IBB-Institute for Biotechnology and Bioengineering,
Centre of Biological Engineering, University of Minho,
Campus de Gualtar, Braga, Portugal
[2]LEPAE, Department of Chemical Engineering, Faculty of Engineering,
University of Porto, Porto, Portugal

Abstract

The provision of safe drinking water (DW) is a top priority issue in any civilized society. Safe DW is a basic need to human development, health and well-being. The main challenge to the DW industry is to deliver a product that is microbiologically and chemically safe, aesthetically pleasing and adequate in quantity and delivery pressure. Normally, the water that leaves a treatment station has quality, but its quality decreases along the travel in the drinking water distribution systems (DWDS). Water industries and governments over the world are working together in order to improve DW quality through the effective treatment, monitoring of its physicochemical and microbiological properties, and the design and the operational management of the distribution networks. Although DW is strictly monitored in developed countries, waterborne outbreaks are still being reported due to microbial contamination. Biofilms contribute notoriously to these events, creating a protective and nutritional reservoir for pathogens growth and survival. Nevertheless, the dynamics of microbial growth in DW networks is very complex, as a large number of interacting processes (physicochemical and biological) are involved. DW biofilms constitute one of the major microbial problems in DWDS that most contributes to the deterioration of water quality. Although biofilm elimination from DWDS is almost impossible, several aspects can be manipulated in order to prevent and control their growth. This book chapter provides a contribution to better understand the important biological and ecological mechanisms involved in biofilm formation in DWDS, with intent to control and prevent their formation, in order to improve DW quality that reaches to consumer's tap.

[*] Email: luciachaves@deb.uminho.pt.

Keywords: Biofilm; Drinking water; Public health

Introduction

The main goal of water companies is to deliver to each consumer microbiological safe drinking water (DW), adequate in quantity and delivery pressure and acceptable in terms of taste, odour and appearance. Studies in a full-scale drinking water distribution system (DWDS) indicated that most bacteria derived from the biofilm of pipeline surfaces. DWDS are known to harbour biofilms, even in the presence of a disinfectant. Biofilms are constituted by a microbial community adapted to conditions of low nutrient concentration and high disinfectant levels. The presence of biofilms in DWDS constitutes one of the currently recognized hazards affecting the microbiological quality of DW and may lead to a number of unwanted effects on the quality of the distributed water [1]. Microbial growth may affect the turbidity, taste, odour and colour of the water, contribute to the increase of the amount of cells in the bulk phase, promote the deterioration of metallic pipes, induce a disinfectant demand and therefore promote disinfectant decay in the distribution system [2]. Also, biofilms can constitute a reservoir of pathogenic microorganisms, which are responsible for several waterborne diseases [1, 3-4].

The development of biofilms in DWDS is influenced by several factors, including microbial quality of intake water, concentration of biodegradable organic matter, amount of available nutrients, sediment accumulation, concentration of residual disinfectants, water residence time, environmental factors (pH, temperature and turbidity of the water), design of network (presence of dead ends, diameter of pipes), hydrodynamics (shear stress at the biofilm-liquid interface), characteristics of material covering the distribution pipes (composition, porosity, roughness) and their conservation state [2]. Recent studies into the microbial ecology and population dynamics of DWDS have found that other important mechanisms play a determinant role in DWDS biofilm formation and on their resistance to disinfectants. Those include the microbial diversity, interspecies interactions, autoaggregation and coaggregation, presence/release of microbial metabolites and molecules (cell-cell signalling), and transfer of genetic material [2]. However, the role of those mechanisms in DWDS biofilm formation remains poorly understood. The purpose of this book chapter is to provide new and relevant information on the role and mechanisms (physicochemical and biological) of biofilm formation in DWDS.

Biofilms in Drinking Water Distribution Systems

Biofilms: Definition and their Impact

In general, a biofilm can be defined as a community of microorganisms that is irreversibly attached to a biotic or abiotic surface and that is enclosed in a matrix of exopolymeric products [5-6]. DW biofilms, particularly, are composed by complex microbial communities functionally organized and embedded in a gelatinous matrix of extracellular polymers excreted by microorganisms (Figure 1a). Extracellular polymers also known as extracellular polymeric substances (EPS) are the key substances keeping biofilm organisms together, gluing them to the surface and providing protection against agents of stress. Any inorganic particle passing nearby (e.g. corrosion products, clays, sand, etc.) may also be incorporated in the biofilms (Figure 1b) increasing its "mechanical strength" [7-8]. According to Characklis and Marshall [9], bacteria are generally dominant in whatever biofilm due to their high growth rates, small size, adaptation capacities and the ability to produce EPS. However, virus, protozoa, fungi and algae may also be present in DW biofilms as reported by several authors [10-14].

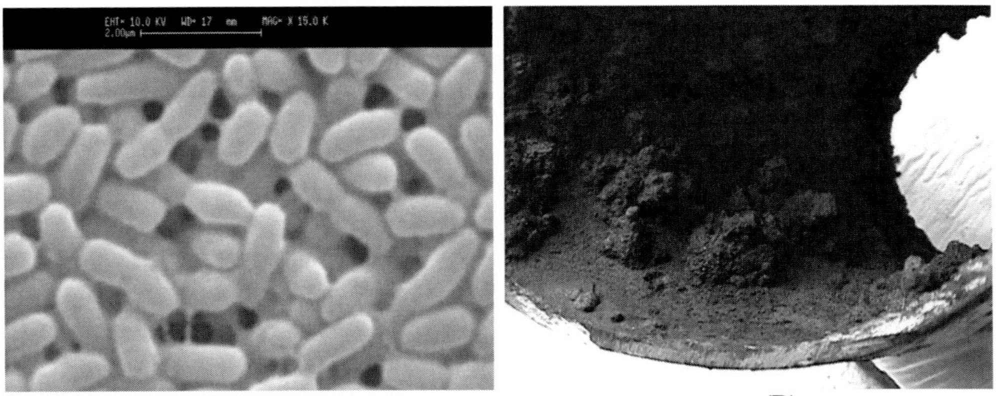

(A) (B)

Figure 1. (a) Scanning electron microscopy (SEM) photomicrographs of 24 hours old biofilms formed by the opportunistic Gram-negative *Burkholderia cepacia* (isolated from laboratorial DWDS) evidencing the presence of an extracellular polymeric matrix (× 15000 magnification; bar = 2 µm). (b) Ductile iron pipe section from a DWDS with biofilm and high amounts of corrosion products. This section of DWDS was obtained as result of a pipe break in the DWDS.

Biofilms are well organized structures where microorganisms are protected from environmental stress and allow complex interactions among different species, i.e. antagonistic or synergistic relationships [15-18]. In biofilms, the way that cells communicate and organize in a social community is controlled by the secretion of signalling molecules in a process called "quorum sensing". This promotes communication between cells and regulates the relationship between cells resulting in a group behaviour instead of an individual performance [19-21].

Relevance of Biofilms in Water Industry

The biofilms in DWDS, also designated as biofouling, are a well-recognized problem in water industry. Biofouling, in general, refers to the undesirable accumulation of biotic matter on a surface. It has been shown to be of considerable hygienic, operational and economical relevance, not only in DWDS but also in other purified water supply systems [1].

Many problems in DWDS are microbial in nature, including biofilm growth, nitrification, microbially-mediated corrosion and the occurrence and persistence of pathogens [22-26]. Biofilms are suspected to be the primary source of microorganisms in DWDS that are fed with treated water and have no pipeline breaches and are of particular concern in older

DWDS [27-28]. Flemming et al., [29] estimated that 95% of the overall biomass is attached to pipe walls, while only 5% is in the water phase. Therefore, the microbial growth in biofilms is highly relevant for water quality since they may directly affect cell density in the bulk phase.

By adopting this sessile mode of life, biofilm-embedded microorganisms enjoy a number of advantages over their planktonic counterparts, namely: the ability to excrete the EPS matrix to capture and concentrate nutrients; resistance to a number of removal strategies such as antimicrobial and antifouling agents as well as shear stress conditions; the possibility of metabolic interactions between bacteria with different physiological requirements; bacterial communication through excreted signalling molecules and the potential for dispersion via detachment, maintaining a persistent bacterial source population that is resistant to antimicrobial agents, while at the same time enabling continuous shedding to promote bacterial spread [1].

The current knowledge of the structure and activities in biofilm communities is still limited, because analysis of microbial physiology and genetics have been largely confined to studies of microorganisms from few lineages for which cultivation conditions have been determined and for some process conditions, not mimicking real environments. The dynamics of the microbial growth in DW networks is very complex, as a large number of interacting processes are involved. DW pipes inner-surfaces are invariably colonized by biofilms, regardless of the presence of a residual disinfectant. In addition to the possibility of causing corrosion, taste and odour problems, biofilms control the microbiological contents of the distributed water and are a potential source of pathogens [30-31]. The interaction of pathogens with other biofilm microorganisms has been a principle of concern in man-made water systems, particularly DWDS. In fact, biofilms formed within potable-water systems contain bacterial pathogens such as *L. pneumophila* and coliforms of intestinal and non-intestinal origin [28,32-33]. Furthermore, protozoa are commonly found within DWDS biofilms and have been associated with pathogen persistence and invasiveness [34-35]. Despite Payment et al., [36] in their work did not find any relationship between biofilm presence in DWDS and occurrence of disease, it has been proved that pathogens such as *L. pneumophila*, *Mycobacterium* spp., *P. aeruginosa*, *Klebsiella* spp., *Burkholderia* spp., *Giardia* and *Cryptosporidium*, among others (Table 1), are transmitted by contaminated water and biofilms are a good candidate as they can act as a protective niche for their survival in DW as shown by several authors [31, 37-38]. The consumption of contaminated DW can cause a wide range of diseases and health-related problems in all people or in more susceptible groups like infants, young children, elderly or sick or immune-compromised people. Waterborne diseases are any illness caused by the utilization of DW contaminated by human or animal faeces, which contain pathogenic microorganisms, or by chemical products. Waterborne pathogens are disease-causing bacteria, protozoa, virus and helminths that are transmitted to people when they consume untreated or inadequately treated water. A list of the most relevant agents can be found in Table 1. If these pathogenic microorganisms are not removed by disinfection and reach the consumer's tap, they may cause outbreaks of disease within the community. The occurrence of outbreaks of waterborne diseases is not limited to developing countries; affluent countries are also affected [39-42]. Such findings demonstrate the essential role of an efficient disinfection plan to control microorganisms in the bulk phase and their biofilms in order to provide high quality DW [1].

Table 1. Pathogens associated to waterborne diseases

Bacteria	Protozoa	Viruses	Helminths
Acinetobacter spp.	*Acanthamoeba castellani*	Adenovirus	*Ascaris lumbricoides*
Aeromonas spp.	*Balantidium coli*	Astrovirus	*Dracunculus medinensis*
Burkholderia pseudomallei	*Blastocystis hominis*	Coxsackie vírus A	*Fasciola* spp.
Campylobacter coli	*Cryptosporidium parvum*	Coxsackie vírus B	*Schistosoma* spp.
Escherichia coli pathogenic	*Cyclospora cayetanensis*	Echovirus	Free-living nematodes other than *Dracunculus medinensis*
E. coli enterohaemorrhagic	*Entamoeba histolytica*	Enterovirus	
Francisella tularensis	*Giardia duodenalis*	Hepatite A virus	
Helicobacter pylori	*Giardia intestinalis*	Hepatite E virus	
Klebsiella spp.	*Giardia lamblia*	Norovirus	
Legionella pneumophila	Microsporidia	Poliovirus	
Leptospira spp.	*Naegleria fowleri*	Rotavirus	
Mycobacterium spp. (non-tubercolous)	*Sarcocytis* spp.	Sapovirus	
Pseudomonas aeruginosa	*Toxoplasma gondii*		
Salmonella typhi			
Salmonella paratyphi			
Salmonella spp.			
Shigella spp.			
Staphylococcus aureus			
Toxic *cyanobacteria*			
Tsukamurella spp.			
Vibrio cholera			
Yersinia enterocolitica			

[1, 33, 235-236].

Biofilm Formation: Physical, Chemical and Biological Processes

The biofilm formation mechanisms were already described extensively, there are several excellent comprehensive reviews on this topic [43-48]. There are a number of mechanisms by which numbers of microbial species are able to come into closer contact with a surface, attach firmly to it, promote cell-cell interactions and grow as a complex structure [47]. Biofilm formation is a dynamic process and comprises a sequence of steps. Currently, processes governing biofilm formation that have been identified include the following steps (Figure 2) [47, 49]:

1) preconditioning of the adhesion surface either by macromolecules present in the bulk liquid or intentionally coated on the surface;
2) Transport of planktonic cells from the bulk liquid to the surface;
3) Adsorption of cells at the surface;
4) Desorption of reversibly adsorbed cells;
5) Irreversible adsorption of bacterial cells at a surface;
6) Production of cell-cell signalling molecules;
7) Transport of substrates to and within the biofilm;
8) Substrate metabolism by the biofilm-bound cells and transport of products out of the biofilm. These processes are accompanied by cell growth, replication, and production of EPS;
9) Biofilm removal by detachment or sloughing.

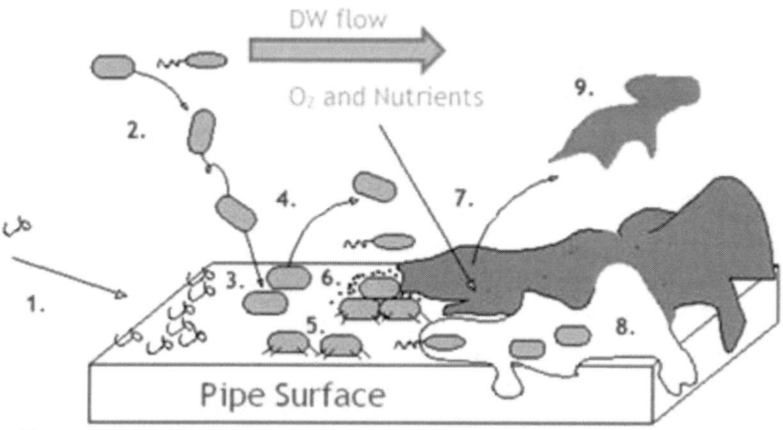

Adapted from Simões et al., [49].

Figure 2. Processes governing DW biofilm formation: (1) Preconditioning the pipe surface by macromolecules (organic and inorganic) present in the water; (2) Transport of planktonic cells from water to pipe surface; (3) Adsorption of cells at the pipe surface; (4) Desorption of reversibly adsorbed cells; (5) Irreversible adsorption of cells; (6) Production of QS molecules; (7) Transport of substrates to and within the biofilm; (8) Substrate metabolism by the biofilm-bound cells and transport of products out of the biofilm, accompanied by cell growth, replication, and production of EPS; (9) Biofilm removal by detachment or sloughing.

Conditioning Film

The first step in biofilm formation (step 1) is the preconditioning of the adhesion surface. The conditioning film is a thin layer of organic molecules and ions covering the adhesion surface that is formed before any microorganisms attach to the surface. These molecules may adhere to the surfaces by physical or chemical adsorption. Physical adsorption is generally a reversible process in which one monolayer is formed, involving nonspecific bonds (London and van der Waals forces). In chemical adsorption, several adsorbed molecular layers are formed involving specific chemical bonds (electrostatic, covalent and hydrogen bonds), dipole interactions, and hydrophobic interactions [50]. The strength of biofilm adhesion is largely dependent on the cohesion of the conditioning as observed by several authors [51-52].

Adhesion

Steps between 2 and 5 correspond to the effective adhesion of microorganisms to surfaces. This is started by the transport of microbial cells to the adhesion surface either by fluid dynamics, gravitational forces and Brownian motion, or by migration through active cell motility (e.g. flagella). Also, the surface electrostatic charge and hydrophobic interactions affect this approaching and the adhesion process. When the cells approach the surface they can interact with each other by the establishment of long and short/intermediate distance forces. The long distance forces are described by the Derjaguin-Landau-Verwey-Overbeek (DLVO) theory and comprise the attractive forces of van der Waals and the repulsive forces of the electrostatic double-layer. The short/intermediate distance forces include hydrophobic interactions, hydrophobic pressure, steric forces, Born repulsion forces and polymer bridges [53-54]. In equilibrium, when favourable, this results in the adhesion of microorganisms.

Biofilm Growth and Maturation

After cellular adhesion to surfaces, the growth and maturation are the following stages of biofilm formation (steps 6-9). The attached microorganisms start growing, they form microcolonies, excreting organic polymers and initiating the formation of the biofilm matrix. Exopolysaccharide synthesis has been shown to be important for the formation of microcolonies [55-56]. As biofilm thickness increases, transport of nutrients from the external liquid media to the inner layers of biofilm and transport of excreted metabolites in the opposite direction are important for biofilm maintenance. Throughout the phase of biofilm growth, bacteria detachment events occur although to a lower extent compared to the growth rate. In the maturation phase, there is the development of a complex and organized consortia of microorganisms embedded in an organic matrix that protects the microorganisms inside from stress factors. It is in this stage that microorganisms produce large amounts of EPS. The structure of a mature biofilm depends on the microbial composition, EPS production, the nutrient availability, hydrodynamic conditions and temperature. In a mature biofilm several processes may occur simultaneously: bacteria detachment into water, attachment of planktonic bacteria, growth and death. However, in this stage these processes are at equilibrium and the number of attached cells per unit surface area is constant in time, although with periodic fluctuations [57-58]. At this phase, the biofilm should reach the highest thickness that is essentially dependent on the hydrodynamic conditions, mass transport and biofilm cohesion.

Detachment

The last phase of biofilm formation (step 9) is the detachment of cells and other components from the biofilm. Hydraulic shear stress provoked by high flow velocities can lead to detachment of bacteria and biofilm aggregates (sloughing), with higher detachment rates at increasing shear [59]. Detachment occurs due to different mechanisms: erosion (the continuous release of single cells or small clusters of cells), sloughing (the rapid detachment of large portions of the biofilm), abrasion (collision of solid particles with the biofilm), and predator grazing. Erosion and sloughing can result from biofilm-associated processes, such as enzyme production [60-61], the excretion of certain signalling molecules [62], cell-cycle-mediated events [63-64], and the excretion of surface modified products (surfactants) by certain bacteria [65], or from external factors such as shear forces [64,66], variations in the nutrient concentration [67], chemical change in EPS due to the presence of chelating agents (Ca^{2+}) that will reduce the cohesive strength of the attached cells [68], abrasion, and predator grazing [69].

Biofilm Structure and Composition

The knowledge of biofilm structure allows a better understanding of how developing biofilms are influenced by the surrounding environment and enables better interpretation of biofilm processes. Over time there has been a shift on perception of the structure of microbial biofilms from that of a homogenous layer of cells in a slime matrix to a much more heterogeneous arrangement. So, several structures have been proposed as biofilm visualizing techniques were improved through the years. The first simplifying assumption that probably extended through the 1980's well into the following decade was that a biofilm could be represented as a simple planar structure, largely 2D, with a relatively constant thickness [70]. In the meantime episcopic differential interference contrast microscope was developed by Keevil and Walker [71] and the heterogeneous mosaic model was proposed for biofilms growing on the inner surfaces of DWDS. These researchers discerned stacks consisting of microcolonies of bacteria held together by EPS and appearing as columns surrounded by a liquid phase in which grazing protozoa could be discerned. Below the stacks there was a layer of cells about 5 µm thick attached to the substratum. These types of structures led Bill Keevil to name this the "heterogeneous mosaic model". Another biofilm structure was proposed by Costerton and co-workers [72-75]. When working with river biofilms supplemented with nutrients, these researchers observed a heterogeneous structure composed of mushrooms with the stalk narrower than the upper surface parts, the whole being penetrated by channels allowing the transportation of water, nutrients and metabolites.

According to Wimpenny and Colasanti [76], who proposed a unifying hypothesis for the microbial biofilm structure based on simple and automaton model, all these conceptual structure models were correct since the final structure was largely dependent on the resource concentration. Thus, the first type was dense relatively uniform biofilm found in habitats where the nutrient levels are generally high (e.g. the human mouth), or periodically extremely high. The second type appeared in water distribution systems where the substrate concentration is very low. The third type was generated in the laboratory using media containing significant nutrient concentration [77]. However, there are reports that indicate the presence of channels in dental plaque biofilms [78] and describe a dense flat biofilm formed under conditions of phosphate starvation [79].

The biofilm structure can be determined by a great variety of environmental parameters (hydrodynamics, nutrient composition, temperature and pH) that, consequently, affect the density, porosity and thickness [76,80-81]. Most environmental biofilms are heterogeneous microbial communities that have different behaviours depending on the conditions (e.g. exopolymers) and the interactions with each other (e.g. chemotaxis, metabolic interactions), hence forming unique biofilms where all resources and energies are optimized.

The structure of DW biofilms on the pipe surface does not follow a standard rule: they may cover the entire inner surface [82-84] or be formed by dispersal aggregates [71]. The surface coverage degree depend of many factors, such as the type of microorganisms, biofilm age, hydrodynamic conditions, presence of inorganic particles, nutrients and temperature.

As result of the application of advanced microscopy, such as confocal laser scanning microscopy and episcopic differential interference contrast microscopy, molecular and electrochemical high-resolution methods have provided insights into the structural organization and function of biofilm communities. Therefore, a mature biofilm is seen as very heterogeneous arrangement, consisting of microcolonies of bacterial cells encased in EPS matrix separated by water channels [44,85]. But although some structural attributes can generally be considered universal, every microbial community is unique [86]. This is due to the fact that a biofilm structure can be influenced by several conditions, such as surface and interface properties, nutrient availability, the microbial community composition, and hydrodynamics, making the exact structure of any biofilm probably a sole feature of the environment in which it develops [87-89]. The water channels that separated the matrix enclosed microcolonies are vital for biofilm maintenance, providing a nutrient flow system inside it [44], that delivers nutrients deep within the complex community [90] and allows the exchange of metabolic products with the bulk fluid layer [91].

Concerning the biofilm composition, water is considered to be the major component of the biofilm, representing from 70 to 99% while bacteria occupy only between 10 and 50% of the total volume of biofilm [92-94]. EPS, the major component of biofilm matrix, are considered the organic substances excreted by attached microorganisms, account for 50 to 90% of the total organic carbon of biofilms [95] and are important keys for the biofilm start-up [94,96]. Their composition and amount are also highly influenced by the type of microorganisms and environmental conditions such as nutrients, temperature, pH and hydrodynamics. For example, the excess of available carbon and the limitation of other nutrients (nitrogen and phosphate), promote exopolysacharides synthesis [80,96]. The EPS determine the structural and functional integrity of microbial biofilms, and contribute significantly to the organization of the biofilm community [97]. EPS are involved in the formation and maintenance of a 3-dimensional, gel-like, highly hydrated and locally charged biofilm matrix, in which the microorganisms are more or less immobilized.

Besides polysaccharides, proteins, nucleic acids or phospholipids, non-cellular materials such as mineral crystals, corrosion products or blood components, may also be found in the biofilm matrix [82]. The biofilm matrix (composed by all inorganic and organic substances surrounding the cells) has several functions. Furthermore, acting as the structural backbone, biofilm matrix protects bacteria from being washed out, from mechanical shocks, from toxic/lethal attacks by antibiotics [98], disinfection chemicals [92,99], UV radiation [100], predators [71] and from desiccation [80,99,101]. As well promotes the storage of nutrients for intake during periods of limitation [102], the retention of extracellular enzymes [103], the horizontal gene transfer [104], and the exchange of signalling molecules and metabolites [62].

Factors Affecting Biofilm Growth in Drinking Water Distribution Systems

The attachment of microorganisms to surfaces and the subsequent biofilm development are very complex processes, affected by several factors, as previously stated. In DWDS these include the nature and concentration of nutrients, sediment accumulation, the type and diversity of microorganisms present and their microbial interactions, concentration of free residual disinfectants, environmental factors (including pH and temperature), water residence time, hydrodynamics conditions, design of network (presence of dead ends, diameter of pipes), characteristics of the material covering the distribution pipes and their age. However, in real systems all these factors work together to influence biofilm accumulation. Thus, the impact of some of them may be insignificant compared with the impact of others and must therefore be considered carefully for each system. During the last decades an extensive research has been done in this topic which resulted into several published reports on the effects of diverse factors in DW biofilm formation [105-111]. The main factors will be briefly described.

Environmental Factors (pH and Temperature)

pH and temperature are considered two important factors affecting life by modifying the electrostatic interactions between surfaces and microorganisms, microbial metabolism, enzymatic activity, kinetics and equilibrium of reactions, and other properties (e.g. diffusivity, solubility) [2]. Also, pH and temperature affect the effectiveness of disinfection. Chlorine residuals present in DWDS are drastically reduced when temperature increases and pH decreases [2].

Disinfectant

Other important variable in biofilm formation is the concentration of disinfectant residual in DWDS. The most used disinfectants are chlorine, chloramines, chlorine dioxide, ozone and UV radiation [112]. From all the disinfectants chlorine is by far the most widely used in DWDS. Chemical disinfection and maintenance of chlorine residual through the distribution systems are almost worldwide strategy to prevent bacterial regrowth during water transportation [113-115]. Even so, regrowth may occur when the chlorine residual decays further down in the distribution system [116-117]. Some studies have demonstrated that chlorine is able to control biofilm form8ation by reducing the rate of biofilm growth, promoting the biofilm detachment and decreasing the activity of microorganisms [105,115,118-119]. However, the presence of residual chlorine is also one of the stress factors that leads to biofilm formation [120]. Nevertheless, some European countries notably the Netherlands, Germany, Austria and Switzerland have taken the approach of distributing high quality DW without the use of residual chlorine. The control of microbial growth in these countries is obtained through limitation of the nutrients essential for growth by more appropriate DW treatments (sedimentation, filtration, UV disinfection, ozone, peroxide), i.e. by the production of biologically stable DW [1].

Hydrodynamics

The hydrodynamic conditions in DWDS are variable. These conditions alternate from laminar to turbulent flow, but stagnant waters also occur in places where the water consumption is low, as well as in reservoirs and buildings. The flow velocity may cause different effects on biofilm accumulation and detachment [96]. Nutrient transport rates within the biofilm increase with the flow velocity until a maximum value is reached, and then decrease as the velocity is further increased. This transport rate promotes bacterial growth within the biofilm. On the other hand, the biofilm density and detachment increase with the flow velocity [96]. As result of wide research on effects of flow velocity on biofilm accumulation controversial results were obtained. Several authors observed that biofilm formation increases with flow velocity [111,121-123], while others achieved the opposite effect [124-126]. A mechanistic explanation about the effects of hydrodynamics on biofilm growth was given by several studies on the biofilm metabolism [127-128]. Higher flow velocities increase the cellular hydrophobicity and will promote cell aggregation and hence biofilm accumulation.

Nutrients

Generally, DWDS are considered oligotrophic environments (with low contents of nutrients like carbon, nitrogen and phosphorous). However, the increase of nutrients in water promotes biofilm formation. Studies have shown a positive relationship between the concentration of nutrients in DW and bacterial regrowth in DWDS [129-130]. Several studies from the DWDS around the world observed that the organic carbon content was the limiting nutrient. An increase in this nutrient promoted bacteria regrowth: in Australia [116], France [131-133], USA [27,134], Singapore [135], Spain [136], Netherlands [129] and in China [137]. Batté et al., [132] observed that the addition of phosphorous did not affect the accumulation of biofilm although phosphorous was being incorporated in biofilm. Other researchers observed that the limiting nutrient was phosphorous in DWDS from Japan [138] and Finland [139]. The detection of the limiting nutrient in DWDS is very important since the addition of phosphate based compounds has been proposed to prevent pipe corrosion and the bacterial regrowth [31,140-141]. The concentration of nutrients such as carbon/nitrogen ratio is important to the production of extracellular polymers and thus affects the adhesion of microorganisms to surfaces [142].

Hydrodynamic conditions and nutrients are the two main parameters that influence biofilm growth in particular the structure, density and thickness [76,81,143]. High shear stress and nutrient limitations led to thin and dense biofilms that will have reduced internal nutrient diffusion [144-145] and increased resistance to removal and cohesion [57,123,125,146]. Under low flow velocities and high nutrients content, the biofilms grow quickly with a low dense structure but with many pores, channels and protuberances [147].

Materials

The variability of materials in DWDS is high. Formerly, the majority of pipelines in DW networks were made of iron-based or cement-based materials. More recently, polymeric materials have been preferred, mainly polyvinyl chloride (PVC), polyethylene (PE), because they are easier to handle and implement. In fact, it is possible to find all this types of materials in the same DWDS. The influence of support materials on biofilm growth is well documented in the literature [10,148-152]. However, there is still controversy about the effects of surface materials on biofilm development when polymeric and metallic materials were compared. Some reports demonstrated that DW biofilms grew less on polymeric materials than on iron matrices [149-150,153-155]. This fact was attributed to iron corrosion products that favour biofilm protection from mechanical and chemical stresses. Other studies reported higher biofilm formation on PVC and PE surfaces than on galvanized steel materials [148,151,156]. While, other works concluded that there was no significant difference in the colonization of the investigated materials (stainless steel, PVC and PE) after decades of operation [157-159]. Lehtola et al., [160] found that biofilms grew faster on PE than on copper pipes, but such differences could not be detected in older piping systems. The main characteristics of materials that have been identified as important on biofilm formation are the roughness and the surface physicochemical properties (chemical composition, solid surface tension, hydrophobicity and surface charge). Another aspect is the leaching of volatile components from pipe materials that can be metabolized by biofilm microorganisms. van der Kooij et al., [161] also observed that the polymeric materials in contact with DW could release biodegradable compounds, thus enhancing biofilm formation. Moreover, corrosion resistance of the materials may be another important factor when choosing the material for the DWDS. Corroded iron pipes may offer numerous bacterial attachment sites and bacteria protection from the effect of flow rate and of disinfectants as well as may release undesirable products to the water [150,162]. Also, the corrosion products may retain nutrients (such as, humic matter) for subsequent utilization by biofilm bacteria [163]. The corrosion on metallic surfaces may be induced by the activity of physiologically diverse microbial species within the biofilms [164].

Microorganisms

The physiological state and the type of microorganisms present in the bulk water will affect the attachment process, since each microorganism has different surface properties, extracellular appendages and abilities to produce EPS. Cell surface hydrophobicity and the presence of extracellular filamentous appendages may influence the rate and the extent of microbial attachment. The hydrophobicity of the cell surface is important in adhesion because hydrophobic interactions tend to increase with the increase in the non-polar nature of one or both surfaces involved, *i.e.,* the microbial cell and the adhesion surface [82]. According to Drenkard and Ausubel [165], the ability of bacteria to attach to each other and to surfaces depends in part on the interaction of hydrophobic domains.

Many microorganisms produce extracellular filamentous appendages. These may, therefore, play a role in the attachment process. In fact, their radius of interaction with the surface is far lower than that of the cell itself. A number of such structures are known to exist - *flagella*, *pili* or *fimbrae*, *prothecae*, *stalks* and *holdfast* [166]. These structures are responsible for motility, involved in the cell-surface interactions and adhesiveness [8]. EPS produced by microorganisms are responsible for binding cells and other particulate materials together (cohesion) and to the surface (adhesion) [80,167-168]. The general composition of bacterial EPS comprises polysaccharides, proteins, nucleic acids, lipids, phospholipids, and humic substances [80,169-170]. According to Tsuneda et al., [171], proteins and polysaccharides account for 75-89% of the biofilm EPS composition, indicating that they are the major components. The type of microorganisms and the interactions established between them has an important role in the biofilm dynamics. The protozoa in DWDS are considered

Additionally, the autochthonous microflora could sustain the growth of protozoa and metazoa that are visible or may have adverse effects on the taste of the DW [190]. Filamentous fungi and microfungi were also observed in DWDS [14,203-204].

Microbial Interactions

Under natural conditions, true monospecies biofilms are rare and in most natural and industrial environments, such as DWDS, biofilms are complex communities. Diversity in microbial communities leads to a variety of complex relationships involving interspecies and intraspecies interactions. Interactions among bacterial species may have a profound influence on the initial stages of biofilm formation and development.

The ecology of a biofilm is a complex function of prevailing growth conditions, hydrodynamic forces, presence of microbial metabolites and molecules (cell-to-cell signalling communications) excreted by the microorganisms and dominant microbial inhabitants in the biofilm [47]. Bacteria have the ability to signal and sense the state of population density in order to change physiological needs under different growth conditions. This phenomenon is commonly called quorum-sensing (QS) [205]. Therefore, QS is a strategy of cell-to-cell communication benefiting the biofilm community by controlling unnecessary overpopulation and competition for nutrients [206]. QS has been demonstrated to play a role in cell attachment and detachment from biofilms [20,82]. Bacteria are considered colonial microorganisms by nature and exploit elaborate systems of intercellular interactions and communications to facilitate their adaptation to changing environments [8,207-209]. The successful adaptation of bacteria to changing natural conditions is dependent on their ability to sense and respond to the external environment and modulate gene expression accordingly [20].

Surfaces provide a niche that promotes the evolution of complex interactions between bacterial cells. Once cells are firmly bound, the activity of the community is dependent on the metabolism and growth of each member species under local surface conditions. Such metabolic activities can include substrate consumption, cellular growth and replication, and synthesis of extracellular polymeric substances [47]. The biological complexity of a system is defined by intra as well as interpopulation cell behaviour. The metabolic activities of those microorganisms that become associated with a surface cause these interfacial chemical gradients to evolve over time and space, creating conditions not normally encountered in the bulk aqueous phase [210].

The microbial heterogeneity found in DW and the existence of interspecies relationships can provide improved strategies for microbial growth control [193,211]. Competition for substrate is considered to be one of the major evolutionary driving forces in the microbial world, and experimental data obtained in laboratory conditions showed how different microorganisms may effectively outcompete others because of better utilization of a given energy source [16,212]. Central to the structure, composition and function of any community is a complex set of interactions. For instance, Hansen et al., [213] found that spatial structure was the key environmental factor for *P. putida* KT2440 and *Acinetobacter* sp. strain C6 to establish a structured community for interspecies interactions. Previously, Møller et al., [212] showed the metabolic synergy between *P. putida* and *Acinetobacter* sp. community members when biodegrading toluene and related aromatic compounds. There is evidence that biofilm community diversity can affect disinfection efficacy and pathogen survival within biofilms [214].

Most of the research into interspecies interactions within biofilms has focused on the beneficial aspects of these relationships. However, not all interactions will be beneficial, since antagonistic interactions may play an important role in the development of microbial communities. The production of antimicrobial molecules, including toxins, bacteriolytic enzymes, antibiotics and bacteriocins seems to be a generic phenomenon for most bacteria [215-216]. Table 2 shows relevant interactions found for several multispecies biofilms from diverse environments.

Table 2. Relevant interspecies interactions in biofilm communities

Interspecies interactions	Strains	Reference
Antagonism	Marine epiphytic bacteria	[237]
	Enteric bacteria	[216]
	Marine pelagic bacteria	[238]
	DW-isolated bacteria	[18]
	Bacillus cereus/*Pseudomonas fluorescens*	[239]
Commensalism	*Pseudomonas* sp. strain GJ1/*Pseudomonas putida* DMP1	[240]
	Acinetobacter sp./*Pseudomonas putida*	[16]
	Lactococcus lactis ssp. *cremoris*/*Pseudomonas fluorescens*	[241]
Competition	*Acinetobacter* sp./*Pseudomonas putida*	[16]
	Klebsiella oxytoca/*Burkolderia cepacia*	[242]
	Marine epiphytic bacteria	[17]
	Denitrifying bacteria	[243]
Mutualism (protocooperation and symbiose)	Soil bacteria	[244]
	Oral bacteria	[245]
	Marine epiphytic bacteria	[214]
	E. coli PHL565/ *P. putida* MT2; *E. coli* PHL565/environmental *E. coli* from DWDS	[246]
	DW-isolated bacteria	[18]
Neutralism	*Pseudomonas* sp.; *Corynebacterium* sp.; *Candida* sp.; *Schizosaccharomyces* sp.; *Saccharomyces* sp.; *Schizosaccharomyces* sp.	[247]
	DW-isolated bacteria	[18]

Coaggregation

Coaggregation, the specific recognition and adherence of genetically distinct bacteria to one another, occurs in a variety of ecosystems [217-219] and was first demonstrated for bacteria from dental plaque [220], where both intergeneric and intrageneric coaggregation occurs [221]. However, coaggregation is a widespread phenomenon that has now been observed amongst bacteria from other biofilm communities in several diverse habitats. Few reports on the coaggregation abilities of freshwater biofilm bacteria have been published [219,222-225], and it has been suggested that coaggregation may also mediate the sequential integration of species of bacteria into freshwater biofilms [226-227].

This mechanism of adhesion is highly specific and is thought to have a role in the development of multispecies biofilms in many different environments [221,228-231] and is now recognized as a mechanism for allowing specific association between collaborating bacterial species. Aggregation conveys advantages to microorganisms including transfer of chemical signals, exchange of genetic information, protection from adverse environmental conditions, metabolic cooperation between different species, as well as cell differentiation in some populations [232]. Coaggregation interactions contribute to the development of biofilms by two routes. The first route is by single cells in suspension specifically recognizing and adhering to genetically distinct cells in the developing biofilm. The second is by the prior coaggregation in suspension of secondary colonizers followed by the subsequent adhesion of this coaggregate to the developing biofilm [229]. In both cases, bacterial cells in suspension specifically adhere to biofilm cells in a process known as coadhesion [51,233].

Coaggregation between pairs of freshwater bacteria is typically mediated by protein "adhesion" on one cell type and a complementary saccharide "receptor" on the other. These protein-saccharide interactions can be blocked by the addition of simple sugars [222,229-230]. Thus, the mechanism mediating adhesion between coaggregating pairs in freshwater biofilm bacteria is very similar to the one verified in oral bacteria. The coaggregation between freshwater bacteria is growth-phase-dependent. It depends on cells being in the optimum physiological state for coaggregation and it is maximized when both partner bacteria are in stationary phase. Maximum expression of coaggregation generates clearly visible flocs of cells in mixtures of the two types of cells [226] and is maintained for up to 48 h into stationary phase, depending on the coaggregating pair. The ability to coaggregate then decreases and eventually is lost completely [223]. The optimum coaggregation between a pair might be dependent upon a change in coaggregation ability of one or both partner bacteria. Moreover, and as suggested by Malik et al., [218], bacterial cell surface properties, namely the hydrophobicity, are other factor thought to play an important role in coaggregation, as well as in cell-substratum interactions. Recently, Min et al., [234] studied the influence of some physicochemical parameters (ionic strength, pH, temperature, and viscosity) on coaggregation ability between the freshwater bacteria. These physicochemical factors are important to consider when developing buffers to detect coaggregation in freshwater environments as well as when the intent is to develop novel approaches to control freshwater biofilm formation by blocking bacterial coaggregation.

Studies on freshwater biofilm bacteria have also demonstrated that coaggregation often occurs between bacteria that are taxonomically distant (intergeneric coaggregation) and occasionally between strains belonging to the same species (intraspecies coaggregation) [222,224]. Intergeneric coaggregation is common between oral bacteria [228], but intraspecies coaggregation has not yet been referred between oral plaque bacteria. Thus, intraspecies coaggregation may well be a characteristic that is unique to freshwater biofilm bacteria.

Simões et al., [230] investigated the intergeneric coaggregation ability among DW-isolated bacteria and the role of this specific mechanism in multispecies biofilm formation. This is the first report demonstrating that *A. calcoaceticus* has a bridging function in DW biofilm formation. This bacterium may facilitate the association of the other species that do not coaggregate directly with each other, increasing the opportunity for metabolic cooperation. Other report by Min and Rickard [231] also explores the role of coaggregation by freshwater bacteria in dual biofilm formation. These authors concluded that coaggregation promotes biofilm integration by facilitating attachment to partner species and likely contributes to the expansion of coaggregating *S. natatoria* populations in dual-species biofilms through competitive interactions. These studies raise the question of whether freshwater bridging organisms such as *A. calcoaceticus* and *S. natatoria* can aid the retention of microbial pathogens or if the ability to coaggregate with many species (i.e. bridge) is a mechanism to outcompete other species in freshwater multispecies biofilms.

Bacteria are affected by the environment they live in and the variety of other species present. Coaggregation can take the form of intra, inter or multigeneric interactions, a combination of which contributes to the overall structure and diversity of bacterial community in freshwater biofilms. The specific mechanism for this remains unknown, but a more complete picture of microbial community diversity and interspecies relationships should facilitate a better understanding of disinfection resistance phenomena and will provide new data to design innovative and effective control strategies that will guarantee microbial safe and high quality DW.

Acknowledgments

The authors acknowledge the financial support provided by the Portuguese Foundation for Science and Technology (SFRH/BPD/81982/2011 – Lúcia C. Simões).

References

[1] Simões, L. C. & Simões, M. (2013). Biofilms in drinking water: problems and solutions. *RSC Advances, 3,* 2520–2533.

[2] Simões, L. C. (2012). Biofilms in drinking water: formation and control. LAP LAMBERT Academic Publishing (ISBN 978-3848413508).

[3] Hall-Stoodley, L. & Stoodley, P. (2005). Biofilm formation and dispersal and the transmission of human pathogens. *Trends in Microbiology, 13*, 7-10.

[4] Huq, A., Whitehouse, C. A., Grim, C. J., Alam, M. & Colwell, R. R. (2008). Biofilms in water, its role and impact in human disease transmission. *Current Opinion in Biotechnology, 19,* 244–247.

[5] Costerton, J. W., Stewart, P. S. & Greenberg, E. P. (1999). Bacterial biofilms: a common cause of persistent infections. *Science, 284*, 1318-1322.

[6] Prakash, B., Veeregowda, B. M. & Krishnappa, G. (2003). Biofilms: A survival strategy of bacteria. *Current Science, 85*, 1299-1307.

[7] Vieira, M. J. & Melo, L. F. (1995). Effect of clay particles on the behaviour of biofilms formed by *Pseudomonas fluorescens*. *Water Science and Technology, 32*, 45-52.

[8] Simões, M., Simões, L. C. & Vieira, M. J. (2010b). A review of current and emergent biofilme control statagies. *LWT – Food Science and Technology, 43*, 573-583.

[9] Characklis, W. G. & Marshall, K. C. (1990). Biofilms: a basis for an interdisciplinary approach. In W. G. Characklis, & K. C. Marshall (Eds.), Biofilms. New York: John Wiley & Sons, Inc.

[10] Momba, M., Kfir, R., Venter, S. N. & Cloete, T. E. (2000). An overview of biofilm formation in distribution systems and its impact on the deterioration of water quality. *Water SA, 26*, 59–66.

[11] Codony, F., Miranda, A. M. & Mas, J. (2003). Persistence and proliferation of some unicellular algae in drinking water systems as result of their heterotrophic metabolism. *Water SA, 29*, 113-116.

[12] Skraber, S., Schijven, J., Gantzer, C. & de RodaHusman, A. M. (2005). Pathogenic viruses in drinking-water biofilms: a public health risk? *Biofilms, 2*, 105–117.

[13] Snelling, W. J., Moore, J. E., McKenna, J. P., Lecky, D. M. & Dooley, J. S. G. (2006). Bacterial-protozoa interactions: an update on the role these phenomena play towards human illness. *Microbes and Infection, 8*, 578-587.

[14] Siqueira, V. M., Oliveira, H. M., Santos, C., Paterson, R., Gusmao, N. & Lima, N. (2011). Filamentous Fungi in drinking water, particularly in relation to biofilm formation. *International Journal of Environmental Research and Public Health, 8*, 456-469.

[15] Nielsen, A. T., Tolker-Nielsen, T., Barken, K. B. & Molin, S. (2000). Role of commensal relationships on the spatial structure of a surface attached microbial consortium. *Environmental Microbiology, 2*, 59-68.

[16] Christensen, B.B., Haagensen, J. A. J., Heydorn, A. & Molin, S. (2002). Metabolic commensalism and competition in a two-species microbial consortium. *Applied and Environmental Microbiology, 68*, 2495-2502.

[17] Rao, D., Webb, J. S. & Kjelleberg, S. (2005). Competitive interactions in mixed species biofilms containing the marine bacterium *Pseudoalteromonas tunicata*. *Applied and Environmental Microbiology, 71*, 1729-1736.

[18] Simões, L. C., Simões, M. & Vieira, M. J. (2007b). Biofilm interactions between distinct bacterial genera isolated from drinking water. *Applied and Environmental Microbiology, 73*, 6192-6200.

[19] Watnick, P. & Kolter, R. (2000). Biofilm, city of microbes. *Journal of Bacteriology, 182*, 2675-2679.7

[20] Daniels, R., Vanderleyden, J. & Michiels, J. (2004). Quorum sensing and swarming migration in bacteria. *FEMS Microbiology Reviews, 28*, 261-289.

[21] Parsek, M. R. & Greenberg, E. P. (2005). Sociomicrobiology: the connections between quorum sensing and biofilms. *Trends in Microbiology, 13*, 27-33.

[22] Regan, J. M., Harrington, G. W., Baribeau, H., Leon, R. D. & Noguera, D. R. (2003). Diversity of nitrifying bacteria in full-scale chloraminated distribution systems. *Water Research, 37*, 197-205.

[23] Beech, I. B. & Sunner, J. (2004). Biocorrosion: towards understanding interactions between biofilms and metals. *Current Opinion in Biotechnology, 15*, 181-186.

[24] Camper, A. K. (2004). Involvement of humic substances in regrowth. *International Journal of Food Microbiology, 92*, 355-364.

[25] Emtiazi, F., Schwartz, T., Marten, S. M., Krolla-Sidenstein, P. & Obst, U. (2004). Investigation of natural biofilms formed during the production of drinking water from surface water embankment filtration. *Water Research, 38*, 1197-1206.

[26] Teng, F., Guan, Y. T. & Zhu, W. P. (2008). Effect of biofilm on cast iron pipe corrosion in drinking water distribution system: Corrosion scales characterization and microbial community structure investigation. *Corrosion Science, 50*, 2816-2823.

[27] LeChevallier, M. W., Babcock, T. M. & Lee, R. G. (1987). Examination and characterization of distribution system biofilms. *Applied and Environmental Microbiology, 53*, 2714-2724.

[28] Wingender, J. & Flemming, H.-C. (2011) Biofilms in drinking water and their role as reservoir for pathogens. *International Journal of Hygiene and Environmental Health, 214*, 417– 423.

[29] Flemming, H.-C., Percival, S. & Walker, J. (2002). Contamination potential of biofilms in water distribution systems. *Water Science and Technology: Water Supply, 2*, 271-280.

[30] Percival, S. L. & Walker, J. T. (1999). Potable water and biofilms: a review of the public health implications. *Biofouling, 42*, 99-115.

[31] Szewzyk, U., Szewzyk, R., Manz, W. & Schleifer, K. H. (2000). Microbiological safety of drinking water. *Annual Reviews in Microbiology, 54*, 81-127.

[32] September, S. M., Els, F. A., Venter, S. N. & Brözel, V. S. (2007). Prevalence of bacterial pathogens in biofilms of drinking water distribution systems. *Journal of Water and Health, 05.2*, 219-227.

[33] WHO (2011). Guidelines for drinking water quality (4th ed.). Geneva, Switzerland: World Health Organization.

[34] Tyndall, R. L. & Domingue, E. L. (1982). Co-cultivation of *Legionella pneumophila* and free-living amoebas. *Applied and Environmental Microbiology, 44*, 954-959.

[35] Thomas, J. M. & Ashbolt, N. J. (2011). Do free-living amoebae in treated drinking water systems present an emerging health risk? *Environmental Science and Technology, 45*, 860–869.

[36] Payment, P., Franco, E. & Siemiatycki, J. (1993). Absence of relationship between health effects due to tap water consumption and drinking water quality parameters. *Water Science and Technology, 27*, 137-143.

[37] Ford, T. E. (1999). Microbiological safety of drinking water: United States and global perspectives. *Environmental Health Perspectives, 107*, 191-206.

[38] Sharma, S., Sachdeva, P. & Virdi, J. S. (2003). Emerging waterborne pathogens. *Applied Microbiology and Biotechnology, 61*, 424-428.

[39] Hrudey, S. E. & Hrudey, E. J. (2004). Safe drinking water. Lessons from recent outbreaks in affluent nations. London: International Water Association Publishing.

[40] Craun, M. F., Craun, G. F., Calderon, R. L. & Beach, M. J. (2006). Waterborne outbreaks reported in the United States. *Journal of Water and Health, 4*, 19–30.

[41] Beaudeau, P., de Valk. H., Vaillant, V., Mannschott, C., Tillier, C., Mouly, D. & Ledrans, M. (2008). Lessons learned from ten investigations of waterborne

gastroenteritis outbreaks, France, 1998–2006. *Journal of Water and Health, 6,* 491–503.

[42] Blasi, M. F., Carere, M., Pompa, M. G., Rizzuto, E. & Funari, E. (2008). Water-related diseases outbreaks reported in Italy. *Journal of Water and Health, 6,* 423–432.

[43] O'Toole, G., Kaplan, H. B. & Kolter, R. (2000). Biofilm formation as microbial development. *Annual Reviews in Microbiology, 54,* 49-79.

[44] Donlan, R. M. & Costerton, J. W. (2002). Biofilms: survival mechanisms of clinically relevant microorganisms. *Clinical Microbiology Reviews, 15,* 167-193.

[45] Hall-Stoodley, L. & Stoodley, P. (2002). Developmental regulation of microbial biofilms. *Current Opinion in Biotechnology, 13,* 228-233.

[46] Chmielewski, R.A.N. & Frank, J. F. (2003). Biofilm formation and control in food processing facilities. *Comprehensive Reviews in Food Science and Food Safety, 2,* 22-32.

[47] Bryers, J. D. & Ratner, B. D. (2004). Bioinspired implant materials befuddle bacteria. *ASM News, 70,* 232-237.

[48] Verstraeten, N., Braeken, K., Debkumari, B., Fauvart, M., Fransaer, J., Vermant, J. & Michiels, J. (2008). Living on a surface: swarming and biofilm formation. *Trends in Microbiology, 16,* 496-506.

[49] Simões, M., Bennett, R. N. & Rosa, E. A. S. (2009). Understanding antimicrobial activities of phytochemicals against multidrug resistant bacteria and biofilms. *Natural Product Reports, 26,* 746-757.

[50] Marshall, K. C. (1996). Adhesion as a strategy for access to nutrients. In M. Fletcher (Ed.), Bacterial adhesion: molecular and ecological diversity. Wiley Series.

[51] Busscher, H. J., Bos, R. & van der Mei, H. C. (1995). Initial microbial adhesion is a determinant for the strength of biofilm adhesion. *FEMS Microbiology Letters, 128,* 229–234.

[52] Bos, R., van der Mei, H. C. & Busscher, H. J. (1999). Physico-chemistry of initial microbial adhesive interactions - its mechanisms and methods for study. *FEMS Microbiology Reviews, 23,* 179-229.

[53] Freter, R., Gingell, D., Jones, G. W., Mutter, P. R. & Dazzo, F. B. (1984). Mechanisms of adhesion. In: K. C. Marshall (Ed.), Microbial adhesion and aggregation, (pp. 5-19). Berlin: Springer, Verlag.

[54] Oliveira, D. R. (1992) Physico-chemical aspects of adhesion. In L. F. Melo, T. R. Bott, M. Fletcher, & B. Capdeville (Eds.), Biofilms - science and technology (pp.45-58). Dordrecht: Kluwer Academic Publishers.

[55] Allison, D. G. & Sutherland, I. W. (1987). The role of exopolysaccharides in adhesion of freshwater bacteria. *Journal of Genetic Microbiology, 133,* 1319-1327.

[56] Watnick, P. & Kolter, R. (1999). Steps in the development of a *Vibrio cholerae* El Tor biofilm. *Molecular Microbiology, 34,* 586-95.

[57] Vieira, M. J., Melo, L. & Pinheiro, M. M. (1993). Biofilm formation: hydrodynamic effects on internal diffusion and structure. *Biofouling, 7,* 67-80.

[58] Bryers, J. D. (2000). Biofilm formation and persistence. In J. D. Bryers (Ed.), Biofilms II: Process analysis and applications, (pp. 45-88). New York: John Wiley & Sons.

[59] Characklis,W. G., Turakhia, M. H. & Zelver, N. (1990). Transport and interfacial transfer phenomena. In W. G. Characklis, & K. C. Marshall (Eds.), Biofilms. New York: John Wiley & Sons, Inc.

[60] Lee, S. F., Li, Y. H. & Bowden, G. W. (1996). Detachment of *Streptococcus mutans* biofilm cells by an endogenous enzymatic activity. *Infection and Immunity, 64*, 1035-1038.

[61] Allison, D. G., Ruiz, B., SanJose, C., Jaspe, A. Gilbert, P. (1998). Extracellular products as mediators of the formation and detachment of *Pseudomonas fluorescens* biofilms. *FEMS Microbiology Letters, 167*, 179-184.

[62] Wuertz, S., Okabe, S. & Hausner, M. (2004). Microbial communities and their interactions in biofilm systems: an overview. *Water Science and Technology, 49*, 327-336.

[63] Allison, D. G., Evans, D. J., Brown, M. R. W. & Gilbert, P. (1990). Possible involvement of the division cycle in dispersal of *Escherichia coli* from biofilms. *Journal of Bacteriology, 172*, 1667-1669.

[64] Gilbert, P., Evans, D. J. & Brown, M. R.W. (1993). Formation and dispersal of bacterial biofilms in vivo and in situ. *Journal of Appllied Bacteriology, 74*, 67S–78S.

[65] Rosenberg, E. & Ron, E. Z. (1999). High- and low-molecular-mass microbial surfactants. *Applied Microbiology and Biotechnology, 52*, 154-162.

[66] Picioreanu, C., van Loosdrecht, M. C. M. & Heijnen, J. J. (2001). Two dimensional model of biofilm detachment caused by internal stress from liquid flow. *Biotechnology Bioengineer, 72*, 205-218.

[67] Sauer, K., Cullen, M. C., Rickard, A. H., Zeef, L. A. H., Davies, D. G. & Gilbert, P. (2004). Characterization of nutrient-induced dispersion in Pseudomonas aeruginosa PAO1 biofilm. *Journal of Bacteriology, 186*, 7312-7326.

[68] Chen, X. & Stewart, P. S. (2002). Role of electrostatic interactions in cohesion of bacterial biofilms. *Applied Microbiology and Biotechnology, 59*, 718-720.

[69] Stewart, P. S. (1993). A model of biofilm detachment. *Biotechnology and Bioengineer, 41*, 111-117.

[70] Wimpenny, J. Manz, W. & Szewzyk, U. (2000). Heterogeneity in biofilms. *FEMS Microbiology Reviews, 24*, 661-671.

[71] Keevil, C. W. & Walker, J. T. (1992). Nomarski DIC microscopy and image analysis of biofilms. *Binary Computing Microbiology, 4*, 93-95.

[72] Lawrence, J. R., Korber, D. R., Hoyle, B. D., Costerton, J. W. & Caldwell, D. E. (1991). Optical sectioning of microbial biofilms. *Journal of Bacteriology, 173*, 6558-6567.

[73] Costerton, J. W., Lewandowski, Z., DeBeer, D., Caldwell, D., Korber, D. & James, G. (1994). Biofilms, the customized microniche. *Journal of Bacteriology, 176*, 2137-2142.

[74] Stoodley, P., DeBeer, D. & Lewandowski, Z. (1994). Liquid flow in biofilm systems. *Applied and Environmental Microbiology, 60*, 2711-2716.

[75] James, G. A., Korber, D. R., Caldwell, D. E. & Costerton, J W. (1995). Digital image analysis of growth and starvation responses of a surface- colonizing *Acinetobacter* sp. *Journal of Bacteriology, 177*, 907-915.

[76] Wimpenny, J. & Colasanti, R. (1997). A unifying hypothesis for the structure of microbial biofilms based on cellular automaton models. *FEMS Microbiology Ecology, 22*, 1-6.

[77] Wimpenny, J. (2000). An overview in biofilms as functional communities. In D. G. Allison, P. Gilbert, H. M. Lappin-Scott, & M. Wilson (eds.), Community structure and co-operation in biofilms (pp. 1-24). Cambridge, UK: Cambridge University Press.

[78] Wood, S. R., Kirkham, J., Marsh, P. D., Shore, R. C., Naltress, B. & Robinson, C. (2000). Architecture of intact natural human plaque biofilms studied by confocal laser scanning microscopy. *Journal of Dental Research, 79*, 21-27.
[79] Huang, C.-T., Xu, K. D., McFeters, G. A. & Stewart, P. S. (1998). Spatial patterns of alkaline phosphatase expression within bacterial colonies and biofilms in response to phosphate starvation. *Applied Environmental Microbiology, 64*, 1526-1531.
[80] Sutherland, I. W. (2001a). The biofilm matrix – an immobilized but dynamic microbial environment. *Trends in Microbiology, 9*, 222-227.
[81] Horn, H., Wasche, S. & Hempel, D. C. (2002). Simulation of biofilm growth, substrate conversion and mass transfer under different hydrodynamic conditions. *Water Science and Technology, 46*, 249-252.
[82] Donlan, R. M. (2002). Biofilms: microbial life on surfaces. *Emerging Infectious Diseases, 8*, 881-890.
[83] Chu, C. W., Lu, C. Y., Lee, C. M. & Tsai, C. (2003). Effects of chlorine level on the growth of biofilm in water pipes. *Journal of Environmental Science and Health Part A-Toxic/Hazardous Substances and Environmental Engineering, 38*, 1377-1388.
[84] Yu, S. L., Yu, X. J., Shi, W. X., Wang, D. & Qiu, X. X. (2007). Bacterial species and variation in the biofilm of water distribution system in Harbin city. *Journal of Water Supply Research and Technology-Aqua, 56*, 445-451.
[85] Lewandowski, Z. (2000). Structure and function of biofilms. In L. V. Evans (Ed.), Biofilms: recent advances in their study and control (pp.1-17). Amsterdam: Harwood Academic Publishers.
[86] Tolker-Nielsen, T. & Molin, S. (2000). Spatial organization of microbial biofilm communities. *Microbial Ecology, 40*, 75-84.
[87] Stoodley, P., Boyle, J. D., Dodds, I. & Lappin-Scott, H. M. (1997). Consensus model of biofilm structure. In J. W. T. Wimpenny, P. S. Handely, P. Gilbert, H. M. Lappin-Scott, & M. Jones (Eds.), Biofilms: community interactions and controls (pp. 1-9). Cardiff, UK: BioLine.
[88] Stoodley, P., Lewandowski, Z., Boyle, J. D. & Lappin-Scott, H. M. (1999). Influence of hydrodynamic conditions and nutrients on biofilm structure. *Journal of Applied Microbiology, 85*, 19S-28S.
[89] Sutherland, I. W. (2001b). Biofilm exopolysaccharides: a strong and sticky framework. *Microbiology, 147*, 3-9.
[90] Stoodley, P., Sauer, K., Davies, D. G. & Costerton, J. W. (2002). Biofilms as complex differentiated communities. *Annual Reviews in Microbiology, 56*, 187-209.
[91] Costerton, J. W. (1995). Overview of microbial biofilms. *Journal of Industrial Microbiology & Biotechnology, 15*, 137-140.
[92] Costerton, J. W., Lewandowski, Z., Caldwell, D. E., Korber, D. R. & Lappin-Scott, H. M. (1995). Microbial biofilms. *Annual Review of Microbiology, 49*, 711-745.
[93] Zhang, X. Q., Bishop, P. L. & Kupferle, M. J. (1998). Measurements of polysaccharides and proteins in biofilm extracellular polymers. *Water Science and Technology, 37*, 345-348.
[94] Watnick, P. & Kolter, R. (2000). Biofilm, city of microbes. *Journal of Bacteriology, 182*, 2675-2679.7
[95] Flemming, H.-C., Wingender, J., Mayer, C., Körstgens, V. & Borchard, W. (2000). Cohesiveness in biofilm matrix polymers. In D. Allison, P. Gilbert, H. M. Lappin-Scott,

& M. Wilson (Eds.), Community structure and cooperation in biofilms, (pp. 87-105). SGM Symposium 59: Cambridge University Press.

[96] Melo, L. F. (2003). Biofilm formation and its role in fixed film processes. In Elsevier (Ed.), The Handbook of Water and Wastewater Microbiology.

[97] Branda, S. S., Vik, A., Friedman, L. & Kolter, R. (2005). Biofilms: the matrix revisited. *Trends in Microbiology, 13*, 20-26.

[98] Stewart, P. S. & Costerton, J. W. (2001). Antibiotic resistance of bacteria in biofilms. *Lancet, 358*, 135-138.

[99] LeChevallier, M. W., Cawthon, C. D. & Lee, R. G. (1988a). Inactivation of biofilm bacteria. *Applied and Environmental Microbiology, 54*, 2492-2499.

[100] Hijnen, W. A. M., Beerendonk, E. F. & Medema, G. J. (2006). Inactivation credit of UV radiation for viruses, bacteria and protozoan (oo)cysts in water: a review. *Water Research, 40*, 3-22.

[101] LeChevallier, M. W., Cawthon, C. D. & Lee, R. G. (1988b). Factors promoting survival of bacteria in chlorinated water supplies. *Applied and Environmental Microbiology, 54*, 649-654.

[102] Wolfaardt, G. M., Lawrence, J. R., Robarts, R. D. & Caldwell, D. E. (1998). In situ characterization of biofilm exopolymers involved in the accumulation of chlorinated organics. *Microbial Ecology, 35*, 213-223.

[103] Flemming, H.-C. (2002). Biofouling in water systems - cases, causes and countermeasures. *Applied Microbiology and Biotechnology, 59*, 629-640.

[104] Molin, S. & Tolker-Nielsen, T. (2003). Gene transfer occurs with enhanced efficiency in biofilms and induces enhanced stabilisation of the biofilm structure. *Current Opinion in Biotechnology, 14,* 255- 261.

[105] Lund, V. & Ormerod, K. (1995). The influence of disinfection processes on biofilm formation in water distribution systems. *Water Research, 29,* 1013-1021.

[106] Camper, A. K., Jones, W. L. & Hayes, J. T. (1996). Effect of growth condition and substratum composition on the persistence of coliforms in mixed-population biofilms. *Applied and Environmental Microbiology, 62*, 4014-4018.

[107] Volk, C. J. & LeChevallier, M. W. (1999). Impacts of the reduction of nutrient levels on bacterial water quality in distribution systems. *Applied and Environmental Microbiology, 65*, 4957–4966.

[108] Norton, C. D. & LeChevallier, M. W. (2000). A pilot study of bacteriological population changes through potable water treatment and distribution. *Applied and Environmental Microbiology, 66*, 268-276.

[109] Chu, C., Lu, C. & Lee, C. (2005). Effects of inorganic nutrients on the regrowth of heterotrophic bacteria in drinking water distribution systems. *Journal of Environmental Management, 74*, 255-263.

[110] Gagnon, G. A., Rand, J. L., O`Leary, K. C., Rygel, A. C., Chauret, C. & Andrews, R. C. (2005). Disinfectant efficacy of chlorite and chlorine dioxide in drinking water biofilms. *Water Research, 39,* 1809-1817.

[111] Lehtola, M. J., Laxander, M., Miettinen, I. T., Hirvonen, A., Vartiainen, T. & Martikainen, P. J. (2006). The effects of changing water flow velocity on the formation of biofilms and water quality in pilot distribution system consisting of copper or polyethylene pipes. *Water Research, 40*, 2151-2160.

[112] Lee, Y. & Nam, S. (2005). Comparative study of chemical disinfection in drinking water supplies. *Water Conditioning and Purification,* 1-6.

[113] LeChevallier, M. W., Welch, N. J. & Smith, D. B. (1996). Full-scale studies of factors related to coliform regrowth in drinking water. *Applied and Environmental Microbiology, 62*, 2201-2211.

[114] Momba, M. N. B., Cloete, T. E., Venter, S. N. & Kfir, R. (1998). Evaluation of the impact of disinfection processes on the formation of biofilms in potable surface water distribution systems. Water Science and Technology, 38, 283-289.

[115] Codony, F., Morato, J. & Mas, J. (2005). Role of discontinuous chlorination on microbial production by drinking water biofilms. Water Research, 39, 1986-1906.

[116] Chandy, J. P. & Angles, M. L. (2001). Determination of nutrients limiting biofilm formation and the subsequent impact on disinfectant decay. *Water Research,* 35, 2677-2682.

[117] Hallam, N. B., West, J. R., Forster, C. F., Powell, J. C. & Spencer, I. (2002). The decay of chlorine associated with the pipe wall in water distribution systems. *Water Research,* 36, 3479-3488.

[118] de Beer, D., Srinivasan, R. & Stewart, P. S. (1994). Direct measurement of chlorine penetration into biofilms during disinfection. *Applied and Environmental Microbiology, 60*, 4339-4344.

[119] Codony, F., Morato, J., Ribas, F. & Mas, J. D. (2002). Effect of chlorine, biodegradable dissolved organic carbon and suspended bacteria on biofilm development in drinking water systems. *Journal of Basic Microbiology, 42*, 311-319.

[120] Kokare, C. R., Chakraborty, S., Khopade, A. N. & Mahadik, K. R. (2009). Biofilm: importance and applications. *Indian Journal of Biotechnology, 8*, 159-168.

[121] Percival, S. L., Knapp, J. S., Wales, D. S. & Edyvean, R. G. J. (1999). The effect of turbulent flow and surface roughness on biofilm formation in drinking water. *Journal of Industrial Microbiology and Biotechnology, 22*, 152-159.

[122] Simões, L. C., Azevedo, N., Pacheco, A., Keevil, C. W. & Vieira, M. J. (2006). Drinking water biofilm assessment of total and culturable bacteria under different operating conditions. *Biofouling, 22*, 91-99.

[123] Paris, T., Skali-Lami, S. & Block, J. C. (2007). Effect of wall shear rate on biofilm deposition and grazing in drinking water flow chambers. *Biotechnology and Bioengineerin, 97*, 1550-1561.

[124] Soini, S. M., Koskinen, K. T., Vilenius, M. J. & Puhakka, J. A. (2002). Effects of fluid-flow velocity and quality on planktonic and sessile microbial growth in water hydraulic systems. *Water Research, 36*, 3812-3820.

[125] Chen, M. J., Zhang, Z. & Bott, T. R. (2005). Effects of operating conditions on the adhesive strength of *Pseudomonas fluorescens* biofilms in tubes. *Colloids and Surfaces B: Biointerfaces,* 43, 61-71.

[126] Tsai, Y. P. (2005). Impact of flow velocity on the dynamic behaviour of biofilm bacteria. *Biofouling, 21*, 267-277.

[127] Liu, Y. & Tay, J. H. (2001). Detachment forces and their influence on the structure and metabolic behaviour of biofilms. *World Journal of Microbiology & Biotechnology, 17,* 111-117.

[128] Liu, Y., Lin, Y. M., Yang, S. F. & Tay, J. H. (2003). A balanced model for biofilms developed at different growth and detachment forces. *Process Biochemistry, 38,* 1761-1765.

[129] van der Kooij, D. (1992). Assimilable organic carbon as an indicator of bacterial regrowth. *Journal American Water Works Association, 84*, 57–65.

[130] Owen, D. M., Amy, G. L., Chowdhury, Z. K., Paode, R., McCoy, G. & Viscosil, K. (1995). NOM characterization and treatability. *Journal American Water Works Association, 87*, 46–63.

[131] Servais, P., Laurent, P. & Randon, G. (1995). Comparison of the Bacterial Dynamics in Various French Distribution-Systems. *Journal of Water Supply Research and Technology-Aqua, 44*, 10-17.

[132] Batté, M., Koudjonou, B., Laurent, P., Mathieu, L., Coallier, J. & Prévost, M. (2003a). Biofilm responses to ageing and to a high phosphate load in a bench-scale drinking water system. *Water Research, 37*, 1351–1361.

[133] Servais, P., Anzil, A., Gatel, D. & Cavard, J. (2004). Biofilm in the Parisian suburbs drinking water distribution system. *Journal of Water Supply Research and Technology-Aqua, 53*, 313-324.

[134] LeChevallier, M. W., Schulz, W. & Lee, R. G. (1991). Bacterial nutrients in drinking water. *Applied and Environmental Microbiology, 57*, 857-862.

[135] Hu, J. Y., Yu, B., Feng, Y.Y., Tan, X. L., Ong, S. L., Ng, W. J. & Hoe, W. C. (2005). Investigation into biofilms in a local drinking water distribution system. *Biofilms, 2,* 19-25.

[136] Frias, J., Ribas, F. & Lucena, F. (2001). Effects of different nutrients on bacterial growth in a pilot distribution system. *Antonie Van Leeuwenhoek, 80,* 129-138.

[137] Bai, X. H., Zhang, L., Zhu, B., Jiang, C. T. & Wang, H. L. (2006). Relationship between water biological stability and the bacteria growth in drinking water distribution system. *Zhongguo Huanjing Kexue/China Environmental Science, 26*, 180-182.

[138] Sathasivan, A. & Ohgaki, S. (1999). Applicat.ion of new bacterial regrowth potential method for water distribution system - A clear evidence of phosphorus limitation. *Water Research, 33*, 137-144.

[139] Lehtola MJ, Miettinen IT, Vartiainen T and Martikainen PJ (2002) Changes in content of microbially available phosphorus, assimilable organic carbon and microbial growth potential during drinking water treatment processes. *Water Research* 36: 3681-3690.

[140] Appenzeller, B. M. R., Batté, M., Mathieu, L., Block, J. C., Lahoussine, V., Cavard, J. & Gate, D. (2001). Effect of adding phosphate in drinking water on bacterial growth in slightly and highly corroded pipes. *Water Research, 35*, 1100–1105.

[141] Appenzeller, B. M. R., Duval, Y. B., Thomas, F. & Block, J. C. (2002). Influence of phosphate on bacterial adhesion onto iron oxyhydroxide in drinking water. *Environmental Science and Technology, 36*, 646-652.

[142] Veiga, M. C., Jain, M. K., Wu, W. M., Hollingsworth, R. I. & Zeikus, J. G. (1997). Composition and role of extracellular polymers in methanogenic granules. *Applied and Environmental Microbiology, 63*, 403-407.

[143] Wasche, S., Horn, H. & Hempel, D. C. (2002). Influence of growth conditions on biofilm development and mass transfer at the bulk/biofilm interface. *Water Research, 36*, 4775-4784.

[144] Zhang, T. C. & Bishop, P. L. (1994). Structure, Activity and Composition of Biofilms. *Water Science and Technology, 29*, 335-344.
[145] Melo, L. F. & Vieira, M. J. (1999). Physical stability and biological activity of biofilms under turbulent flow and low substrate concentration. *Bioprocess Engineering, 20,* 363-368.
[146] van Loosdrecht, M. C. M., Picioreanu, C. & Heijnen, J. J. (1997). A more unifying hypothesis for biofilm structures. *FEMS Microbiology Ecology, 24*, 181-183.
[147] van Loosdrecht, M. C. M,, Eikelboom, D., Gjaltema, A., Mulder, A., Tijhuis, L. & Heijnen, J. J. (1995). Biofilm structures. *Water Science and Technology, 32*, 35-43.
[148] Schwartz, T., Hoffmann, S. & Obst, U. (1998). Formation and bacterial composition of young, natural biofilms obtained from public bank-filtered drinking water systems. *Water Research, 32*, 2787-2797.
[149] Kerr, C. J., Osborn, K. S., Robson, G. D. & Handley, P. S. (1999). The relationship between pipe material and biofilm formation in a laboratory model system. *Journal of Applied Microbiology, 85,* 29S – 38S.
[150] Niquette, P., Servais, P. & Savoir, R. (2000). Impacts of pipe materials on densities of fixed bacterial biomass in a drinking water distribution system. *Water Research, 34*, 1952-1956.
[151] Cloete, T. E., Westaard, D. & van Vuuren, S. J. (2003). Dynamic response of biofilm to pipe surface and fluid velocity. *Water Science and Technology, 47*, 57-59.
[152] Yu, J., Kim, D. & Lee, T. (2010). Microbial diversity in biofilms on water distribution pipes of different materials. *Water Science Technology, 61*,163-171.
[153] Hem, L. J. & Skjevrak, I. (2002). Potential water quality deterioration of drinking water caused by leakage of organic compounds from materials in contact with the water. In Copenhagen: 20th NoDig Conference.
[154] Momba, M. N. B. & Kaleni, P. (2002). Regrowth and survival of indicator microorganisms on the surfaces of household containers used for the storage of drinking water in rural communities of South Africa. *Water Research, 36*, 3023-3028.
[155] Chang, Y. C., Puil, M. L., Biggerstaff, J., Randall, A. A., Schulte, A. & Taylor, J. S. (2003). Direct estimation of biofilm density on different pipe material coupons using a specific DNA-probe. *Molecular and Cellular Probes, 17*, 237–243.
[156] Bachmann, R. T. & Edyvean, R. G. J. (2006). AFM study of the colonisation of stainless steel by *Aquabacterium commune. International Biodeterioration & Biodegradation, 58*, 112-118.
[157] Pedersen, K. (1990). Biofilm development on stainless-steel and PVC surfaces in drinking-water. *Water Research, 24*, 239-243.
[158] Zacheus, O. M., Iivanainen, E. K., Nissinen, T. K., Lehtola, M. J. & Martikainen, P. J. (2000). Bacterial biofilm formation on polyvinyl chloride, polyethylene and stainless steel exposed to ozonated water. *Water Research, 34*, 63-70.
[159] Wingender, J. & Flemming, H.-C. (2004). Contamination potential of drinking water distribution network biofilms. *Water Science and Technology, 49*, 277-286.
[160] Lehtola, M. J., Miettinen, I. T., Keinanen, M. M., Kekki, T. K., Laine, O., Hirvonen, A., Vartiainen, T. & Martikainen, P. J. (2004). Microbiology, chemistry and biofilm development in a pilot drinking water distribution system with copper and plastic pipes. *Water Research, 38*, 3769-3779.

[161] van der Kooij, D., Vrouwenvelder, H. S. & Veenendaal, H. R. (1995). Kinetic aspects of biofilm formation on surfaces exposed to drinking water. *Water Science and Technology, 32*, 61-65.

[162] LeChevallier, M. W., Lowry, C. D., Lee, R. G. & Gibbon, D. L. (1993). Examining the Relationship between Iron Corrosion and the Disinfection of Biofilm Bacteria. *Journal American Water Works Association, 85*, 111-123.

[163] Butterfield, P. W., Camper, A. K., Biederman, J. A. & Bargmeyer, A. M. (2002a). Minimizing biofilm in the presence of iron oxides and humic substances. *Water Research, 36*, 3898-3910.

[164] Beech, I. B. (2004). Corrosion of technical materials in the presence of biofilms-current understanding and state-of-the art methods of study. *International Biodeterioration & Biodegradation, 53*, 177-183.

[165] Drenkard, E. & Ausubel, F. M. (2002). *Pseudomonas* biofilm formation and antibiotic resistance are linked to phenotypic variation. *Nature, 416*, 740-743.

[166] Harbron, R. S. & Kent, C. A. (1988). Aspects of cell adhesion. In L. F. Melo, T. R. Bott, & C. A. Bernardo (Eds.), (Vol. 145, pp. 125-140). NATO ASI series.

[167] Characklis, W. G. & Wilderer, P. A. (1989). Glossary. In W. G. Characklis, & P. A. Wilderer (Eds.), Structure and function of biofilms, (pp. 369-371). Chichester: John Wiley & Sons, Inc.

[168] Allison, D. G. (2003). The biofilm matrix. *Biofouling, 19*, 139-150.

[169] Jahn, A. & Nielsen, P. H. (1998). Cell biomass and exopolymer composition in sewer biofilms. *Water Science and Technology, 37*, 17-24.

[170] Wingender, J., Neu, T. R. & Flemming, H.-C. (1999). What are bacterial extracellular polymeric substances? In J. Wingender, T. R. Neu, & Flemming, H.-C. (Eds.) Microbial extracellular polymeric substances – characterization, structure and function (pp. 1-19). Berlin, Germany: Springer Berlin Heidelberg.

[171] Tsuneda, S., Aikawa, H., Hayashi, H., Yuasa, A. & Hirata, A. (2003). Extracellular polymeric substances responsible for bacterial adhesion onto solid surface. *FEMS Microbiology Letters, 223*, 287-292.

[172] Kalmbach, S., Manz, W. & Szewzyk, U. (1997). Dynamics of biofilm formation in drinking water: phylogenetic affiliation and metabolic potential of single cells assessed by formazan reduction and in situ hybridization. *FEMS Microbiology Ecology, 22*, 265–279.

[173] Sibille, I., Sime-Ngando, T., Mathieu, L. & Block, J. C. (1998). Protozoan bacterivory and *Escherichia coli* survival in drinking water distribution systems. *Applied and Environmental Microbiology, 64,* 197-202.

[174] Brown, M. R. W. & Barker, J. (1999). Unexplored reservoirs of pathogenic bacteria: protozoa and biofilms. *Trends in Microbiology, 7*, 46-50.

[175] Berry, D., Xi, C. & Raskin, L. (2006). Microbial ecology of drinking water distribution systems. *Current Opinion in Biotechnology, 17*, 297-302.

[176] Keevil, C. W. (2003). Rapid detection of biofilms and adherent pathogens using scanning confocal laser microscopy and episcopic differential interference contrast microscopy. *Water Science and Technology, 47*, 105-116.

[177] EPA. Health Risks from Microbial Growth and Biofilms in Drinking Water Distribution Systems. US Environmental Protection Agency Office of Ground Water

and Drinking Water Standards and Risk Management Division 1200 Pennsylvania Ave., NW Washington DC; 2002.

[178] van der Kooij, D. (2000). The unified biofilm approach: a framework for addressing biological phenomena in distribution systems. International Distribution Research Symposium. Denver, CO.

[179] Melo, L. F. & Bott, T. R. (1997). Biofouling in water systems. *Experimental Thermal and Fluid Science, 14,* 375-381.

[180] Payment, P. & Robertson, W. (2004). The microbiology of piped distribution systems and public health. In R. Ainsworth (Ed.), Safe piped waste: managing microbial water quality in piped distribution systems. WHO Drinking-water Quality Series, London, UK: IWA Publishing.

[181] Sartory, D. P. (2004). Heterotrophic plate count monitoring of treated drinking water in the UK: a useful operational tool. *International Journal of Food Microbiology, 92,* 297–306.

[182] Oliver, J. D. (2000). The public health significance of viable but nonculturable bacteria. In R. R. Colwell, & D. J. Grimes (Eds.), Nonculturable microorganisms in the environment (pp. 277-300). Washington, DC: ASM Press.

[183] Kalmbach, S., Manz, W. & Szewzyk, U. (2000). Development of a new method to determine the metabolic potential of bacteria in drinking water biofilms, probe active counts (PAC). In H.-C. Flemming, U. Szewzyk, & T. Griebe (Eds), Biofilms, investigative methods and applications (pp. 107–121). Lancaster, UK: Technomic Publishing.

[184] Batté, M., Mathieu, L., Laurent, L. & Prévost, M. (2003b). Influence of phosphate and disinfection on the composition of biofilms from drinking water systems as measured by fluorescence in situ hybridization. *Canadian Journal of Microbiology, 49,* 741-753.

[185] Martiny, A. C., Jorgensen, T. M., Albrechtsen, H.-J., Arvin, E. & Molin, S. (2003). Long-term succession of structure and diversity of a biofilm formed in a model drinking water distribution system. *Applied and Environmental Microbiology, 69*, 6899–6907.

[186] Schmeisser, C., Stockigt, C., Raasch, C., Wingender, J., Timmis, K. N., Wenderoth, D. F., Flemming, H.-C., Liesegang, H., Schmitz, R. A., Jaeger, K.-E. & Streit, W. R. (2003). Metagenome survey of biofilms in drinking-water networks. *Applied and Environmental Microbiology, 69*, 7298–7309.

[187] Williams, M. M. & Braun-Howland, E. B. (2003). Growth of *Escherichia coli* in model distribution system biofilms exposed to hypochlorous acid or monochloramine. *Applied and Environmental Microbiology, 65*, 5463–5271.

[188] Williams, M. M., Santo Domingo, J. W., Meckes, M. C., Kelty, C. A. & Rochon, H. S. (2004). Phylogenetic diversity of drinking water bacteria in a distribution system simulator. *Journal of Applied Microbiology, 96*, 954-964.

[189] Hoefel, D., Monis, P. T., Grooby, W. L., Andrews, S. & Saint, C. P. (2005). Culture-independent techniques for rapid detection of bacteria associated with loss of chloramine residual in a drinking water system. *Applied and Environmental Microbiology, 71*, 6479-6488.

[190] Eichler, S., Christen, R., Holtje, C., Westphal, P., Botel, J., Brettar, I., Mehling, A. & Hofle, M. G. (2006). Composition and dynamics of bacterial communities of a drinking water supply system as assessed by RNA- and DNA-based 16S rRNA gene fingerprinting. *Applied and Environmental Microbiology, 72*, 1858-1872.

[191] Poitelon, J.-B., Joyeux, M., Welté, B., Duguet, J.-P., Prestel, E., Lespinet, O. & DuBow, M. S. (2009). Assessment of phylogenetic diversity of bacterial microflora in drinking water using serial analysis of ribosomal sequence tags. *Water Research, 43*, 4197-4206.

[192] Camper, A. K., Brastrup, K., Sandvig, A., Clement, J., Spencer, C. & Capuzzi, A. J. (2003). Effect of distribution system materials on bacterial regrowth. *Journal of American Water Works Association, 95*, 107-121.

[193] Simões, L. C., Simões, M. & Vieira, M. J. (2010a). Influence of the diversity of bacterial isolates from drinking water on resistance of biofilms to disinfection. *Applied and Environmental Microbiology, 76*, 6673-6679.

[194] Block, J.-C., Sibille, I., Gatel, D., Reasoner, D. J., Lykins, B. & Clark, R. M. (1997). Biodiversity in drinking water distribution systems: a brief review. In D. Sutcliffe (Ed.), The microbiological quality of water (pp. 63-71). London, UK: Freshwater Biological Association.

[195] Martiny, A. C., Albrechtsen, H.-J., Arvin, E. & Molin, S. (2005). Identification of bacteria in biofilm and bulk water samples from a nonchlorinated model drinking water distribution system: detection of a large nitrite-oxidizing population associated with *Nitrospira* spp. *Applied and Environmental Microbiology, 71*, 8611-8617.

[196] Bachmann, R. T. & Edyvean, R. G. J. (2005). Biofouling: an historic and contemporary review of its causes, consequences and control in drinking water distribution systems. *Biofilms, 2*, 197-227.

[197] LeChevallier, M. W. (1990). Coliform regrowth in drinking water: a review. *Journal of the American Water Works Association, 82*, 74–86.

[198] Le Dantec, C., Duguet, J. P., Montiel, A., Dumoutier, N., Dubrou, S. & Vincent, V. (2002). Chlorine disinfection of atypical *Mycobacteria* isolated from a water distribution system. *Applied and Environmental Microbiology, 68*, 1025-1032.

[199] Lee, S. H. & Kim, S. J. (2002). Detection of infectious enteroviruses and adenoviruses in tap water in urban areas in Korea. *Water Research, 36*, 248-256.

[200] Park, S. R., Mackay, W. G. & Reid, D. C. (2001). *Helicobacter* sp. recovered from drinking water biofilm sampled from a water distribution system. *Water Research, 35*, 1624-1626.

[201] Sen, K. & Rodgers, M. (2004). Distribution of six virulence factors in *Aeromonas* species isolated from US drinking water utilities: a PCR identification. *Journal of Applied Microbiology, 97*, 1077-1086.

[202] Karanis, P., Sotiriadou, I., Kartashev, V., Kourenti, C., Tsvetkova, N. & Stojanova, K. (2006). Occurrence of *Giardia* and *Cryptosporidium* in water supplies of Russia and Bulgaria. *Environmental Research, 102*, 260-271.

[203] Doggett, M. S. (2000). Characterization of fungal biofilms within a municipal water distribution system. *Applied and Environmental Microbiology, 66*, 1249-1251.

[204] Gonçalves, A. B., Paterson, R. R. M. & Lima, N. (2006). Survey and significance of filamentous fungi from tap water. *International Journal of Hygiene and Environmental Health, 209*, 257-264.

[205] Swift, S., Throup, J. P., Williams, P., Salmond, G. P. & Stewart, G. S. (1996). Quorum sensing: a population-density component in the determination of bacterial phenotype. *Trends in Biochemical Science, 21*, 214-219.

[206] Davey, M. E. & O`Toole, G. A. (2000). Microbial biofilms: from ecology to molecular genetics. *Microbiology and Molecular Biology Reviews, 64*, 847-867.

[207] Davies, D. G., Parsek, M. R., Pearson, J. P., Iglewski, B. H., Costerton, J. W. & Greenberg, E. P. (1998). The involvement of cell-to-cell signals in the development of a bacterial biofilm. *Science, 280*, 295-298.

[208] Sauer, K. & Camper, A. K. (2001). Characterization of phenotypic changes in *Pseudomonas putida* in response to surface-associated growth. *Journal of Bacteriology, 183*, 6579-6589.

[209] Fuqua, C. & Greenberg, E. P. (2002). Listening in on bacteria acyl-homoserine lactone signalling. *Nature Reviews in Molecular Cell Biology, 3*, 685-695.

[210] Geesey, G. G. (2001). Bacterial behaviour at surfaces. *Current Opinion in Microbiology, 4*, 296-300.

[211] Rasmussen, T. B., Skindersoe, M. E., Bjarnsholt, T., Phipps, R. K., Christensen, K. B., Jensen, P. O., Anderson, J. B., Koch, B., Larsen, T. O., Hentzer, M., Eberl, L., Høiby, N. & Givskov, M. (2005). Identity and effects of quorum-sensing inhibitors produced by *Penicilium* species. *Microbiology, 151*, 1325-1340.

[212] Møller, S., Steinberg, C., Andersen, J. B., Christensen, B. B., Ramos, J. L., Givskov, M. & Molin, S. (1998). In situ gene expression in mixed-culture biofilms: evidence of metabolic interactions between community members. *Applied and Environmental Microbiology, 64*, 721-732.

[213] Hansen, S. K., Rainey, P. B., Haagensen, J. A. J. & Molin, S. (2007). Evolution of species interactions in a biofilm community. *Nature, 445*, 533-536.

[214] Burmølle, M., Webb, J. S., Rao, D., Hansen, L. H., Sørensen, S. J. & Kjelleberg, S. (2006). Enhanced biofilm formation and increased resistance to antimicrobial agents and bacterial invasion are caused by synergistic interactions in multispecies biofilms. *Applied and Environmental Microbiology, 72*, 3916-3923.

[215] Riley, M. A. (1998). Molecular mechanism of bacteriocin evolution. *Annual Review of Genetics, 32*, 255-278.

[216] Tait, K. & Sutherland, I. W. (2002). Antagonistic interactions amongst bacteriocin-producing enteric bacteria in dual species biofilms. *Journal of Applied Microbiology, 93*, 345-352.

[217] Kolenbrander, P. E. (2000). Oral microbial communities: biofilms, interactions, and genetic systems. *Annual Review of Microbiology, 54*, 413–437.

[218] Malik, A., Sakamoto, M., Hanazaki, S., Osawa, M., Suzuki, T., Tochigi, M. & Kakii, K. (2003). Coaggregation among nonflocculating bacteria isolated from activated sludge. *Applied and Environmental Microbiology, 69*, 6056–6063.

[219] Rickard, A. H., McBain, A. J., Ledder, R. G., Handley, P. S. & Gilbert, P. (2003a). Coaggregation between freshwater bacteria within biofilm and planktonic communities. *FEMS Microbiology Letters, 220*, 133–140.

[220] Gibbons, R J. & Nygaard, M. (1970). Interbacterial aggregation of plaque bacteria. *Archives of Oral Biology, 15*, 1397-1400.

[221] Kolenbrander, P. E., Andersen, R. N., Clemans, D. L., Whittaker, C. J. & Klier, C. M. (1999). Potential role of functionally similar coaggregation mediators in bacterial succession. In H. N. Newman, & M. Wilson (Eds.), Dental plaque revisited: oral biofilms in health and disease (pp. 171-186). Cardiff, UK: Bioline.

[222] Buswell, C. M., Herlihy, Y. M., Marsh, P. D., Keevil, C. W. & Leach, S. A. (1997). Coaggregation amongst aquatic biofilm bacteria. *Journal of Applied Microbiology, 83*, 477-484.

[223] Rickard, A. H., Stephen, A. L., Buswell, C. M., High, N. J. & Handley, P. S. (2000). Coaggregation between aquatic bacteria is mediated by specific-growth-dependent lectin-saccharide interations. *Applied and Environmental Microbiology, 66,* 431-434.

[224] Rickard, A. H., Leach, S. A., Hall, L. S., Buswell, C. M., High, N. J. & Handley, P. S. (2002). Phylogenetic relationships and coaggregation ability of freshwater biofilm bacteria. *Applied and Environmental Microbiology, 68*, 3644–3650.

[225] Rickard, A. H., Gilbert, P. & Handley, P. S. (2004). Influence of growth environment on coaggregation between freshwater biofilm bacteria. *Journal of Applied Microbiology, 96*, 1367-1373.

[226] Rickard, A. H., Thomas, J., Leach, S. A., Buswell, C. M., High, N. J. & Handley, P. S. (1999). Coaggregation amongst aquatic and oral bacteria is mediated by lectin-saccharide interactions. In J. Wimpenny, P. Gilbert, J. Walker, M. Brading, & R. Bayston (Eds.), Biofilms: the good, the bad and the ugly (pp. 343-354). Cardiff, UK: Bioline.

[227] Handley, P. S., Rickard, A. H., Leach, S. A., Buswell, C. M. & High, N. J. (2001). Coaggregation-is it a universal phenomenon? In P. Gilbert, D. Allison, M. Brading, J. Verran, & J. Walker (Eds.), Biofilm community interactions: chance or necessity? (pp. 1–10). Cardiff, UK: Bioline.

[228] Kolenbrander, P. E. & London, J. (1993). Adhere today, here tomorrow: oral bacterial adherence. *Journal of Bacteriology, 175*, 3247–3252.

[229] Rickard, A. H., Gilbert, P., High, N. J., Kolenbrander, P. E. & Handley, P. S. (2003b). Bacterial coaggregation: an integral process in the development of multispecies biofilms. *Trends in Microbiology, 11*, 94–100.

[230] Simões, L. C., Simões, M. & Vieira, M. J. (2008a). Intergeneric coaggregation between drinking water bacteria: evidence for the role *Acinetobacter calcoaceticus* as a bridging bacterium. *Applied and Environmental Microbiology, 74*, 1259-1263.

[231] Min, K. R. & Rickard, A. H. (2009). Coaggregation by the freshwater bacterium *Sphingomonas natatoria* Alters Dual-Species Biofilm Formation. *Applied and Environmental Microbiology, 75*, 3987-3997.

[232] Wimpenny, J. & Colasanti, R. (2004). A simple cellular automaton model for coaggregation. *Biofilms, 1*, 369-375.

[233] Bos, R., van der Mei, H. C., Meinders, J. M. & Busscher, H. J. (1994). Quantitative method to study co-adhesion of microorganisms in a parallel plate-flow chamber – basic principles of the analysis. *Journal of Microbiological Methods, 20*, 289–305.

[234] Min, K. R. Zimmer, M. N. & Rickard, A. H. (2010). Physicochemical parameters influencing coaggregation between the freshwater bacteria *Sphingomonas natatoria* 2.1 and *Micrococcus luteus* 2.13. *Biofouling, 26*, 931–940.

[235] Ashbolt, N. J. (2004). Microbial contamination of drinking water and disease outcomes in developing regions. *Toxicology, 198*, 229-238.

[236] Karanis, P., Kourenti, C. & Smith, H. (2007). Waterborne transmission of protozoan parasites: a worldwide review of outbreaks and lessons learnt. *Journal of Water and Health, 5*, 1-38.

[237] Burgess, J. G., Jordan, E. M., Bregu, M., Mearns-Spragg, A. & Boyd, K. G. (1999). Microbial antagonism: a neglected avenue of natural products research. *Journal of Biotechnology, 70*, 27-32.

[238] Bhattarai, H. D., Lee, Y. K., Cho, K. H., Lee, H. K. & Shin, H. W. (2006). The study of antagonistic interactions among pelagic bacteria: a promising way to coin environmental friendly antifouling compounds. *Hydrobiology, 568*, 417-423.

[239] Simões, M., Simões, L. C., Pereira, M. O. & Vieira, M. J. (2008c). Antagonism between *Bacillus cereus* and *Pseudomonas fluorescens* in planktonic systems and in biofilms. *Biofouling, 24*, 339-349.

[240] Cowan, S. E., Gilbert, E., Liepmann, D. & Keasling, J. D. (2000). Commensal interactions in a dual-species biofilm exposed to mixed organic compounds. *Applied and Environmental Microbiology, 66*, 4481-4485.

[241] Kives, J., Guadarrana, D., Orgaz, B., Rivera-Sen, A., Vazquez, J. & SanJose, C. (2005). Interactions in biofilms of *Lactococcus lactis* spp. *cremoris* and *Pseudomonas fluorescens* cultured in cold UHT milk. *Journal of Dairy Science, 88,* 4165-4171.

[242] Komlos, J., Cunningham, A. B., Camper, A. K. & Sharp, R. R. (2005). Interaction of *Klebsiella oxytoca* and *Burkholderia cepacia* in dual-species batch cultures and biofilms as a function of growth rate and substrate concentration. *Microbial Ecology, 49*, 114–125.

[243] Andersson, S., Dalhammar, G. & Rajarao, G. K. (2009). Persistence and competition of denitrifying bacteria in a dual-strain biofilm subjected to a natural wastewater flora. *Vatten, 65*, 87-92.

[244] Wolfaardt, G. M., Lawrence, J. R., Robarts, R. D., Caldwell, S. J. & Caldwell, D. E. (1994). Multicellular organization in a degradative biofilm community. *Applied and Environmental Microbiology, 60*, 434-446.

[245] Palmer, R. J., Karmerzak, K., Hansen, M. C. & Kolenbrander, P. E. (2001). Mutualism versus independence: strategies of mixed-species oral biofilms in vitro using saliva as the sole nutrient source. *Infection and Immunity, 69*, 5794-5804.

[246] Castonguay, M.-H., van der Schaaf, S., Koester, W., Krooneman, J., van der Meer, W., Harmsen, H. & Landini, P. (2006). Biofilm formation by *Escherichia coli* is stimulated by synergistic interactions and co-adhesion mechanisms with adherence-proficient bacteria. *Research in Microbiology, 157*, 471-478.

[247] Yu, A., Vinarov, Z., Robysheva, N., Smirnov, V. N. & Sokolov, D. P. (2002). Studies of the stability of microbial association use in industrial biofiltering of gaseous discharges. *Applied Biochemistry and Microbiology, 38,* 445-449.

In: Biofilms in Bioengineering
Editors: Manuel Simões and Filipe Mergulhão

ISBN: 978-1-62948-161-6

Chapter VIII

Biofilms and Biocorrosion

C. K. Gunsch[1,*] ***and A. S. Danko***[2,†]
[1]Duke University, Department of Civil and Environmental Engineering, Durham, NC, US
[2]University of Porto, Department of Mining Engineering, Geo-Environmental and Resources Research Center, Porto, Portugal

Abstract

Biocorrosion or microbiologically influenced corrosion (MIC) is an electrochemical process in which microorganisms initiate, facilitate or accelerate abiotic corrosion reactions. This process has been shown to be mediated by both prokaryotic and eukaryotic microorganisms. Generally, microorganisms form biofilms directly on a pipe or other structure and microenvironments develop which facilitate the growth of microorganisms which participate in the biocorrosive reactions. While many microorganisms have been shown to be involved in biocorrosion, it is widely accepted that sulfate reducing bacteria (SRB) play a key role in most cases. SRBs have been shown to participate in several reactions including cathodic depolarization and production of H_2S, both of which are known to contribute to MIC. Biocorrosion can occur in both aerobic and anaerobic environments. Generally, bacteria will grow at the surface of the biofilm and develop an oxygen gradient allowing for SRB growth inside the biofilm. The aerobic microorganisms consume oxygen and produce by-products, both of which are critical to MIC occurrence. Biocorrosion has been found to occur on a variety of materials including carbon steel, iron, manganese, aluminum and magnesium alloys. It has been well documented in both fresh and marine water environments. Many methods have been developed to control biocorrosion including mechanical cleaning devices, oxidizing and non-oxidizing biocides as well as a variety of coatings.

Keywords: Aerobic; Anaerobic; Biocorrosion; Biofilm; Sulfate reducing bacteria

* Tel: (919) 660-5208; E-mail: ckgunsch@duke.edu.
† Tel: +351-22-508-1426; E-mail: asdanko@fe.up.pt.

Introduction

Corrosion of metals is an abiotic electrochemical reaction occurring in aqueous environments when metal ions go into solution (anodic reaction), leaving electrons that combine with oxygen to produce hydroxyl ions (cathodic reaction). In general, metals which are difficult to extract (e.g., gold and silver) tend to be the least susceptible to corrosion. This factor is attributed to thermodynamic tendency of the metal to revert to the oxides from which it came [1]. Therefore, metals which have not been tremendously refined are likely to be less affected by corrosion. In addition, factors which facilitate metal reversion to oxides such as those which overcome atomic bonding of the metal and encourage metal ions to leave the surface will promote corrosion.

Microorganisms are known to provide environments that initiate, facilitate or accelerate an already occurring corrosion reaction [2-4]. The microbial component of these reactions is known as biocorrosion, microbial corrosion or microbiologically influenced corrosion (MIC). Microorganisms play a critical role because of the diverse interactions they have in the environment surrounding the metal surface. The nature of the interaction between the microorganisms, aqueous solution and metal will dictate the extent as well as the rate at which corrosion will occur.

Microorganisms

Prokaryotic and eukaryotic microorganisms have both been shown to be involved in biocorrosion [5-35]. Their small size enables them to grow in virtually any compartment including inaccessible crevices and pits which offer protection from the fluid shear stress of the aqueous medium flowing over the metal surface. Furthermore, several attributes including their high surface area to volume ratio, metabolic products and rapid growth rates all play important roles in accelerating metal corrosion processes by as much as 100,000- fold [36].

Microbes interact with the metal in biofilm structures which occur on the metal surface. These biofilms which are mainly composed of cells and extrapolymeric substances (EPS) are porous and in many cases non-uniform both in terms of time and space. Their physical attributes enable them to adsorb and entrap materials including metabolic products (many of which are corrosive) as well as debris [37-41]. In addition, the biofilm structure allows for the development of microenvironments which enable synergistic microbial relationships. The development of microenvironments and the microbial community fingerprint in the biofilm play key roles in the development and progression of biocorrosion [23, 32].

Metal and Aqueous Solution

The type of metal as well as its composition are two other important factors to consider for biocorrosion. In practice, most metallic surfaces have deposits such as corrosion products, scaling or biofilms, all of which can alter corrosion rates or, in some instances, lead to localized corrosion. Several factors have been shown to increase corrosion such as heat treatments [42], manufacturing defects, deformation of grain structure [43] while the use of alloying protects against biocorrosion [42]. Furthermore, the nature of the aqueous solution in terms of pH as well as ionic composition is critical in determining the electrochemical behavior of a metal in the field. Many lab studies rely on media that contain compounds which may affect biocorrosion differently than the situations encountered in natural and industrial environments. Therefore, caution should be exercised when utilizing lab data to explain field results.

Biocorrosion of Carbon and Stainless Steel in Anaerobic Environments

Corrosion in anaerobic environments is generally thought to be primarily caused by sulfate reducing bacteria (SRB) although some studies have shown that methanogens, nitrate reducing bacteria (NRB) and metal reducing bacteria (MRB) are also involved [25, 29, 44-46].

Common SRBs found to participate in corrosion include those from the genus *Desulfovibrio* with *Desulfovibrio vulgaris* and *Desulfovibrio desulfurocans* being the two most commonly studied. Both sessile and planktonic SRB have been shown to be involved in the biocorrosion process [47-49]. Synergistic activities of different types of bacteria have shown to enhance the microbial effects within biofilm structures. An example of which is anaerobic areas forming within aerobic biofilms allowing SRB bacteria to grow [50].

Interestingly, one study found that the presence of O_2 had a significant impact on corrosion rates of iron [51]. Indeed, during successive aeration and dearation cycles, biocorrosion rates increased significantly when O_2 was introduced into the system versus when oxygen was absent. Tubercles were suggested to play an important role (Figure 3).

There is an ongoing debate as to whether SRB can utilize oxygen as a terminal electron acceptor. Complete and partial genome sequences to *D. vulgaris* (catalase and superoxide dismutase) and *Desulfitobacterium hafiense* (catalase) showed enzymes capable of dealing with the toxic effects of oxygen [52, 53].

There are several proposed mechanisms for biocorrosion by SRBs and these involve both direct and indirect mechanisms. Such mechanisms include cathodic polarization [54], production of hydrogen sulfide (H_2S) and other metabolites such as acetate [55-58]. SRB affect corrosion directly through the production of these metabolic intermediates. In addition, they can also be affected by their surroundings, which could change the environment from corrosive to non-corrosive. These factors include pH, ionic composition, and oxygen levels [59].

Cathodic depolarization is a process by which the electrons, which are lost from the ionization of the metal (anodic reaction), reduce protons from water forming hydrogen. This hydrogen remains on the metal surface and in dynamic equilibrium. SRB then consume the hydrogen using it as an electron donor reducing sulfate and forming H_2S.

$4\ Fe \rightarrow 4\ Fe^{2+} + 8\ e^-$ (Anodic Reaction)
$8\ H_2O \rightarrow 8\ H^+ + 8\ OH^-$ (Electrolytic dissociation of water)
$8\ H^+ + 8\ e^- \rightarrow 8\ H$ (Cathodic Reaction)
$SO4^{2-} + 8\ H \rightarrow S^{2-} + 4\ H_2O$ (Cathodic depolarization by SRB)

$Fe^{2+} + S^{2-} \rightarrow FeS$ (Corrosion Product)
$3\ Fe^{2+} + 6\ OH^{-} \rightarrow 3\ Fe(OH)_2$ (Corrosion Product)

The overall reaction is as follows:

$$4\ Fe + SO4^{2-} + 4\ H_2O \rightarrow FeS + 3\ Fe^{2+} + 8\ OH^{-}$$

The cathodic depolarization process due to the activity of SRB is shown in Figures 1 and 2.

The H_2S, formed by the dissimilatory reduction of sulfate, is free to react with the iron forming iron sulfides (FeS) and hydrogen. The hydrogen that is produced can be used by SRB or other microorganisms that have hydrogenases. The level of hydrogenase activity is often associated with biocorrosion and is more important than the total number of cells [62]. In addition, hydrogenase activity is not necessarily dependent on viable cells within the biofilm matrix. Indeed, hydrogenase activity can retain its activity for months even without viable cell counts or production of H_2S [63].

The H_2S that is produced can also be transformed biologically to elemental sulfur or chemically oxidized when it reaches the aerobic headspace of sewers and other microaerophilic niches. This can be problematic because elemental sulfur is known to be severely corrosive to carbon steel and has been found near areas where pitting has been observed in carbon steel biocorrosion [1]. The rate of chemical oxidation was found to be pH and temperature dependent and also dependent on the form of hydrogen sulfide with HS^{-} since it is more readily oxidized when compared to H_2S [64]. It should be noted there are many different forms of sulfide and that this is pH dependent. The two principal forms are H_2S and HS^{-} with H_2S as the predominant compound when the pH is lower than 7 and the hydrogen sufide ion (HS^{-}) the predominant compound above that pH value. At extremely alkaline conditions (pH > 14), the sulfide (S^{-2}) ion predominates.

Another important practical consideration is that in nature, the surfaces of carbon steels contain chemical and biological deposits (such as oxides, sulfides, hydroxyides, and biofilms). Therefore, biocorrosion is based on the breaking down of the passive protective layers from the aggressive characteristics of SRB metabolic products [1].

In addition, chloride and other cations and anions in the medium have also been shown to enhance or inhibit biocorrosion. For example, different iron concentrations can lead to the formation of different physical forms iron sulfides that coat the surface of the alloys [65, 66]. When these protective layers are broken, corrosion rates are significantly increased [67]. In addition, sulfate has also been shown to affect biocorrosion rates. Biocorrosion was not observed in the absence of sulfate and was only initiated when sulfate was added to the medium [68]. Alternatively, high concentrations of sulfate were found to be inhibitory to sulfate reduction [69].

Recent evidence has suggested that direct electron transfer from the metallic ions can also be used by specific SRBs rather than utilizing the hydrogen that is formed from cathodic depolarization [70]. The rate of sulfate reduction from these newly isolated microorgansims is much faster than the rate of hydrogen produced from cathodic depolarization. These new isolates are related to *Desulfobacterium* and *Desulfovibrio* species based on phylogenetic and physiological analyses. The authors hypothesized that these SRB utilize a cell-surface-associated redox active component, a process that is known from iron (+2) oxidizing bacteria [70, 71].

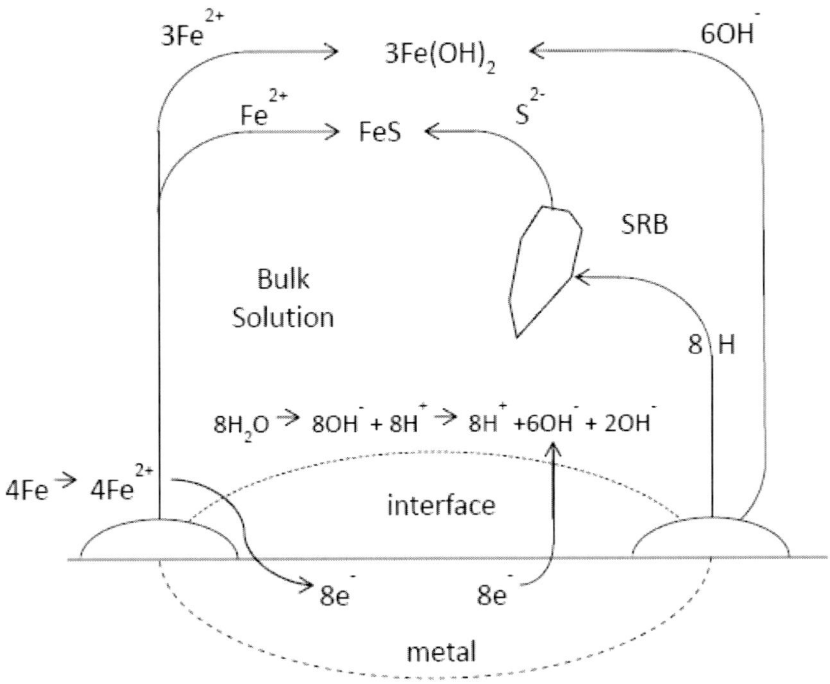

Figure 1. Influence of SRB on corrosion of ferrous metals (adapted from [60]).

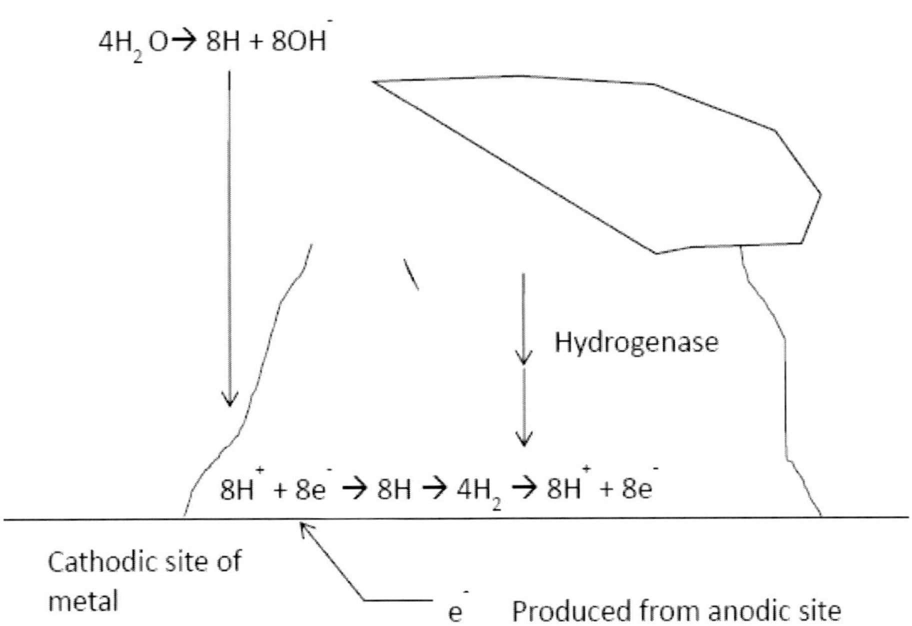

Figure 2. Cathodic depolarization of the surface due to utilization of hydrogen by the hydrogenase of SRB (adapted from [61]).

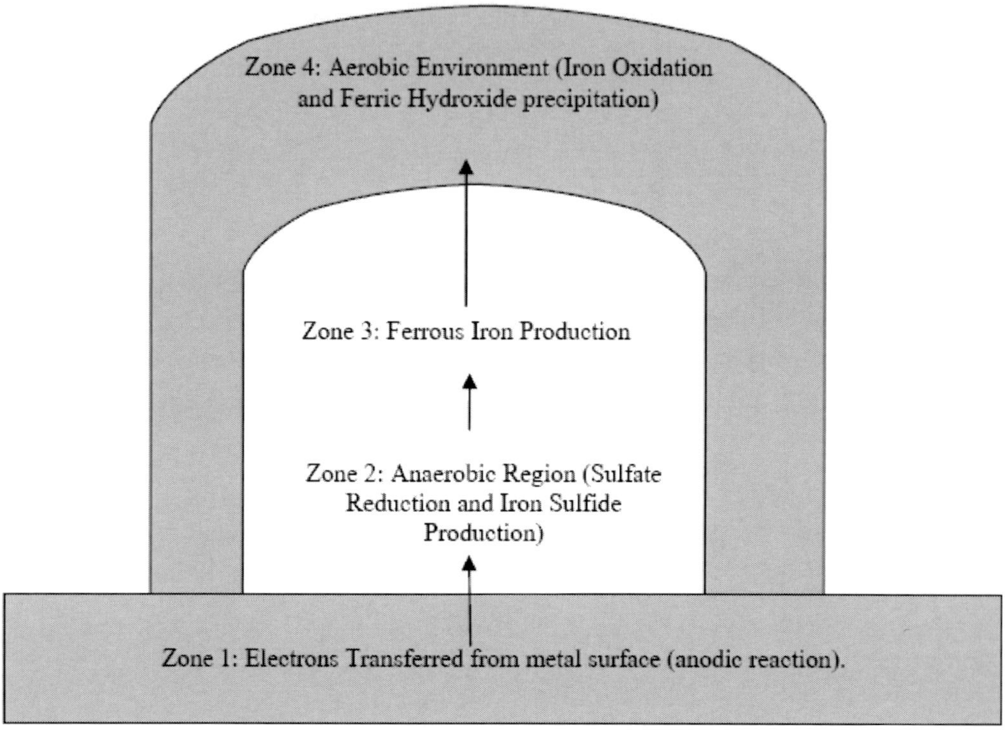

Figure 3. General redox zones and electron transfer reactions in a tubercle (Adapted from [113]).

The presence of nitrate is also a concern in corrosion. This is because nitrate can serve as an electron acceptor for zero valent metals such as iron in an abiotic reaction in conditions with limited oxygen [72]. Microorganisms that can utilize nitrate can also be involved in biocorrosion. *Paracoccus denitrificans* was able to utilize cathodically produced H_2 [73]. In addition, anaerobically grown, hydrogenase positive *Escherichia coli* hastened corrosion in the presence of H_2 and nitrate [74]. Even though, NRBs can facilitate biocorrosion, their contribution is generally considered to be relatively minor compared to SRB [72].

Many different types of MRB that utilize dissimilatory metal reduction have also been shown to be involved in biocorrosion. These include microorganisms such as *Shewanella* sp. and *Pseudomonas* sp. that are capable of iron and manganese reduction [75-78]. These microbes were found to hasten the corrosion of iron and its alloys by the dissolution of corrosion resistant oxide films on the metal surface. This results in further corrosion by the loss of the passive protective layers or replacement by less stable reduced metal films [79]. The type of oxide film has also been found to affect the rate of reaction [80].

Methanogens, just like SRB, NRB, and MRB, are also able to utilize hydrogen. However, it has been shown that methanogens are outcompeted in environments where SRB are present since SRB have higher oxidation rates and affinity for H_2 and acetate [81-83]. However, methanogenic bacteria are able to coexist with SRB and their presence is dependent on H_2 production rates, sulfate availability, and relative population sizes[84, 85]. Previous work with pure cultures of methanogens has been contradictory as some studies suggested their involvement while others have not [86-90]. However, a methanogenic isolate has also been found to utilize the electrons directly from iron as previously described [70]. In this way, these methanogens are also able participate in the corrosion process.

Corrosion also occurs in marine environments. This is because the chlorides present in sea water are efficient electrolytes. Many of the same mechanisms that have been discussed previously in this section are also applicable to marine environments. This includes the influence of environmental parameters such as pH, temperature, ionic composition, and oxygen levels and the activities of SRB, such as production of metabolites and EPS [91, 92, 93]. Another aspect concerns the synergistic effects of sulfides in the presence of chloride or at the biofilm/metal interface [59, 94]. Additionally, the concentration of iron was shown to influence the thickness and characteristics of the ferrous hydroxide layer [95].

Although most research has focused on iron and steel, corrosion also occurs on copper and copper-nickel alloys found in heat exchangers. Even though copper is well known to have anti-microbial properties, biofilms have been found colonizing these materials in marine environments. Two mechanisms occur that can influence the rate of corrosion on copper alloys: the replacement of existing alloys with copper hydroxychlorides [96], and the formation of imperfect oxide layers in the presence of sulfides [97]. In either case, the protection of the underlying alloys is severely compromised. Corrosion of copper in marine environments has also been shown to be influenced by the production of EPS [98].

Iron and Manganese Oxidation

Corrosion caused by iron oxidation is the result of three distinct mechanisms which are usually interconnected [99]. First, soluble iron species as well as iron bearing scale can oxidize leading to pipe mass loss. Second, scale can accumulate and form tubercules which provide microenvironments ideal for bacterial growth. Lastly, soluble or particulate iron corrosion by-products can also be released. Microorganisms have been reported in iron tubercules and, it is generally accepted that their presence is the major contributor to biocorrosion in cast iron pipes [100-103]. The major iron-oxidizing genera responsible for the oxidation of ferrous ions to ferric ions include *Sphaerotilus*, *Crenothrix*, and *Leptothrix* as well as the stalked microorganism *Gallionella* [104-111].

Differential aeration zones are created within the tubercular structure which leads to the growth of microorganisms capable of working in a synergistic fashion. Metal-oxidizing bacteria provide the perfect niches for the accumulation of chloride ions which in turn leads to the formation of acidic ferric chloride [112]. The general electron transport in a tubercule is shown in Figure 1. Generally, the outer area of the tubercule acts as the cathode due to the continuous supply of oxygen from the fluid. The anode is found at the bottom of the tubercule where oxygen is limited and conditions are generally anaerobic. While the identity of the aerobic microorganism varies on a case by case basis, SRB are the main microorganism which grows in the anaerobic region of the tubercule and play a critical role in corrosion acceleration of cast iron steels.

While there are reports that show biofilms can actually protect from biocorrosion [114], in most cases, biofilms are associated with detrimental effects on cast iron steel pipe structure. In addition to the promotion of biocorrosion via tubercules discussed earlier, there are many other mechanisms by which microorganisms can contribute to corrosion. Bacteria are known to affect iron speciation by Fe(III) reduction or oxidation of Fe(II) which in turn affects material integrity [115-119]. Some microorganisms create localized pH gradients [102] while others can produce corrosive by-products such as hydrogen sulfide [102, 120] or iron phosphate [121, 122]. All of these factors contribute to biocorrosion.

Aluminum and Aluminum Alloy Oxidation in Fuel and Water Systems

In addition to water distribution systems, there are many other environments where biofilms can lead to biocorrosion. Fuel storage is one common application often plagued with issues related to MIC. Fungi and bacteria are known to be capable of degrading fuel [123-129] and are commonly found in fuel storage tanks. The occurrence of biocorrosion in fuel storage tanks has important environmental implications as a reduction in a tanks integrity may lead to fuel releases. In addition, the presence of microorganisms in the fuel can lead to product quality issues [130].

One area which has received much attention is the deterioration of aluminum alloys used in gas tanks in the aviation industry. The promotion of microbial growth in fuel requires the presence of water. Microenvironments have been shown to develop which retain water to promote growth utilizing the parafinnic hydrocarbons as their primary carbon source. Slime growth as well as tubercule and biofilm formations have all been reported on aluminum metal surfaces. In the case of tubercule formation, biocorrosion mechanisms are similar to that described for iron oxidation. Symbiotic relationships develop between aerobic and anaerobic bacteria. Aerobic bacteria occur on the outside of the tubercule and consume oxygen while SRB grow inside and consume the nutrients produced such as acetic and butyric acid [2, 131]. In general, tubercules form over a deep pit. A range of microorganisms have been isolated from jet fuels and fuel storage systems including *Hormoconis resinae* [132, 133] and some species of the genera *Aspergillus*, *Penicillium* and *Fusarium* [127, 130, 134-137].

Corrosion attack generally occurs at the bottom of the fuel tank where water contamination is most likely to be present. Microbial growth will only take place if macronutrients (carbon, nitrogen, oxygen, hydrogen and phosphorous) as well as micronutrients (metals and cofactors) are present. Generally the presence of nitrogen and phosphorous are the limiting factors on biofilm development although they both tend to be present as fuel impurities as nitrates and phosphates, respectively. However, following the establishment of microbial growth, the absence of nitrogen can actually lead to the increased production of extracellular organic acids [138]. These substances are corrosive in their own right and contribute to the challenge in controlling corrosion in aluminum based fuel storage environments.

There are five main factors which affect corrosion. First, fungal by-products include organic acids which create a localized increase in protons. Lower pH values have been shown to be achieved underneath microbial attachment areas suggesting that biofilm formation plays a critical role in the promotion of biocorrosion [139]. Second, microorganisms change the chemical characteristics of the fluid by consuming oxygen, producing metabolites and cellular lysis products. The decrease in oxygen in the fluid leads to a protection of the metal (cathode), however the production of acidic by-products counteracts the oxygen depletion and leads to pitting corrosion. In this case, information on redox potential can provide some insights as to determine if localized corrosion can be expected. Third, surface energy is inversely proportional to incubation time. The implications are that anions present in the fluid can breakdown the metal film. This phenomenon has been reported in environments with high chloride concentrations [140]. Fourth, the process of microbial adhesion has been shown to promote localized pit corrosion. There are numerous reports showing that the presence of biofilms is linked to localized production of organic acids which leads to localized corrosion [133, 139, 141-146]. Finally, fungi have been shown to consume corrosion inhibitors such as nitrogen and phosphorous which can lead to an increase in chloride/inhibitor ratio and ultimately lead to the catastrophic failure of several aluminum alloys [147]. Others have reported on the consumption of oxygen, lubricating oils and protective coatings which all contribute to corrosion [148, 149].

Much less information is available on bacterial biofilms and the role they play in the corrosion of aluminum and aluminum alloys. Most published work utilizes *Pseudomonas* spp. as well as *Serratia marcescens* [150-153]. The reported mechanisms are similar to those discussed previously with fungi in terms of their effect on redox potential, creation of localized pH decrease where biofilms develop as well as media modification related to bacterial metabolism.

Magnesium and Magnesium Alloy Oxidation

Magnesium alloys are believed to be safe for medical applications and are being investigated for used in implants [154]. Furthermore, magnesium alloys which contain neither aluminum nor zinc are the most resistant to stress corrosion cracking making them ideal for numerous applications. However, their susceptibility to corrosion has to date restricted the use of magnesium alloys [155]. Magnesium is an essential element for many biochemical functions [156]. Intracellular requirements are estimated to be 10 to 20 μM and it is has been found bound to ATP, nucleic acids and ribosomes [157]. Because of all of these reasons, its use as an alloy commonly promotes the formation of biofilms leading to MIC [158]. Much research has gone towards utilizing a sacrificial magnesium anode to protect the underlying structural material [159-161]. Song et al., [162] described the magnesium corrosion reactions as shown below. As the $Mg(OH)_2$ deposits onto the surface, it forms a passive film. However, some of the magnesium ions goes back into the liquid thereby increasing the bulk liquid ionic concentration as well as the pH.

$Mg \rightarrow Mg^{2+} + 2e^-$ (Anode)

$2H_2O + 2\,e^- \rightarrow H_2 + 2OH^-$ (Cathode)

$Mg^{2+} + 2\,OH^- \rightarrow Mg(OH)_2$ (Protective Film Formation)

Although the mechanisms of magnesium corrosion are fairly well established, very little information has been published discussing the role biocorrosion plays in the overall failure of magnesium and magnesium alloys. Specifically, the mechanisms of bacterial colonization have been overlooked for the most part. Nandakumar et al., [157] studied the development of *Pseudomonas* sp. biofilms on the magnesium alloy AZ31B and evaluated the effect of pH and total magnesium concentration on total viable count and biofilm surface area. The bulk medium magnesium concentration was found to increase from 0 to 12-13 ppm after 48 hours and up to 25 ppm after 6 days exposure. The pH in the medium increased to >9 after 48 hours. The initial colonization was found to be higher on magnesium alloys as compared to stainless steel after a 48 hour exposure, however it became gradually smaller as the experiment proceeded while it kept increasing in the case of stainless steel [163]. Additional research is clearly needed to identify the specific mechanisms involved in magnesium related biocorrosion.

Prevention of Biocorrosion

The methods used to inhibit biofilm growth are largely application specific and require an understanding of how each biofilm contributes to biocorrosion locally [164]. There are several types of deposits which can contribute to biocorrosion including scaling (e.g., calcium carbonates, sulfates or silicates) and slime. As discussed earlier, the presence of these compounds can facilitate microbial growth and accelerate MIC. The build-up of scale is directly correlated to temperature, pH, water chemistry as well as hydrodynamic conditions found in the system. Therefore, these are optimal parameters to target to minimize scale formation. Sulfuric acid, inorganic phosphate salts as well as phosphonates have all been used for this purpose [164]. Slime residues are the second type of deposits which can contribute to biocorrosion and these can be either chemical or biological in nature. In general, high flow rates will either prevent or help remove slime while low flow rates will promote their formation [164]. In general, biocorrosion can be prevented by cleaning processes, the use of biocides and altering either coating characteristics or cathodic protection [164].

Cleaning Processes

Deposits can be removed either mechanically, chemically or by a combination of both methods. Mechanical cleaning has been shown to be effective at removing both biofilms as well as metal oxides[164]. In general, rubber spheres are used for heat-exchangers while mechanical pigs are using in production pipes. Injection lines tend to be blasted with sand, grit or water. These methods are used in conjunction with biocides to minimize the amount of localized pitting as well as the formation of differential aeration cells[164]. Chemical cleaning generally involves either chelating agents or mineral or organic acids. Acids in general are not as effective for eliminating biological growth but are useful for scale removal.

Biocides

Biocides are used specifically to inhibit biofilm formation. Both organic and inorganic biocides exist which can target either prokaryotic (i.e., bactericidal) or eukaryotic (i.e., fungicidal or algicidal) microorganisms. Chlorine, bromine, ozone and hydrogen peroxide are the most common oxidizing biocides and are very effective at inhibiting microbial growth [165-170]. However, caution must be exercised because they can have deleterious effects on structural metals. They have been shown to interact with other inhibitors used in water treatment, contribute to metal corrosion as well as affect the integrity of other components such as rubber or plastic coatings [164]. Because of this, the application of oxidizing biocides tends to be intermittent rather than continuous [168, 170, 171]. Non-oxidizing biocides including glutaraldehyde, quartenary ammonium compounds, methylene-bisthiocyanate and tetrakis(hydroxymethyl) phosphonium sulfate have also been shown to be quite effective to combat biofilm formation [171, 172]. This type of biocide presents a number of advantages over their oxidizing counterparts due to their persistence. However, in general, a combination of oxidizing and non-oxidizing biocide is used.

Coating Characteristics

Many different types of coatings have been developed which are composed of non-toxic compounds such as silicone, epoxy resins as well as fluorinated compounds [173-178]. These coatings decrease the surface tension making biofilm attachment more difficult. Coatings are very useful in preventing MIC as they provide a continuous protection against biofilm formation. However, if there is any portion of the coating which is removed overtime, this creates a region of preferential attack for microbial growth.

Cathodic Protection

Cathodic protection is an electrochemical way of corrosion control where oxidation primarily occurs at the anode thereby protecting the metallic cathode. This can be achieved in two ways. First, this can be accomplished by using a more active metal (Zn, Mg) than the metal to be protected (Fe). This can also be achieved by applying an external current which directs the corrosion current away from the metal that needs protection (at the cathode) and towards an inert anode [164, 179, 180]. The use of this technique becomes prohibitively expensive as the amount of current applied increases. A side effect of cathodic protection is an increase in calcareous deposits. This phenomenon occurs because the additional current contributes to a release in hydroxyl ions which decreases the solubility of both calcium and magnesium ions [164]. This development is beneficial from an economical perspective because the presence of scale reduces the current requirement. The scale also inhibits the growth of aerobic organisms. However, the growth of anaerobic organisms such as SRB is increased under these conditions.

Conclusion

Biocorrosion is a process in which prokaryotic and eukaryotic microorganisms initiate, facilitate or accelerate abiotic corrosion reactions. This process is electrochemical by nature and is mediated through a variety of electron transfer reactions. Biofilms play an important role in the proliferation of biocorrosion on metal surfaces as their structure enables the formation of microenvironments which in turn promotes the growth of microorganisms which fill different oxidation-reduction niches. SRBs in particular play an important role and are critical to the development of cathodic depolarization and production of H_2S. A variety of methods have been devised to control the proliferation of microorganisms on metal surfaces to hinder biocorrosion including mechanical cleaning, biocides, various coatings and cathodic protection.

Acknowledgments

The assistance of Liliana Pereira is appreciated. ASD wishes to thank the Portuguese Science and Technology Foundation (FCT) under the Ciência 2008 Program.

References

[1] Videla HA. Manual of Biocorrosion. Boca Raton, Florida: CRC Press; 1996.

[2] Rao TS. Microbiologically influenced corrosion: Basics and case studies. *Corrosion Reviews* 2009:333-65.

[3] Videla HA, Herrera LK. Understanding microbial inhibition of corrosion. A comprehensive overview. *International Biodeterioration & Biodegradation* 2009;63: 896-900.

[4] Vukovic M, Cvetkovski V, Conic V. Mechanisms of microbiologically induced corrosion of metals in the environments containing sulphate-reducing bacteria. *Corrosion Reviews* 2009;27:1-22.

[5] Belov D, Sokolova T, Smirnov V, Kuzina O, Kostyukova L, Kartashov V. Corrosion of aluminum and its alloys under the effect of microscopic fungi. *Protection of Metals* 2008;44:737-42.

[6] Bento FM, Beech IB, Gaylarde CC, Englert GE, Muller IL. Degradation and corrosive activities of fungi in a diesel-mild steel-aqueous system. *World Journal of Microbiology & Biotechnology* 2005;21:135-42.

[7] Bento FM, Englert GE, Gaylarde CC, Muller IL. Influence of aqueous phase on electrochemical biocorrosion tests in diesel/water systems. *Materials and Corrosion-Werkstoffe Und Korrosion* 2004;55:577-85.

[8] Brehm U, Gorbushina A, Mottershead D. The role of microorganisms and biofilms in the breakdown and dissolution of quartz and glass. *Palaeogeography Palaeoclimatology Palaeoecology* 2005;219:117-29.

[9] Gadd GM. Metals, minerals and microbes: geomicrobiology and bioremediation. *Microbiology-Sgm*; 156:609-43.

[10] Gadd GM. Heterotrophic solubilization of metal-bearing minerals by fungi. In: CotterHowells JD, Campbell LS, ValsamiJones E, Batchelder M, editors. Environmental Mineralogy: Microbial Interactions Anthropogenic Influences, Contaminated Land and Waste Management 2000. p. 57-75.

[11] Gall JC, Duringer P, Krumbein W, Paicheler JC. Impact of microbial ecosystems on sedimentation. *Palaeogeography Palaeoclimatology Palaeoecology* 1994;111:17-28.

[12] Klintworth R, Reher HJ, Viktorov AN, Bohle D. Biological induced corrosion of materials II: New test methods and experiences from Mir station. *Acta Astronautica* 1999;44:569-78.

[13] Klintworth R, Reher HJ, Viktorov AN, Bohle D, European Space A. Biological induced corrosion of materials: New test methods and experiences from MIR station. 7th International Symposium on Materials in Space Environment 1997. p. 513-22.

[14] Krumbein WE, Urzi CE, Gehrmann C. Biocorrosion and biodeterioration of antique and medieval glass. *Geomicrobiology Journal* 1991;9:139-60.

[15] Lugauskas A, Prosycevas I, Ramanauskas R, Griguceviciene A, Selskiene A, Pakstas V. The Influence of Micromycetes on the Corrosion Behaviour of Metals (Steel, Al) under Conditions of the Environment Polluted with Organic Substances. *Materials Science-Medziagotyra* 2009;15:224-35.

[16] Morton LHG, Surman SB. Biofilms in biodeterioration - A review. *International Biodeterioration & Biodegradation* 1994;34:203-21.

[17] Savelyev Y, Rudenko A, Robota L, Koval E, Savelyeva O, Markovskaya L, et al., Novel polymer materials for protecting crew and structural elements of orbital station against microorganisms attack throughout long-term operation. *Acta Astronautica* 2009;64:36-40.

[18] Smirnov VF, Belov DV, Sokolova TN, Kuzina OV, Kartashov VR. Microbiological corrosion of aluminum alloys. *Applied Biochemistry and Microbiology* 2008;44:192-6.

[19] Torres ES, Cravo WB, de Franca FP. Performance of foamy pigs on the removal of biofilms from AISI-1020 carbon steel pipelines. *Corrosion Reviews* 2004;22:193-208.

[20] Torres ES, de Franca FP. Kinetics of biofilm formation as a function of dissolved oxygen concentration on AISI-1020 carbon steel coupons. *Corrosion Reviews* 2002;20:115-28.

[21] Warscheid T, Braams J. Biodeterioration of stone: a review. *International Biodeterioration & Biodegradation* 2000;46:343-68.

[22] Bethencourt M, de Lomas JG, Corzo A, Villahermosa D, Matres V. Effect of biofilm in the corrosion of austenitic stainless steels in wastewater treatment plants. *Revista De Metalurgia*;46:37-51.

[23] Bohus V, Toth EM, Szekely AJ, Makk J, Baranyi K, Patek G, et al., Microbiological investigation of an industrial ultra pure supply water plant using cultivation-based and cultivation-independent methods. *Water Research*; 44:6124-32.

[24] Cetin D, Aksu ML. Corrosion behavior of low-alloy steel in the presence of *Desulfotomaculum* sp. *Corrosion Science* 2009;51:1584-8.

[25] Herrera LK, Videla HA. Role of iron-reducing bacteria in corrosion and protection of carbon steel. *International Biodeterioration & Biodegradation* 2009;63:891-5.

[26] Korenblum E, Valoni E, Penna M, Seldin L. Bacterial diversity in water injection systems of Brazilian offshore oil platforms. *Applied Microbiology and Biotechnology*; 85:791-800.

[27] Li J, Li JY, Yuan WS, Du YL. Biocorrosion characteristics of the copper alloys BFe30-1-1 and HSn70-1AB by SRB using Atomic Force Microscopy and Scanning Electron Microscopy. *International Biodeterioration & Biodegradation*; 64:363-70.

[28] Li J, Xu ZY, Li JY, Jiao D. Characteristics of the Microbiologically Influenced Corrosion of 304 Stainless Steel in Reclaimed Water Enviroment. *Acta Physico-Chimica Sinica;* 26:2638-46.

[29] Mori K, Tsurumaru H, Harayama S. Iron corrosion activity of anaerobic hydrogen-consuming microorganisms isolated from oil facilities. *Journal of Bioscience and Bioengineering*; 110:426-30.

[30] Motamedi S, Marandi R, Mazhar SF. Studying and identifying the biocorrosion of water supply pipes in copper mine. In: Donati ER, Viera MR, Tavani EL, Giaveno MA, Lavalle TL, Chiacchiarini PA, editors. *Biohydrometallurgy*: A Meeting Point between Microbial Ecology, Metal Recovery Processes and Environmental Remediation 2009. p. 133-6.

[31] Nercessian D, Duville FB, Desimone M, Simison S, Busalmen JP. Metabolic turnover and catalase activity of biofilms of *Pseudomonas fluorescens* (ATCC 17552) as related to copper corrosion. *Water Research*; 44:2592-600.

[32] Pavissich JP, Vargas IT, Gonzalez B, Pasten PA, Pizarro GE. Culture dependent and independent analyses of bacterial communities involved in copper plumbing corrosion. *Journal of Applied Microbiology*; 109:771-82.

[33] Pedersen K. Analysis of copper corrosion in compacted bentonite clay as a function of clay density and growth conditions for sulfate-reducing bacteria. *Journal of Applied Microbiology*; 108:1094-104.

[34] Rajasekar A, Anandkumar B, Maruthamuthu S, Ting YP, Rahman P. Characterization of corrosive bacterial consortia isolated from petroleum-product-transporting pipelines. *Applied Microbiology and Biotechnology*; 85:1175-88.

[35] Stadler R, Wei L, Furbeth W, Grooters M, Kuklinski A. Influence of bacterial exopolymers on cell adhesion of *Desulfovibrio vulgaris* on high alloyed steel: Corrosion inhibition by extracellular polymeric substances (EPS). *Materials and Corrosion-Werkstoffe Und Korrosion*; 61:1008-16.

[36] Costello JA. Corrosion of metals by microorganisms - A literature survey. *International Biodeterioration Bulletin* 1969;5:101-&.

[37] Beech IB, Sunner JA, Hiraoka K. Microbe-surface interactions in biofouling and biocorrosion processes. *International Microbiology* 2005;8:157-68.

[38] Bergel A. Recent advances in electron transfer between biofilms and metals. In: Schippers A, Sand W, Glombitza F, Willscher S, editors. *Biohydrometallury*: From the Single Cell to the Environment 2007. p. 329-34.

[39] Jan-Roblero J, Posadas A, de la Serna JZD, Garcia R, Hernandez-Rodriguez C. Phylogenetic characterization of bacterial consortia obtained of corroding gas pipelines in Mexico. *World Journal of Microbiology & Biotechnology* 2008;24:1775-84.

[40] Landoulsi J, El Kirat K, Richard C, Feron D, Pulvin S. Enzymatic approach in microbial-influenced corrosion: A review based on stainless steels in natural waters. *Environmental Science & Technology* 2008;42:2233-42.

[41] Sheng XX, Ting YP, Pehkonen SO. The influence of ionic strength, nutrients and pH on bacterial adhesion to metals. *Journal of Colloid and Interface Science* 2008;321:256-64.

[42] Borenstein SW. Why does microbiologicall influenced corrosion occur at or adjacent to austenitic stainless steel weldments? *Corrosion* 1991;91.

[43] Wagner PA, Ray RI. Surface Analytical Techniques for Microbiologically Influenced Corrosion - A review. In: Kearns JR, Little BJ, editors. *Microbiologically Influenced Corrosion Testing* 1994. p. 153-69.

[44] Torrado JR, Calixto DM, Sarmiento AE, Panqueva JH. Evaluation of molybdate and nitrite on SRB related with corrosion in industrial systems. *International Biodeterioration & Biodegradation* 2008;62:5-.

[45] Boopathy R, Daniels L. Pattern of organotin inhibition of methanogenic bacteria. *Applied and Environmental Microbiology* 1991;57:1189-93.

[46] Duncan KE, Gieg LM, Parisi VA, Tanner RS, Tringe SG, Bristow J, et al., Biocorrosive Thermophilic Microbial Communities in Alaskan Worth Slope Oil Facilities. *Environmental Science & Technology* 2009;43:7977-84.

[47] Hamilton WA. Sulphate-reducing bacteria and the role in anaerobic corrosion. *Annual Rev. Microbiol.* 1985;39:195-217.

[48] Pankhania IP. Hydrogen metabolism in sulphate-reducing bacteria and their role in anaerobic corrosion. *Biofouling* 1988;1:27-47.

[49] Lee W, Lewandowski Z, Nielsen PH, Hamilton WA. Role of sulphate-reducing bacteria in corrosion of mild steel: a review. *Biofouling* 1995;8:165-94.

[50] Videla HA, Characklis WG. Biofouling and microbiologically induced corrosion. *International Biodeterioration & Biodegradation* 1992;29:195-212.

[51] Hardy JA, Brown JL. The corrosion of mild steel under anaerobic biofilms. *Corrosion* 1984;40:650-4.

[52] Beech IB, Sunner J. Biocorrosion: towards understanding interactions between biofilms and metals. *Current Opinion in Technology* 2004;15:181-6.

[53] Heidelberg JF, Seshadri R, Haveman SA, Hemme CL, Paulsen IT, Kolonay JF, et al., The genome sequence of the anaerobic, sulfate-reducing bacterium *Desulfovibrio vulgaris* Hildenborough. *Nature Biotechnology* 2004;22:554-9.

[54] Von Wolzogen Kuhr CAH, Van Der Vlugt LS. De grafiteering van Gietijzer als electrobiochemich Process in anaerobe Gronden. *Water* 1934;18:147-65.

[55] Pak K-R, Lee H-J, Lee H-K, Kim Y-K, Oh Y-S, Choi S-C. Involvement of organic acids during corrosion of iron coupon by *Desulfovibrio desulfuricans. J. Microbiol. Biotechnol.* 2003;13:937-41.

[56] Costello JA. Cathodic depolarization by sulfate-reducing bacteria. *South Af. J. Sci.* 1974;70:202-4.

[57] Lee W, Lewandowski Z, Nielsen PH, Hamilton WA. Role of sulfate-reducing bacteria in corrosion of mild steel: a review. *Biofouling* 1995;8:165-94.

[58] Widdel F. Microbial corrosion. In: Finn RK, Praeve P, Schlingmann M, Crueger W, Esser K, Thauer R, et al., editors. *Biotechnology Focus*. Munich: Hanser 1992. p. 277-318.

[59] Videla HA. Electrochemical interpretation of the role of microorganisms in corrosion. In: Houghton RN, Smith RN, Rossmore HOW, editors. Biodeterioration 7. London: Elsevier Applied Science; 1988. p. 359-71.

[60] Nelson EE. Microbiological corrosion in waterfloods. *Corrosion* 1961;18:247.

[61] Flemming HC, Geesey GG. Biofouling and Biocorrosion in Industrial Water Systems. Berlin: Springer-Verlag; 1991.

[62] Bryant RD, Jansen W, Boivin J, Laishley EJ, Costerton JW. Effect of hydrogenase and mixed sulfate-reducing baterial populations on the corrosion of steel. *Applied and Environmental Microbiology* 1991;57:2804-9.

[63] Beech IB. Sulfate-reducing bactera in biofilms on metallic materials and corrosion. *Microbiology Today* 2003;30:115-7.

[64] Nielsen AH, Vollertsen J, Hvitved-Jacobsen T. Chemical sulfide oxidation of wastewater - effects of pH and temperature. *Water Science and Technology* 2004;50:185-92.

[65] Rickard DT. The chemistry of iron sulfide formation at low temperatures. *Acta Universitatis Stockholmiensis* 1969;20:67-95.

[66] Morse JW, Millero FJ, Cornwell JC, Rickard D. The chemistry of the hydrogen sulfide and iron sulfide systems in nature waters. *Earth-Science Reviews* 1987;24:1-42.

[67] Mara DD, Williams DGA. The mechanisms of sulfide corrosion by sulfate-reducing bacteria. In: Walter AH, Van Der Plas EHH, editors. *Biodeterioration of Materials.* London: Applied Science Publishers; 1972. p. 103-13.

[68] Fonseca I, Feio MJ, Lino AR, Reis MA, Rainha VL. The influence of the mectia on the corrosion of mild steel by *Desulfovibrio desulfuricans*: an electrochemical study. *Electrochimica Acta* 1998;43:213-22.

[69] Mohanty SS, Das T, Mishra SP, Chaudhury R. Kinetics of SO42- reduction under different growth media by sulfate reducing bacteria. *BioMetals* 2000;13:73-6.

[70] Dinh HT, Kuever J, Mussmann M, Hassel AW, Stratmann M, Widdel F. Iron corrosion by novel anaerobic microorganisms. *Nature* 2004;427:829-32.

[71] Appia-Ayme C, Guiliani N, Ratouchniak J, Bonnefoy V. Characterization of an operon encoding two c-type cytochromes, an aa3-type cytochrome oxidase, and rusticyanin in *Thiobacillus ferrooxidans* ATCC 33020. *Applied and Environmental Microbiology* 1999;65:4781-7.

[72] Kielemoes J, De Boever P, Verstraete W. Influence of denitrification on the corrosion of iron and stainless steel powder. *Environmental Science and Technology* 2000; 34:663-71.

[73] Till BA, Weathers LJ, Alvarez PJJ. Fe^0-supported autotrophic denitrification. *Environmental Science and Technology* 1999;32:634-9.

[74] Mara DD, Williams DJ. Corrosion of mild steel by nitrate reducing bacteria. *Chemical & Industry* 1971;22:566-7.

[75] Obuekwe CO, Westlake DWS, Cook FD, Costerton W. Surface change in mild steel coupons from the action of corrosion-causing bacteria. *Applied and Environmental Microbiology* 1981;41:766-74.

[76] Obuekwe CO, Westlake DWS, Plambeck JA, Cook FD. Corrosion of mild steel in cultures of ferric iron reducing bacterium isolated from crude oil. I. Polarization characteristics. *Corrosion* 1981;37:461-7.

[77] Obuekwe CO, Westlake DWS, Plambeck JA, Cook FD. Corrosion of mild steel in cultures of ferric iron reducing bacterium isolated from crude oil. II. Mechanism of anodic depolarization. *Corrosion* 1981;37.

[78] Dawood Z, Brözel VS. Corrosion enhancing potential of *Shewanella putrefaciens* isolated from industrial cooling waters. *Journal of Applied Microbiology* 1998;84:929-36.

[79] Beech IB, Gaylarde CC. Recent advances in the study of biocorrosion- An overview. *Revista de Microbiología* 1999;30:177-90.

[80] Little B, Wagner P. Myths related to microbially influenced corrosion. *Materials Performance* 1997;36:40-4.

[81] Lovley DR. Minimum threshold for hydrogen metabolism by methanogenic bacteria. *Applied and Environmental Microbiology* 1985;49:1530-1.

[82] Robinson JA, Tiedje JM. Competition between sulfate-reducing and methangenic bacteria for H_2 under resting and growing conditions. *Archives of Microbiology* 1984;137:26-32.

[83] Kristjansson JK, Schönheit P, Thauer RK. Different Ks values for hydrogen of methanogenic bacteria and sulfate reducing bacteria: an explanation for the apparent inhibition of methanogenesis by sulfate. *Archives of Microbiology* 1982:278-32.

[84] Lovley DR, Dwyer DF, Klug MJ. Kinetic analysis of competition between sulfate reducers and methanogens for hydrogen in sediments. *Applied and Environmental Microbiology* 1982;43:1373-9.

[85] Lovley DR, Phillips EJP. Competitive mechanism for inhibition of sulfate reduction and methane production in the zone of ferric iron reduction in sediments. *Applied and Environmental Microbiology* 1987;53:2636-41.

[86] Daniels L, Belay N, Rajagopal BS, Weimer PJ. Bacterial methanogenesis and growth from CO_2 with elemental iron as the sole source of electrons. *Science* 1987;237:509-11.

[87] Deckena S, Blotovogel K-H. Growth of methanogenic and sulfate-reducing bacteria with cathodic hydrogen. *Biotechnology Letters* 1990;12:615-20.

[88] Deckena S, Blotovogel K-H. Fe^0-oxidation in the presence of methanogenic and sulfate-reducing bacteria and its possible role in anaerobic corrosion. *Biofouling* 1992;5:287-93.

[89] Belay N, Daniels L. Elemental metals as electron source for biological methane formation from CO_2. *Antoine van Leeuwenhoek* 1990;57:1-7.

[90] Boopathy R, Daniels L. Effect of pH on anaerobic mild steel corrosion by methangenic bacteria. *Applied and Environmental Microbiology* 1991;57:2104-8.

[91] Videla HA. Corrosion of mild steel induced by sulfate-reducing bacteria. A study of the passivity breakdown of biogenic sulfides. In: Dexter SC, editor. Biologically Induced Corrosion, NACE-8 International Conferece Corrosion Series. Houston, Texas: NACE International; 1986. p. 162-71.

[92] Videla HA, Gomez de Saravia SG, de Mele MFL, Hernandez G, Hartt W. The influence of microbial biofilms on cathodic protection at different temperatures. Corrosion/93. Houston, Texas: NACE International; 1993.

[93] Zinkevich V, Bogdarina I, Kang H, Hill MAW, Tapper R, Beech IB. Characterisation of exopolymers produced by different isolates of marine sulphate-reducing bacteria. *International Biodeterioration & Biodegradation* 1996:163-72.

[94] Hamilton WA. Sulfate-reducing bacteria and their role in corrosion. In: Flemming HC, Geesy GG, editors. *Biofouling and Biocorrosion in Industrial Water Systems*. Berlin: Springer-Verlag; 1991. p. 187-93.

[95] Videla HA, Swords CL, De Mele MFL, Edyvean RG, Watkins P, Beech IB. The role of iron in SRB influenced corrosion of mild steel. Corrosion/98. Houston, Texas: NACE International; 1998.

[96] De Mele MFL, Brankevich G, Videla HA. A bioelectrochemical study of 70/30 copper-nickel alloy behavior in artificial chloride solutions and natural seawater. *Corrosion Reviews* 1990;9:269-75.

[97] Gomez de Saravia SG, De Mele MFL, Videla HA. Interaction of biofilms and inorganic passive layers in the corrosion of Cu/Ni alloys in chloride environments. *Corrosion* 1990;46:302-6.

[98] Jolley JG, Geesey GG, Hankins MR, Wright RB, Wichlacz PL. In Situ, Real-Time FT-IR/CIR/ATR Study of the Biocorrosion of Copper by Gum Arabic, Alginic Acid, Bacterial Culture Supernatant and Pseudomonas atlantica Exopolymer Applied Spectroscopy 1989;43:1062-7.

[99] McNeill LS, Edwards M. Iron pipe corrosion in distribution systems. *Journal American Water Works Association* 2001;93:88-+.

[100] Allen MJ, Taylor RH, Geldreich EE. The occurrence of microorganisms in water main encrustations. *Journal American Water Works Association* 1980;72:614-25.

[101] Emde KME, Smith DW, Facey R. Initial investigation of microbially influenced corrosion (MIC) in a low temperature water distribution system. *Water Research* 1992;26:169-75.

[102] Tuovinen OH, Button KS, Vuorinen A, Carlson L, Mair DM, Yut LA. Bacterial, chmical and mineralogical characteristics of tubercles in distribution pipelines. *Journal American Water Works Association* 1980;72:626-35.

[103] Victoree.Ht. Control of water quality in transmission and distribution mains. *Journal American Water Works Association* 1974;66:369-70.

[104] Gino E, Starosvetsky J, Kurzbaum E, Armon R. Combined Chemical-Biological Treatment for Prevention/Rehabilitation of Clogged Wells by an Iron-Oxidizing Bacterium. *Environmental Science & Technology*; 44:3123-9.

[105] Rancourt DG, Thibault PJ, Mavrocordatos D, Lamarche G. Hydrous ferric oxide precipitation in the presence of nonmetabolizing bacteria: Constraints on the mechanism of a biotic effect. *Geochimica Et Cosmochimica Acta* 2005;69:553-77.

[106] Tuhela L, Carlson L, Tuovinen OH. Biogeochemical transformations of Fe and Mn in oxic groundwater and well water environments. *Journal of Environmental Science and Health* Part a-Environmental Science and Engineering & Toxic and Hazardous Substance Control 1997;32:407-26.

[107] Wiklander L. The precipitation of iron in field drains. K Lantbruks Akad Handl Och Tidskr [Stockholm] 1945;84:358-67.

[108] El Gheriany IA, Bocioaga D, Hay AG, Ghiorse WC, Shuler ML, Lion LW. Iron Requirement for Mn(II) Oxidation by Leptothrix discophora SS-1. *Applied and Environmental Microbiology* 2009;75:1229-35.

[109] Takeda M, Kondo K, Yamada M, Koizumi JI, Mashima T, Matsugami A, et al., Solubilization and structural determination of a glycoconjugate which is assembled into the sheath of Leptothrix cholodnii. *International Journal of Biological Macromolecules*; 46:206-11.

[110] de Vet W, Dinkla IJT, Muyzer G, Rietveld LC, van Loosdrecht MCM. Molecular characterization of microbial populations in groundwater sources and sand filters for drinking water production. *Water Research* 2009;43:182-94.

[111] Suzuki I, Sahabi DM, Takeda M, Koizumi J. Bacterial diversity and mechanism of drinking water treatment using bio-filtration for removal of iron and manganese from groundwater. *Journal of Bioscience and Bioengineering* 2009;108:S90-S.

[112] Chamritski IG, Burns GR, Webster BJ, Laycock NJ. Effect of iron-oxidizing bacteria on pitting of stainless steel. *Corrosion* 2004;60:658-69.

[113] Torrabadelia J, Serra R. Actividad bacteriana en el interior de las redes de distribucion de agua y su relacion con los fenomenos de incrustacion y corrosion. *Doc. Invest. Hidrol.* 1972;13:111-35.

[114] Oconnor JT, Hash L, Edwards AB. Deterioration of water quality in distribution systems. *Journal American Water Works Association* 1975;67:113-6.

[115] Nemati M, Webb C. A kinetic model for biological oxidation of ferrous iron by *Thiobacillus ferrooxidans*. *Biotechnology and Bioengineering* 1997;53:478-86.

[116] Chapelle FH, Lovley DR. Competitive-exclusion of sulfate reduction by Fe(III)-reducing bacteria - A mechanism for producing discrete zones of high-iron ground-water. *Ground Water* 1992;30:29-36.

[117] Okereke A, Stevens SE. Kinetics of iron oxidation by Thiobacillus ferrooxidans. *Applied and Environmental Microbiology* 1991;57:1052-6.

[118] Kovalenko TV, Karavaiko GI, Piskunov VP. Effect of Fe-3+ ions in the oxidation of ferrous iron by Thiobacillus ferrooxidans at various temperatures. *Microbiology* 1982;51:142-6.

[119] Denisov GV, Kovrov BG, Kovaleva TF. Effect of pH and temperature of the medium on the rate of oxidation of Fe-2+ to Fe-3+ by a culture of *Thiobacillus ferrooxidans* and the coefficient of efficient of biosynthesis. *Microbiology* 1981;50:696-700.

[120] Dearaujojorge TC, Coutinho C, Deaguiar LEV. Sulfate-reducing bacteria associated with biocorrosion - A review. Memorias Do Instituto Oswaldo Cruz 1992;87:329-37.

[121] Iverson WP. Possible source of a phosphorus compound produced by sulfate-reducing bacteria that cause anaerobic corrosion of iron. *Materials Performance* 1998;37:46-9.

[122] Hamilton WA. Sulfate-reducing bacteria and anaerobic corrosion. *Annual Review of Microbiology* 1985;39:195-217.

[123] Hutchins SR, Sewell GW, Kovacs DA, Smith GA. Biodegradation of aromatic-hydrocarbons by aquifer microorganisms under denitrifying conditions. *Environmental Science & Technology* 1991;25:68-76.

[124] Pasteris G, Werner D, Kaufmann K, Hohener P. Vapor phase transport and biodegradation of volatile fuel compounds in the unsaturated zone: A large scale lysimeter experiment. *Environmental Science & Technology* 2002;36:30-9.

[125] Stapleton RD, Sayler GS, Boggs JM, Libelo EL, Stauffer T, Macintyre WG. Changes in subsurface catabolic gene frequencies during natural attenuation of petroleum hydrocarbons. *Environmental Science & Technology* 2000;34:1991-9.

[126] Zajic JE, Suppliss.B, Volesky B. Bacterial degradation and emulsification of No 6 Fuel Oil. *Environmental Science & Technology* 1974;8:664-9.

[127] Itah AY, Brooks AA, Ogar BO, Okure AB. Biodegradation of International Jet A-1 Aviation Fuel by Microorganisms Isolated from Aircraft Tank and Joint Hydrant Storage Systems. *Bulletin of Environmental Contamination and Toxicology* 2009;83:318-27.

[128] Mariano AP, Tomasella RC, Di Martino C, Maciel Filho R, Regali Seleghim MH, Contiero J, et al., Aerobic biodegradation of butanol and gasoline blends. *Biomass and Bioenergy* 2009;33:1175-81.

[129] Zanaroli G, Di Toro S, Todaro D, Varese GC, Bertolotto A, Fava F. Characterization of two diesel fuel degrading microbial consortia enriched from a non acclimated, complex source of microorganisms. *Microbial Cell Factories*;9.

[130] Ferrari MD, Neirotti E, Albornoz C. Occurrence of heterotrophic bacteria and fungi in an aviation fuel handling system and its relationship with fuel fouling. *Revista Argentina de Microbiologia* 1998;30:105-14.

[131] Matias PM, Pereira IAC, Soares CM, Carrondo MA. Sulphate respiration from hydrogen in Desulfovibrio bacteria: a structural biology overview. *Progress in Biophysics & Molecular Biology* 2005;89:292-329.

[132] Ayllon ES, Rosales BM. Electrochemical test for predicting microbiologically influenced corrosion of aluminum and AA 7005 alloy. *Corrosion* 1994;50:571-5.

[133] McNamara CJ, Perry TD, Leard R, Bearce K, Dante J, Mitchell R. Corrosion of aluminum alloy 2024 by microorganisms isolated from aircraft fuel tanks. *Biofouling* 2005;21:257-65.

[134] Hedrick HG, Reynolds RJ, Crum MG. Identification and viability of microorganisms from jet fuel samples. Raper, Kenneth B Developments in Industrial Microbiology Vol 9 Proceedings of the Twenty-Fourth General Meeting of the Society for Industrial Microbiology Xi + 482p Illus American Institute of Biological Sciences: Washington, DC, USA 1968:415-25.

[135] Hedrick HG, Reynolds RJ, Crum MG. Identification and viability of microorganisms from jet fuel samples. *Develop. Ind. Microbiol.* 1968;9:415-25.

[136] Koval EZ, Bugaenko LP. Cladosporium resinae and its degradation of gasoline and other petroleum products. *Mikologiya i Fitopatologiya* 1977;11:250-4.

[137] Sheridan JE, Soteros JJ. Survey of fungi in jet aircraft fuel systems in New Zealand. *International Biodeterioration Bulletin* 1974;10:105-7.

[138] Rivers C. The growth, metabolism, and corrosive effects of *Cladosporium resinae*. PhD Tehsis, Corrosion and Protection Centre, University of Manchester, UK 1973.

[139] Parbery DG. The role of *Cladosporium resinae* in the corrosion of aluminum alloys. *International Biodeterioration Bulletin* 1968;4.

[140] Scott JA, Hill EC. Microbial aspects of subsonic and pupersonic aircraft. *Proc. Symp. Microbiol.*, The Institute of Petroleum, London 1971.

[141] Ayllon ES, Rosales BM. Corrosion of AA-7075 aluminum-allow in media contaminated with Cladosporium resinae. *Corrosion* 1988;44:638-43.

[142] Parbery DG. Role of Cladosporium Resinae in Corrosion of Aluminium Alloys. *International Biodeterioration Bulletin* 1968;4:79-&.

[143] Simmons RB, Rose LJ, Crow SA, Ahearn DG. The occurrence and persistence of mixed biofilms in automobile air conditioning systems. *Current Microbiology* 1999;39:141-5.

[144] Yang SS, Lin JY, Lin YT. Microbiologically induced corrosion of aluminum alloys in fuel-oil/aqueous system. *J. Microbiol. Immunol. Infect.* 1998;31:151-64.

[145] Yang S-S, Chen C-Y, Wei C-B, Lin Y-T. Microbial corrosion of aluminum alloy. *Chinese Journal of Microbiology and Immunology* (Taipei) 1996;29:186-96.

[146] Yang S-S, Lin J-Y, Lin Y-T. Microbiologically induced corrosion of aluminum alloys in fuel-oil/aqueous system. *Chinese Journal of Microbiology and Immunology* (Taipei) 1998;31:151-64.

[147] Samuels BW, Sotoudeh K, Foley RT. Inhibition and acceleration of aluminum corrosion. *Corrosion* 1981;37:92-7.

[148] Videla HA, Herrera LK. Microbiologically influenced corrosion: looking to the future. *International Microbiology* 2005;8:169-80.

[149] Busalmen JP, Vazquez M, de Sanchez SR. New evidences on the catalase mechanism of microbial corrosion. *Electrochimica Acta* 2002;47:1857-65.

[150] Rajasekar A, Ting YP. Microbial Corrosion of Aluminum 2024 Aeronautical Alloy by Hydrocarbon Degrading Bacteria Bacillus cereus ACE4 and Serratia marcescens ACE2. *Industrial & Engineering Chemistry Research*; 49:6054-61.

[151] Busalmen JP, Valcarce MB, de Sanchez SR. Importance of surface chemistry in bacterial adhesion to metals and biocorrosion. *Corrosion Reviews* 2004;22:277-305.

[152] Leitz M, Franke H. Observation of low aluminium corrosion rates during biofilm formation. *Optical Fibers and Sensors for Medical Applications Ii* 2002;4616:229-36.

[153] Stoodley P, Jacobsen A, Dunsmore BC, Purevdorj B, Wilson S, Lappin-Scott HM, et al., The influence of fluid shear and AlCl3 on the material properties of Pseudomonas aeruginosa PAO1 and Desulfovibrio sp. EX265 biofilms. *Water Science and Technology* 2001;43:113-20.

[154] Kuwahara H, Al-Abdullat Y, Ohta M, Tsutsumi S, Ikeuchi K, Mazaki N, et al., Surface reaction of magnesium in Hank's solutions. *Magnesium Alloys* 2000 2000;350-3:349-58.

[155] Makar GL, Kruger J. Corrosion studies of rapidly solidified magnesium alloys. *Journal of the Electrochemical Society* 1990;137:414-21.

[156] Hughes MN, Poole RK. Metals and microorganisms. *Metals and microorganisms* 1989:i-xii, 1-412.

[157] Nandakumar K, Sreekumari KR, Kikuchi Y. Antibacterial properties of magnesium alloy AZ31B: in-vitro studies using the biofilm-forming bacterium Pseudomonas sp. *Biofouling* 2002;18:129-35.

[158] Videla HA, Characklis WG. Biofouling and microbially influenced corrosion. *International Biodeterioration & Biodegradation* 1992;29:195-212.

[159] Fukumoto S, Sugahara K, Yamamoto A, Tsubakino H. Improvement of corrosion resistance and adhesion of coating layer for magnesium alloy coated with high purity magnesium. *Materials Transactions* 2003;44:518-23.

[160] Parthiban GT, Parthiban T, Ravi R, Saraswathy V, Palaniswamy N, Sivan V. Cathodic protection of steel in concrete using magnesium alloy anode. *Corrosion Science* 2008;50:3329-35.

[161] Tsyrul'nyk OT, Heneha BY. Mechanism of the sacrificial protection of stressed low-alloy steel. *Materials Science* 2004;40:597-604.

[162] Song G, Atrens A, Stjohn D, Nairn J, Li Y. The electrochemical corrosion of pure magnesium in 1 N NaCl. *Corrosion Science* 1997;39:855-75.

[163] Sreekumari KR, Nandakumar K, Kikuchi Y. Bacterial attachment to stainless steel welds: Significance of substratum microstructure. *Biofouling* 2001;17:303-16.

[164] Videla HA. Prevention and control of biocorrosion. *International Biodeterioration & Biodegradation* 2002;49:259-70.

[165] Guiamet PS, De Saravia SGG. Laboratory studies of biocorrosion control using traditional and environmentally friendly biocides: An overview. *Latin American Applied Research* 2005;35:295-300.

[166] Piriou P, Dukan S, Levi Y, Guyon F, Villon P. Modelling free and fixed bacterial biomass behaviour in drinking water distribution systems. *Revue des Sciences de l'Eau* 1996;9:381-406.

[167] Borshchevskii AM, Velikova TD, Baturova LP. Effect of ozonation on the corrosion of carbon-steel in the tap water of St Petersburg. *Russian Journal of Applied Chemistry* 1993;66:1699-701.

[168] de Saravia SGG, Guiamet PS, Videla HA. Prevention and protection of the effects of biocorrosion and biofouling minimizing the environmental impact. *Revista De Metalurgia* 2003:49-54.

[169] Dlamini ET, Williams AL, Dashek WV, Swank WT, Vose JM. Protein contents of ozone-fumigated and air-fumigated pinus sobus needles. *Biodeterioration Research* 4 1994:379-90.

[170] Viera MR, Guiamet PS, de Mele MFL, Videla HA. Use of dissolved ozone for controlling planktonic and sessile bacteria in industrial cooling systems. *International Biodeterioration & Biodegradation* 1999;44:201-7.

[171] Cheung CWS, Beech IB. The use of biocides to control sulphate-reducing bacteria in biofilms on mild steel surfaces. *Biofouling* 1996;9:231-49.

[172] Cetin D, Bilgic S, Donmez G. Biocorrosion of low alloy steel by Desulfotomaculum sp and effect of biocides on corrosion control. *Isij International* 2007;47:1023-8.

[173] Al-Darbi MM, Muntasser ZM, Tango M, Islam MR. Control of microbial corrosion using coatings and natural additives. *Energy Sources* 2002;24:1009-18.

[174] Arnould C, Volcke C, Lamarque C, Thiry PA, Delhalle J, Mekhalif Z. Titanium modified with layer-by-layer sol-gel tantalum oxide and an organodiphosphonic acid: A coating for hydroxyapatite growth. *Journal of Colloid and Interface Science* 2009;336:497-503.

[175] Gu XN, Zheng YF, Lan QX, Cheng Y, Zhang ZX, Xi TF, et al., Surface modification of an Mg-1Ca alloy to slow down its biocorrosion by chitosan. *Biomedical Materials* 2009;4.

[176] Wan D, Yuan SJ, Neoh KG, Kang ET. Poly(glycidyl methacrylate)-Polyaniline Bilayer-Modified Mild Steel for Combating Biocorrosion in Seawater. *Journal of the Electrochemical Society* 2009;156:C266-C74.

[177] Xu XH, Lu P, Guo MQ, Fang MZ. Cross-linked gelatin/nanoparticles composite coating on micro-arc oxidation film for corrosion and drug release. *Applied Surface Science*; 256:2367-71.

[178] Yuan SJ, Pehkonen SO, Ting YP, Neoh KG, Kang ET. Antibacterial Inorganic-Organic Hybrid Coatings on Stainless Steel via Consecutive Surface-Initiated Atom Transfer Radical Polymerization for Biocorrosion Prevention. *Langmuir*; 26:6728-36.

[179] Maruthamuthu S, Ponmariappan S, Mohanan S, Palaniswamy N, Palaniappan R, Rengaswamy NS. Control of metallic corrosion through microbiological route. *Indian Journal of Experimental Biology* 2003;41:1023-9.

[180] Yuan SJ, Xu FJ, Pehkonen SO, Ting YP, Neoh KG, Kang ET. Grafting of Antibacterial Polymers on Stainless Steel Via Surface-Initiated Atom Transfer Radical Polymerization for Inhibiting Biocorrosion by *Desulfovibrio desulfuricans*. *Biotechnology and Bioengineering* 2009;103:268-81.

In: Biofilms in Bioengineering
Editors: Manuel Simões and Filipe Mergulhão

ISBN: 978-1-62948-161-6

Chapter IX

Medical Biofilms

Randall Wolcott MD[1,*] and Scot Dowd PhD[2]
[1]Medical Director; Southwest Regional Wound Care Center, Lubbock, Texas, US
[2]Director; Research and Testing Laboratory of the Southwest, Texas, US

Abstract

When Medical Biofilm is viewed as a single disease similar to the way we categorize Cancer, the result is breathtaking. In a general review of the literature looking at infections related to biofilm the findings suggest over 17 million people in the U.S. develop a biofilm infection annually. Also a very large number of people are dying from, or with, their biofilm disease. Ventilator acquired pneumonia is associated with 67,000 deaths per year, chronic obstructive pulmonary disease (COPD) with 121,000 deaths per year, chronic wounds with 50,000 deaths per year and so on down the line until the total of deaths associated with biofilm annually surpasses 500,000 people. This finding alone creates a clear urgency to educate, research, and cure what we now understand is an overwhelming human disease.

Medical microbiology is at a crossroads. It is almost unimaginable to think, given Medicine's current technological capabilities, that the most important responsibility of Medical microbiology - the identification of infecting microorganisms - is relegated to a methodology that is over 150 years old. It is the failure of clinical cultures along with tenaciously holding on to the "one organism one infection" dogma that has prevented Medicine from fully appreciating biofilm's role in chronic infections.

Appreciating the full scope of the devastation produced by Medical Biofilm will take time. Biofilm's role in normal health and well-being through healthy commensals is just now being investigated. If our commensal biofilms are sick then we are sick. The vast carnage of chronic infections produced by pathogenic biofilm has not been fully elucidated. Furthermore, if even a few of the major diseases that behaved like chronic infections, such as atherosclerosis, amyloid formation in the pancreas and brain, chronic inflammation producing cancer, turn out to be related to biofilm, then the importance of Medical Biofilm becomes incalculable.

[*] E-mail: randy@randallwolcott.com.

It is clear that chronic infections from the various host niches will be different yet biofilm principles should translate to each one. Therefore what we've learned from environmental biofilm, industrial biofilm and all the other biofilm disciplines have proven to be applicable to specific chronic infections. With this understanding along with the development and deployment of new technologies we should make rapid progress in managing biofilm-based disease. Early translational work has shown this to be true and if the translation of biofilm principles to human chronic infections holds up then it is not unreasonable to estimate that Medical Biofilm as a disease can be controlled in our lifetime.

Keywords: Biofilm; Biofilm treatment; Chronic infection; Inflammation; Topical antibiotics

Introduction

Medical Biofilm's role in human infectious disease cannot be overstated. A comment during a National Institutes of Health NIH committee meeting suggested that 80% [1]of human infectious disease was related to biofilm, whereas the Centers for Disease Control (CDC) suggests 65%. [2] However, both these estimates are missing the mark, thus failing to fully convey the immensity of Medical Biofilm. Humans, as a species, reside in a biosphere dominated by bacteria which is estimated to be 10^{30} microorganisms for the planet. We mainly live cooperatively and synergistically with biofilms through commensal mechanisms. It is estimated that in looking at the human and the human microbiome as a single entity that only 10% of the cells are contributed by the human genome [3]. What makes us is not just our 23 paired chromosomes but rather our how our cells develop and function in relation to our commensal bacteria. In fact, it is clear that we would lack vitamins, components of host immunity, some neurologic development, and other important components of who we are without the benefits of commensal bacteria. Yet, in this vast sea of bacteria reside individual pathogens, as well as synergistic groups of bacteria acting as pathogens (functional equivalent patho groups), unrelentingly attempting to compromise host defenses. Therefore it is not surprising that these rogue bacteria can breach (planktonic phenotype) and/or maintain (biofilm phenotype) a replicative foothold in almost any tissue of the human body. Del Pollo and Patel described this problem by identifying a number of chronic infections associated with biofilm (Table 1) [4]. We then tried to further develop the concept by treating Medical Biofilm in a similar manner as how malignancies are categorized [5]. Cancer is an inclusive category of all malignancies with the common property of cells of any tissue type outside of host control. Although malignancies are quite different, that is they are diagnosed differently and treated differently, there are principles that translate throughout the entire disease of cancer. Similarly, Medical Biofilm can be grouped together under a common theme of human infections produced by biofilm phenotype bacteria. Although these chronic infections are quite different, for example different sites will select for different species of bacteria, the findings at a molecular, immunological, microbiological level should translate across all chronic infections. When Medical Biofilm is viewed in this comprehensive way, the result is breathtaking. We undertook a general review of the literature looking at infections related to biofilm [6]. Using the numbers provided by authors of individual articles, the findings suggest over 17 million people in the U.S. develop a biofilm infection each year.

Table 1. Medical biofilm diseases

Probable		Possible	Generally Accepted	
Nasal polyps	Epididymitis		Otis externus	Periodontal disease
Conjunctivitis	Urinary stones	Atherosclerosis	Otitis medius	Osteonecrosis of the jaw (bisphosphonate therapy)
Iritis	Chronic pancreatitis	Diabetes (pancreatic beta amyloid formation)	Cholesteatoma	Mastitis
Chronic bronchitis	Ascending cholangitis		Mastoiditis	Ventilator acquired pneumonia
Asthma	Atopic dermatitis	Alzheimer's (beta amyloid formation)	Chronic tonsillitis	Cystic fibrosis
Duodenal ulcer	Psoriasis		Chronic rhinosinusitis	Gastric ulcers (H-pylori)
Crohn's	Burns	Multiple sclerosis	Osteomyelitis	COPD
Ulcerative colitis	Surgical site infections	Chronic inflammation induced cancers	Chronic meningitis	Peritoneal dialysis catheters
Diverticulitis	Folliculitis		Ventricular shunt infections	Urinary catheters
Proctitis	Abscess	Autoimmune diseases	Amniotic fluid sludge	Intravenous catheters
Pelvic inflamm. disease	Chronic vaginosis	Altered commensals (autism, mood, obesity)	Bacterial endocarditis	Infected medical devices
Chronic prostatitis	Postop. endophthalmitis		Wounds	Gall stones
Interstitial cystitis				

More importantly, the literature suggested that very large numbers of people were dying from, or with, their biofilm disease. Ventilator acquired pneumonia was associated with 67,000 deaths per year, COPD with 121,000 deaths per year, chronic wounds with 50,000 deaths per year and so on down the line until the total of deaths associated with biofilm annually surpassed 500,000 people. This finding alone presses Medical Biofilm to the forefront of medicine and creates a clear urgency to educate, research, and cure what we now understand is an overwhelming human disease.

It is interesting to note how physicians have come to treat most chronic infections. Every discipline in medicine encounters chronic infections in the systems which they specialize. The ears, nose and throat (ENT) specialist treating chronic tonsillitis, the gastrointestinal (GI) doctor treating Crohn's disease, or the wound specialist treating chronic wounds independent of each other have developed strategies which are remarkably and sadly similar. Chronic infections are treated with benign neglect, treating the lesions with trial and error therapies, and antibiotics are used for exacerbations. If the host does not contain the chronic infection or the tissue to which it is adhered degrades sufficiently, then the "treatment" becomes rather draconian. The treatment for tonsillitis is removal of the tonsils, for endocarditis it is the replacement of the heart valve, and for chronic wounds it is removal of the limb. The final common pathway for the definitive solution of biofilm disease that every discipline in medicine has developed over time is to remove or replace body parts. A clear understanding of Medical Biofilm will hopefully change this practice.

Biofilm Basics

Medical microbiology has focused on the mobile single cell (planktonic) mode of bacterial life. It is now recognized that up to 99% of bacteria in the environment prefer to live as sessile communities [7]. This protects biofilm against predators (protozoans, bacteriophages, white blood cells), UV light, dehydration, nutrient restriction, antibiotics, biocides, and a number of other stresses [8, 9]. Therefore, biofilm's community mode of life is much more robust and more successful in dealing with hostile environments such as the host and its defenses (Table 2).

Biofilm phenotype is defined as bacteria (usually multiple species) that attach to a surface, produce a self-secreted sanctuary of extracellular polymeric substance (EPS), and demonstrate tremendous heterogeneous gene expression among its members. When a biofilm microcolony is established in the host, it begins to solidify its defenses against complement [10], antibodies [11], and white blood cells [12, 13]. The cells continue to divide until they reach a critical density (quorum). This quorum accumulates enough cell-to-cell signaling molecules that direct the differentiation of the microcolony into a mature biofilm with fully functional community defenses [9]. These defenses include slow penetration, depletion of harmful substances, an altered microenvironment with slow growth or no growth, upregulated horizontal gene transfer, and many others [14–17].

In vitro experiments have demonstrated three important differences between bacteria grown in biofilm phenotype versus planktonic phenotype: biofilms are resistant to antibiotics (50-1,000 times) [15], biofilms are highly resistant to biocides (hydrogen peroxide, acids, bleach) [17], and biofilms evade the host immune system (white blood cells, antibodies, complement) [11, 12]. These properties of biofilm may lead to consistent clinical findings among most human biofilm infections. These common findings include only modest response of chronic infections to antibiotics, which plateau, and clinical symptoms reemerge once the antibiotic is withdrawn. Also, there is the inability to develop vaccines or other augmented host responses that can prevent or clear biofilm disease [14].

Also, with Medical Biofilm, sepsis is uncommon. This is best illustrated by endocarditis. The vegetation is an organized biofilm on the heart valve, which constantly sloughs bacteria into the blood, yet clinical symptoms of sepsis in patients with endocarditis are only 36% [18]. It may be conjectured that successful human biofilm disease resists killing its host. Whether it is resident in heart valves, kidneys, or colons, sepsis seems to be selected against by biofilm communities.

Table 2. Planktonic *vs* biofilm explanations for clinical observations

Observation	Planktonic Concept	Biofilm Concept
In vitro	Planktonic bacterium (seed) expresses proteins and structures for motility and attachment (flagella, fimbria). Its function is to spread the colony to a different location.	Biofilm (vegetation) is a complex colony of bacteria that can express different proteins (variable phenotype) to fulfill different roles to help the community survive. Biofilm is stationary, protecting its location with multiple defenses.
In vivo	Planktonic bacteria are susceptible to antibiotics, biocides and the immune system.	Biofilms are resistant to antibiotics and biocides. Once biofilm is established it cannot be eradicated by the immune system.
Acute wounds heal in 2-4 weeks; chronic wounds in the same area heal in 4-6 months	Acute wounds have intact defenses and destroy planktonic bacteria, not allowing biofilm formation.	Some bacteria may evade host defenses and establish a biofilm. Biofilm possesses defenses against the host immune system and interferes with wound healing.
Antibiotics are ineffective in chronic wounds but effective in acute wounds	Rapidly growing bacteria show a 4- to 5-log reduction in viability.	The biofilm phenotypes quickly adapt and demonstrate only a grudging 1- to 2-log reduction, even at 50 to 1000 times the MIC.
Autograft or Allograft fails on wounds	Planktonic cells are easily cleared by neutrophils, antibodies, and common wound bed preparations and, therefore, do not explain graft failure.	Laying unprotected cells over biofilm adds a second surface and a food source. This leads to the rapid deterioration of the graft and increased exudate, inflammation, and odor of the wound.
Negative wound cultures	Most clinical bacteria in planktonic phenotype are easily cultured.	Wounds have bacteria on their surface yet can culture negative when a biofilm phenotype is present (i.e., viable but not culturable).
Wounds "stuck" in chronic inflammatory state	Planktonic bacteria are easily cleared by host inflammatory response and survive best if inflammation is avoided.	Biofilms are impervious to the host inflammatory response and can even feed off the exudate produced by inflammation. Biofilms often promote inflammation.

The biofilm story is truly fascinating. The fossil records suggest that biofilm communities (stromatolites) emerged approximately 3.25 billion years ago [19]. Even more amazing is the fact that over the vast billions of years of time to the present day, biofilm has propagated essentially unchanged. This remarkable achievement is most likely due to biofilm societies' "unselfish" properties of inclusiveness (co-aggregation), tolerance (diversity), sharing (increased horizontal gene transfer), and open communications (quorum sensing). It seems that biofilm, from the beginning, got it right by pursuing a strategy of "survival of the most cooperative."

Co-aggregation, relies on specific surface components to produce a cell to cell recognition between two distinct bacterial species causing the species to loosely bind [20]. Co-aggregation is much more than diversity in that it is a process that selects for specific species to co-locate within the biofilm, thus taking full advantage of their synergies. These surface structures allow for a loose bond that brings both (or more) bacteria together to the potential infection site or they can also quickly link up prior to EPS formation at the site. Either way the bacteria are selected to spatially reside in close proximity [21].

It must be pointed out that biofilm communities are mainly heterogeneous with exclusive regions of one species swirled into regions of another species [22]. There is no evidence of bacteria species being uniformly dispersed throughout a polymicrobial biofilm. This creates important clinical phenomenon such as we demonstrated in a topologic study of wound biofilm, which showed marked difference in the proportion of different biofilm members in different regions of the wound [23]. All the different members were present at each sampling site, yet the proportions were vastly dissimilar, thus demonstrating a unique species as the predominant cornerstone community at different sampling sites. Chronic wound biofilm shows a definite architecture with a clustering of microbial species and a definite spatial relationship between the clusters [22].

Certain species have a clear advantage to attach to different substrates, thus establishing a biofilm in a host setting. Therefore, the diverse epitopes on different host tissues, as well as host environments will select for these early pioneers. For example, *S. aureus* has multiple adhesins for specific epitopes for extracellular matrix proteins such as fibronectin and fibrinogen, making it very successful in attaching to skin, connective tissue and muscle regions. Once the pioneers are attached and early biofilm is formed, then secondary settlers will be established (recruited)—some through co-aggregation mechanisms or others by chance.

Biofilms possess multiple molecular mechanisms to allow for diversity. Diversity allows for survivability due to different capabilities of individual microbes. Synergies between these microbes and most importantly the shared defenses of the biofilm colony_showed that beta-lactamase producing bacteria could cause an accumulation of beta-lactamase in the biofilm matrix, thus protecting non beta-lactamase producing bacteria. The more different species in a biofilm, the higher the probability that at least one member will survive any given stress to the community. It seems that small numbers of each species survive (persisters) and cannot be completely eradicated; thus if even one species survives with enough reserves to start reconstituting the biofilm, these persisters can reestablish their contribution to the reconstituted biofilm [24].

This property of biofilm, being able to reconstitute itself from near annihilation, raises two very important clinical points. First, it is impossible to eradicate a bacterial species from a biofilm infection in a host setting. The suppressed species can reemerge as long as the biofilm is present. The second point is that a biofilm infection is a cooperative community; therefore, no single bacteria can be pointed out as the “cause” of the infection. This brings about great trouble for the concept of "pathogen," so commonly found in planktonic microbiology. Even minor species of bacteria considered to be non-pathogens can provide support to other bacteria which may allow the group to be “pathogenic.” So, it is impossible to deselect an individual bacterial species as not important in a biofilm infection.

Horizontal gene transfer is of key importance in managing biofilm disease. The individual bacteria comprising the biofilm are spatially organized to increase the efficiency of horizontal gene transfer. The inter-bacterial distance is very precise, especially in the middle third, which optimizes the transfer of mobile genetic material such as antibiotic resistance factors (mecA cassette, van gene, ESBL, etc.) making biofilm much more efficient in disseminating these resistance factors. Therefore, biofilm produces, in the environment of a chronic infection, the most potent mechanism for the transfer of antibiotic resistance.

There are many different molecules that function in quorum sensing pathways. The signaling molecules of gram positives tend to be short peptides from 7-12 amino acids in length. Gram negative bacteria commonly use N-acyl homoserine lactone (AHL) for their quorum sensing pathways. More recently, a family of auto inducers most commonly represented by auto inducer 2 (AI-2) has been found to be active in quorum sensing pathway in most gram negatives and gram positive bacteria. The effect of AI-2 on anaerobic bacteria is unknown [25].

Quorum sensing was first identified in bacteria and is now understood as a decentralized system which coordinates gene expression according to the density of the bacterial populations in a particular biofilm. This allows for the heterogeneous development of bacteria in the different regions of the biofilm. Quorum sensing is clearly responsible for phenotypic development of bacteria in each region, as well as the tolerance necessary to develop the diversity seen in biofilm disease.

Quorum sensing, as the name suggests, is the communication between individual bacterium within the biofilm society and is based on concentration. There are many different broad molecular themes in quorum sensing such as transcription factor (Rpo), Sigma factor, and the most common being two component systems. A two component system is made up of a sensor protein that binds to a specific ligand and then activates response regulator proteins. Usually, there are several intermediate molecules in the molecular cascade which eventually result in specific gene expression. The advantage of a two component system is that multiple signals can be integrated during the cascade process to either upregulate or downregulate specific gene expression. This gives the community tight control yet the flexibility to respond as necessary.

The agr locus is one of the most studied two component systems. There are several promoters, but promoter P2 is responsible for upregulating the transcription of the four main regulatory proteins which comprise the agr two component system. AgrD encodes for the quorum sensing molecule autoinducing peptide (AIP). This is an autoinducing peptide which is approximately 7 to 9 amino acids in different species of the *Staphylococcus* genus. A second agr product is agrC, which is the receptor for the agr system and binds to AIP. Once agrC binds with its AIP ligand, agrC then activates agrA through a phosphorylation process. The phosphorylated agrA then upregulates promoter P2 which further upregulates the agrACDB expression. Thus, AIP becomes a true auto inducing peptide upregulating the expression of more agr products [26].

The increase in promoters also upregulates RNA III transcription which produces an intracellular effector molecule which stimulates expression of virulence factors. It is important to note that the virulence factors produced by this upregulation of the agr system is temporally regulated to mid and late exponential phase of the growth cycle. When the biofilm structure is more established, virulence factors tend to be down regulated [27].

Gram negative bacteria, as represented by *Pseudomonas aeruginosa*, utilize acyl homoserine lactone (AHL) as the quorum sensing molecule for the most global biofilm related operon regulated by the las-rhl system. The las system is responsible for promoting expression of rhamnolipid, elastase and multiple biofilm characteristics such as the rapid upregulation of alginate production and secretion. The las system is also important in upregulating the rhl system.

Rhl is a second layer of regulatory control on expression of the operon. Rhl seems to regulate genes which promote cell survival in biofilm phenotype including rpoS. Even though acyl homoserine lactones are associated with both virulence factors and biofilm behavior such as EPS formation, these pathways are divergent. Smith et al., showed that virulence factors can be deselected in *Pseudomonas aeruginosa* by mutations in LasR while still maintaining dense mucoid phenotype biofilm in the cystic fibrotic airway.[28]] Smith postulates that the numerous virulence factors that are deselected are necessary in the acute phase to establish the biofilm but unnecessary, and possibly detrimental, to the biofilm after prolonged colonization. These findings clearly demonstrate a divergence between the two different infection strategies that a single species such as *Pseudomonas aeruginosa* can utilize to pursue a host infection. The first strategy utilizing virulence factors to breach host defenses and tissues more closely correlates with planktonic phenotype behavior. Whereas in a chronic infection the bacterial society literally jettisons virulence factors in favor of dense colony defenses and alternative strategies to obtain nutrition.

A broader quorum sensing mechanism is illustrated by auto inducer 2 (AI-2) which carries the chemical name furanosyl borate diester signal. AI-2 is a diffusible molecule that acts on luxS gene which is widely located in both gram negative and gram positive bacteria, which indicates that AI-2 is capable of interspecies communication. It is now clear that AI-2 is a coregulator along with AHL systems, as well as the oligopeptide systems in gram positives. This makes it difficult to exactly pinpoint AI-2's role in colony control. Our recent evaluation of clinical wound samples demonstrates the presence of AI-2 in chronic wounds of quite divergent polymicrobial diversity [29]. The value of this at the clinical level is that disruption of AI-2 signaling, as with inhibiting any other quorum sensing signal, may degrade the biofilm community and allow for better clinical outcomes.

The discussion of quorum sensing pathways can be complex and tedious, and only a few examples were offered to illustrate the character of the quorum sensing systems. The systems described, although steeped in jargon, become important, especially when looking at quorum sensing inhibitors which will be developed in detail in the treatment section of this chapter.

Extracellular polymeric substance, that is polymeric sugars assembled within the bacterium and then excreted externally then modified and crosslinked with bacterial DNA, is often thought of as the main component of the matrix binding the biofilm community tightly to the surface, while also providing the main defense for the entire community. Katsuyama in a two part article describes a much different reality for Medical Biofilm [30, 31]. His experiments showed that coagulase positive *S. aureus* first polymerized fibrinogen to fibrin deeply embedding itself in the host proteins. Then bacterial derived polymeric substances were excreted to complete the biofilm matrix. The incorporation of host proteins such as fibrin, collagen, and other linear proteins along with DNA from host and bacterial sources is well known in Medical Biofilm [32–35].

The biofilm matrix is composed of host components as well as bacterial components which produces incredible general colony defenses. These defenses include reaction with chemicals, delaying diffusion, sequestering important molecules (quorum sensing molecules, beta-lactamase, synergistic molecules), providing discrete anatomical structure for differentiation and co-location. Yet biofilm's most potent defense is its ability to recover after a catastrophic event.

It can be argued that the most potent property of Medical Biofilm is its ability to reconstitute itself after a devastating insult. Even after the vast majority of biofilm is physically removed and only small retained fragments are available, these small fragments have the ability to become metabolically active and reconstitute the entire colony within the protection of the biofilm matrix. This property of biofilm is a main contributor to the recalcitrance of chronic infection. But as discussed in the treatment section, the property of biofilm reconstitution can be exploited for therapeutic gain.

In the laboratory, bacterial derived DNA is most often used for crosslinking the polymeric sugars and adding significant stability to the structure of the matrix [36–39]. In Medical Biofilm, neutrophil derived DNA may also participate in cross-linking [40]. Bacterial derived proteins, such as biofilm accumulation protein (BAP), are also a significant part of biofilm grown in the laboratory yet host proteins appear to play a much bigger role in Medical Biofilms [41, 42].

The biofilm matrix is enigmatic in that a bacterial species or a specific group of species depending on host conditions and specific stressors can produce dramatically different matrix compositions. This makes strategies to degrade the EPS difficult, in that the matrix may be primarily host or bacterial derived proteins, diverse sources of DNA or bacterial derived polymeric sugars, but no matter the composition of the matrix, it must always have the property of allowing fluid movement through the biofilm, as well as the ability of nutrients to pass through the biofilm. These are properties that can be exploited in treatment.

Biofilm nutrition offers great insight into the overall strategy that biofilm uses to produce infection. Biofilm must establish and maintain a permanent attachment to the host tissue. Therefore, it is precluded from utilizing virulence factors to necrose the tissue to which it is bound. In fact, multiple different molecular mechanisms have been identified by which bacteria maintain a permanent attachment to the tissue that they infect.

For example, *S. aureus* uses a heme sensing system to rapidly identify an internal host environment and upregulate adhesions which are specific for host extracellular matrix proteins [43]. Once attached to these epitopes, *S. aureus* uses secretory systems to commandeer the surrounding eukaryotic cells and block synthesis, but more importantly to block apoptotic pathways, thus rendering the eukaryotic cells senescent [44, 45]. Senescent cells are alive but unable to respond to appropriate signals and, more importantly, unable to slough, thus creating a stable base for biofilm attachment.

Another example is *Escherichia coli*, which has been demonstrated to work through a TypeIII secretory system to eject effector proteins into the eukaryotic cell to which it's attached [46]. These effector proteins polymerize cytoskeleton actin into a firm pedestal on which the *E. coli* can remain attached. Other effector proteins are secreted effectively blocking apoptotic pathways, blocking the turnover of the epithelial cell, thus establishing a long term attachment to the host. Many other bacterial species have been investigated and found to have very similar molecular mechanisms for maintaining attachments on host tissues [47].

The biofilm once firmly attached on a senescent host site must now obtain nutrition through a completely different pathway than tissue necrosis. There is emerging evidence that microbes in any phenotype express a large array of different virulence factors to specifically dampen and/or reorient both the innate and adaptive immune responses [47, 48]. Studies of multiple different microorganisms have demonstrated that bacteria can, very specifically and intentionally, upregulate proinflammatory cytokines in order to manipulate the host innate immune response. There are several examples, including *Shigella* spp. which expresses a plasma virulence gene that increases the expression of proinflammatory cytokines [49]. Quorum sensing molecules from *P. aeruginosa* can directly act on host cells to induce the expression of proinflammatory cytokines [50] and also, *S. aureus* (as well as other *Staphylococcus* species) expresses modulines and super antigens (enterotoxin, toxic shock syndrome, toxin 1, etc.), which can induce massive and unsustainable releases of proinflammatory cytokines [51] Another unexpected source of proinflammatory cytokines is the Type III secretory system of some pathogens injecting protein effectors into host immune cells, damping phagocytosis and reducing the lethal actions of these neutrophils thereby, increasing the release of proinflammatory cytokines [52].

It seems counterintuitive that bacteria would ever intentionally "activate" the host innate immunity. However, our hypothesis is that microorganisms in biofilm phenotype benefit from this hyperinflammatory state. That is, by hijacking the host immune system, biofilm has created several important mechanisms to maintain its replicative niche. These mechanisms include collateral damage (hyperinflammatory products damaging host cells), degrading the effectiveness of host immunity and most importantly, creating a source of sustainable nutrition (the plasma that leaks from inflamed vessels).

Biofilm Inflammation

Biofilms continually release planktonic "seeds" of bacteria from the safety of the biofilm matrix which can bait the immune system (via pathogen associated molecular patterns recognition mechanisms) which continually upregulates host inflammatory response [53]. The biofilm exploits this hyperinflammatory state by harvesting nutrients from the host exudates that accompanies the inflammatory response. In this way, the sacrifice of a few bacteria promotes the survival of the community through continuous nutrient acquisition. Suppressing the host response with immunosuppressants or physically removing the bacteria load through debridement may break this cycle and suppress the biofilm.

Lipopolysaccharide (LPS) is a major component of gram negative bacteria and has been found to decrease neutrophils' response to Interleukin 8 and also inhibits the phosphatidal serine receptor that resides on neutrophils [54]. Through these pathways, LPS is a potent inducer for continued chemotaxis of neutrophils to the side of the biofilm as well as a potent source to block clearance by macrophages. Thus, the biofilm does not need a specific pathogen with powerful virulence factors to maintain an inflammatory response, but only requires an accessory population of Gram negative bacteria to release LPS. Biofilms have been shown to constantly release LPS in membrane vesicles, a previously unknown function, into the surrounding environment [55, 56].

Along with these above mechanisms, biofilm has also been demonstrated to release its own genetic material (bacterial DNA) into the surrounding chronic infection [57]. Bacterial DNA also stimulates inflammatory recruitment in host tissue [58]. Thus, unlike the single pathogen, biofilms have an arsenal of generalized tools that can stimulate, maintain and ultimately control a hyperinflammatory state. All the while, the heart of the biofilm encased in the matrix remains intact and unperturbed by the host response.

Manipulating Neutrophils

Individual bacterial populations such as *P. aeruginosa* have developed virulence factors such as pyocyanin that prevents the removal of neutrophils [59]. However, wound biofilm can use generalized mechanisms to produce similar results. The literature indicates that excessive neutrophils accumulate at the site of chronic wounds[60]] and many other chronic infections [61–64]. Interleukin 8, a common proinflammatory cytokine, in and around the region of a chronic infection, promotes chemotaxis of neutrophils by binding to their CXCR1 and CXCR 2 receptors [65]. Also, as stated above, bacteria utilize LPS and other factors to degrade the CXCR 1 and 2 receptors [66, 67]. In addition, LPS can interfere with the neutrophils' membrane associated phosphatidal serine. In a healthy immune response, neutrophils begin apoptosis after engulfing the pathogens. They express proteinaceous surface molecules such as phosphatidal serine, which macrophages interpret as a request for phagocytosis [68–70]. Macrophages recognize, engulf, and then degrade these functional terminal neutrophils, preventing necrotic disintegration of the neutrophil *in situ* [71, 72].

A localized accumulation of neutrophils at the site of the chronic infection is a healthy host response, but only if this accumulation is accompanied by an orderly elimination of the neutrophils by macrophages in order to prevent the excessive release of destructive proteases [73]. When bacterial products such as lipopolysaccharide interfere with phosphatidal serine, the macrophages may not recognize the need to clear terminal neutrophils. These neutrophils then degrade in the presence of the chronic wound and release destructive enzymes such as elastase, which produces host cell damage.

A second consequence of neutrophils being drawn into a chronic infection, trapped and then induced into a delayed apoptotic process [74, 75], is to produce a nonphysiologic increase in host inflammation termed hyperinflammation. Because neutrophils are inefficiently cleared at the site of infection, there is a resulting abundance of proteolytic enzymes such as elastase, matrix metalloprotease 8 as well as inflammatory mediators [76, 77]. Interestingly, elastase then degrades neutrophils CXCR 1 receptors, fragments of which stimulate TLR2 receptors to produce proinflammatory cytokines [78, 79].

This completes the cycle that produces the consistent and chronic biochemical pattern which is associated with most biofilm disease, and has been clearly identified in wounds. The result of these unregulated proteases and proinflammatory cytokines is to produce sustained inflammation.

Wound inflammation is perpetuated, on the one hand, because of neutrophils disintegration in situ producing excess proteases and proinflammatory cytokine. On the other hand, there is clearly a downregulation of tissue derived inhibitors of this inflammatory response. There is a communication breach in the immune system at the site of chronic infections which produces a major protease/antiprotease and balance [80] which is generally seen throughout biofilm disease [81].

For example *P. aeruginosa* has the ability to chemotactically recruit neutrophils, commandeer the incoming neutrophils, "generate an inflammatory milieu that turns neutrophils proteases against their comrades, disabled neutrophil mediated bacterial killing and impair the neutrophils response to host survival signals"[79]. This is a redundant and comprehensive hijacking of the host's most important component of innate immunity.

General Mechanisms to Maintain Chronic Inflammation: Proteases

Chronic wounds are highly proteolytic. Although bacteria can produce numerous proteases, chronic wounds seem to possess mainly host derived proteases [82]. Studies have demonstrated that these proteases are mainly elastase, collagenase (MMP 8 and MMP 1), and gelatinase (MMP 2 and MMP 9) mostly originate from the innate immune response. This elevated protease activity seems to be related to the presence of biofilm. In a vicious cycle, effete neutrophils release proteolytic enzymes (mainly elastase and MMP 8). Macrophages are less able to recognize and remove neutrophils, thereby promoting neutrophil necrotic disintegration. The soluble fragments of host elastase-degraded CXCR 1 receptors can stimulate TLR2, producing additional proinflammatory cytokines to upregulate inflammation and recruit additional neutrophils. This perpetual cycle produces sustained elevated levels of inflammation which discourages healing [83].

The matrix metalloproteases, along with elastase, induce necrosis in host cell types producing collateral damage which impairs the wound repair process [84]. This highly destructive host milieu is even more remarkable for what it is missing, mainly tissue inhibitors of metalloproteases (TIMP). It is unclear if this decrease in TIMP is secondary to bacterial subversion of host wound bed cells responsible for its production, or the physical barrier created by the biofilms tight adherence to the wound bed's surface. It is clear that the decreased presence of TIMP is just as important to the elevated presence of proteases which creates the imbalance between protease and antiprotease seen in chronic infections. The pathological consequence of unopposed protease activity is sustained host tissue damage and chronic inflammation [85, 86].

Interestingly, many of these neutrophil released proteases may directly benefit wound biofilm. One example is *Staphylococcus epidermitis*, which requires activated "accumulation-associated protein" (AAP) to initiate biofilm formation. That is, host proteases have been found to proteolytically process AAP, which activates the bacteria's adhesion function [87, 88]. Rohde et al., concludes that, "It is therefore reasonable to assume that in vivo effector mechanisms of innate immunity can directly induce protein dependent *S. epidermitis* cell aggregation and biofilm formation, thereby enabling the pathogen to evade clearance by phagocytosis" [87].

We have explored one pathway of how neutrophils are usurped by biofilm to produce inflammation and release excessive proteases at the site of infection. It is possible that macrophages, dendritic cells, mast cells, and other immune cells that regulate the inflammatory response may also be commandeered by biofilm. Macrophages and dendritic cells can abandon their regulatory role, be confused by a potent invader and pursue a "delayed-type hypersensitivity" approach of "slash and burn" to confront the challenge [89]. When chronic wounds begin to heal they sometimes exhibit this behavior. This is observed clinically as the lysing of layers of the wound bed, tendon and fragments of bone, and the subsequent removal of these infected fragments by wound contraction whereby the material moves to the surface of the wound bed and is expelled. It is likely that chronic infections block or manipulate this potentially useful host defense of sloughing the surface [47].

The exact biochemistry and cellularity observed in a wound will depend on the species constituting the biofilm, expressed virulence factors, host immune competence and yet to be recognized factors. However, a successful pathogenic biofilm will need to maintain a constant source of nutrient rich plasma exudate by promoting hyperinflammation and prevent wound healing by maintaining high protease levels. Traditionally, a single pathogen is thought to be the only mechanism capable of causing infectious disease. But it may be possible that species of what may be considered traditionally "nonpathogenic" bacteria can maintain a chronic infection by working synergistically with other members.

An increase in genetic diversity within the biofilm is commonly associated with increased ability to withstand environmental stress. Based on what is understood about biofilm to date, a chronic infection that demonstrates a single predominant pathogenic species should be less robust and easier to suppress than genetically diverse biofilm. A genetically homogenous infection would be evidenced clinically by infection responding quickly and completely to a single treatment such as antibiotics. In contrast, wound biofilms require multiple simultaneous treatments that must be changed frequently during the course of healing [90]. The recalcitrance of some wounds to single treatment strategies may be explained therefore, by their diversity and generalized mechanisms of pathogenesis described earlier in this chapter.

Several microorganisms such as *S. aureus* and *P. aeruginosa* may possess all the functions necessary to initiate and sustain a chronic infection on their own. A functionally equivalent diverse population may require several cooperating species to achieve the same degree of infection. One species within the population may attach to host epitopes; a second species may self-secrete or organize host components into a protective matrix; and a third species may perpetually release lipopolysaccharide or seeding dispersal, producing a perpetual inflammatory state. If the appropriate species can coexist, then a genetically diverse biofilm with a pathology that is functionally equivalent or superior to a single successful pathogen may be able to form a robust and recalcitrant chronic infection.

Pathogenic biofilms cause the host immune system to violate its very important laws of universality, tolerance and appropriateness. The principle of universality is mainly exhibited in the adaptive immune system with its random sequence generators. These almost endless sequences of amino acids result in a vast number of T-cell receptors, B-cell receptors and antibodies that can recognize a "universal" number of antigens. However, a biofilms' colony defenses can prevent penetration of white blood cells and antibodies [10, 91, 92]. This renders the host immunity "universally" irrelevant in chronic infections.

Tolerance is not only applied to self antigens but to commensal and mutualistic bacteria. Our mutualism with certain bacterial species is permissible in very specific host locations and is complex, but is clearly governed by the principle of tolerance. Biofilm does not promote the process of tolerance. In fact, biofilm inhibits the host from down regulating its immune response or developing any sort of tolerance to the prolonged presence of an infectious biofilm. Pathogen associated molecular patterns, super antigens, degradation of CXCR 1 receptors, seeding dispersal and generalized bacterial factors (LPS, bacterial DNA fragments) perpetuate the destructive level of host inflammation.

And finally, the lack of appropriateness is what sets biofilm disease apart from other infections. Acute infections that have rapidly growing bacteria tend to destroy host tissue through bacterial derived virulence factors. However, chronic biofilm infections can destroy tissue by using host originated products such as proteases inappropriately. Biofilm may regulate host protease production in a chronic infection to produce the destruction of host tissue. Regardless, host immunity is acting inappropriately to destroy host cells.

Many infectious disease experts believe that the destructive level of proinflammatory cytokines and proteases in chronic infections are present because of abnormal immune responses. Suggested causes of these dysfunctions include abnormal host genetics, repetitive trauma, malnutrition or poor perfusion. Currently recognized immune "abnormalities", such as dysfunctional dendritic cells [93], aberrant T-cells [94], reduced TMP 4 and excessive neutrophils, may be better explained by the presence of bacterial pathogens or functionally equivalent pathotypes that organize into biofilms which commandeer the host immune response.

Biofilm Infection

The literature shows a significant dichotomy in the strategies which bacteria pursue at the molecular level. Yet, the literature on bacterial infection is somewhat confusing. The information relative to infected organisms show that in multiple studies when microorganisms establish an infection, there is a significant upregulation in multiple virulence factors [95] while another group of studies of the same microorganism will clearly show that the infecting microorganism will downregulate virulence factors (pyocyannin) [96, 97].

These findings have been contradictory and placed scientists at odds until it became clear that bacteria were pursuing two fundamentally different modes of infection. In one mode of infection, bacteria adopted a planktonic strategy up regulating virulence factors, destroying host tissue via virulence factors and using that tissue for nutrition to sustain itself. A second mode, not as well understood, is microorganisms pursuing a biofilm phenotype strategy forming a colony, down regulating virulence factors, inflaming host tissues to allow the percolation of plasma through the biofilm for its nutrient source. Neither strategy is exclusive of the other.

The first clear model of biofilm infection was put forward by Costerton, Stewart and Greenberg describing biofilm in a cystic fibrosis model [98]. Peg Dyrckx captured this brilliantly in the accompanying illustration which demonstrates the metamorphosis of a planktonic bacterium into a biofilm infection. Kim et al., documents the recent emergence of medicine's nascent understanding of biofilm stating, "We need to understand that bacteria pursue two strategies one to breach and one to maintain on the surface" [99]. It is of critical importance that we fully grasp the importance of this paradigm shift.

For example, using a burn mouse model, Schaber and Rumbaugh demonstrated that a biofilm develops within hours of seeding with planktonic bacteria [100]. The spread of the biofilm infection took place at the capillary basement membrane of the peri-wound vasculature. This is counterintuitive as it would seem more beneficial for the bacteria to select poorly perfused tissue in order to avoid renewable host immunity. However, biofilms are well fortified against most immune strategies, so they may select capillary basement membranes as a source of nutrients.

Clinicians have long understood and taught that there are two types of infections affecting the host. Acute infections are best explained by the properties of planktonic bacteria. That is, individual bacteria breach the host's primary defenses, multiply within the host, secrete virulence factors which kills host tissue, then produce microbial derived proteases which breakdown that tissue into a nutrient source to perpetuate the infection. This is a predatory type of infection which is very rapid and culminates either in death of the host or the host being able to regain control resulting in the eradication of the microorganism. Clinically, acute infections are seen as infections that occur rapidly, tend to progress aggressively yet respond to appropriate antibiotics rapidly and completely. Acute infections are usually measured in days or weeks and once they demonstrate no clinical effect are felt to be "cured."

Chronic infections, on the other hand, are quite different. Of the chronic infections involving the different tissues of the body many have been clearly correlated with biofilm phenotype bacteria. Biofilm characteristics tend to produce a more parasitic type of behavior. The bacteria breach primary host defenses but rather than pursuing a nutrient source of killing host tissue and degrading it, biofilm seeks to establish a sustainable niche. Therefore, biofilm quickly attaches to host epitopes, secretes an extracellular polymeric substance, incorporates host components, downregulates virulence factors, and secretes soluble effectors to commandeer host immunity and maintain the biofilm host interface.

Diagnostics

Dentistry, industry and environmental scientists have long understood the importance of biofilm phenotype bacteria on their disciplines. In fact, most nonmedical disciplines have been dealing with the biofilm problem for many years and yet it is just in the last decade that medicine is turning its attention to biofilm phenotype microorganisms. It is not that biofilm infection in humans is an insignificant problem, it is just that even with the immense technology brought to bear on human disease, the biofilm connection was just never made. The delay in fully recognizing the importance of biofilm may lie in the beginnings of medical microbiology.

Robert Koch, utilizing the very limited technologies of his days, understood that infectious disease was very complex, what he termed chaos [101]. To make sense out of this chaos Koch managed to develop a technique termed "pure culture" still the basis for medical microbiology today. "But others found it nearly impossible. No matter how ingenious the machinery, how careful the researchers, they kept ending up with beakers of mixed bacteria. The inability to get anything but mixed cultures let many scientist to believe that bacteria have to be in mixed groups in order to thrive…" [102]. We can easily see that "pure culture" was a work around for the limitations of the technology available in 1860.

Koch developed the method of pure culture to identify the one organism producing the infection. For acute episodic infectious diseases such as anthrax, cholera and tuberculosis, pure culture was adequate. For polymicrobial infections, however, this was not the case. The fatal flaw was assuming that all infectious disease was produced by a single organism and that postulation has carried over even to the modern era. Even though for years many chronic infections have been identified as polymicrobial, the pillar of Koch's postulate, one microbial organism produces one disease, prevented the adoption of biofilm methods.

Traditional culture based methods of pathogen identification utilize samples from the chronic infection and typically can only grow limited number of the microbes present (less than 5%). Cultures cannot grow "viable but not culturable" biofilm phenotype microorganisms but rather have a tendency to selectively amplified microorganisms which do grow [103]. Herein lies the fundamental issue associated with chronic infections-the need for a paradigm shift in clinical analysis to a broader and more thorough method which elucidates the complete microbial bioburden.

Polymerase chain reaction (PCR) is now a well understood culture free molecular method to amplify specific target DNA sequences [104]. Briefly, the PCR primers can be designed and selected to identify segments of DNA unique to the pathogen genome or for sequences of mobile genetic information such as the mecA cassette (target) that is suspect in the chronic infection. The microbial DNA possibly including the target segment is extracted and purified directly from a physical sample [105]. The sample is placed into a PCR device where it undergoes a series of enzyme-mediated target DNA amplification steps until enough of the specific target sequence is generated to be detected [106]. This process has been improved upon by quantitatively detecting these pathogen specific DNA targets through appropriate normalization in order to determine organisms contribution to the chronic infection [107]. Clinical diagnostic laboratories can use a custom wound pathogen PCR panel to perform quantitative PCR (qPCR). This panel is used to generate information on the percentage of pathogens in the wound as well as primary antibiotic resistance of selected organisms (e.g. mecA cassette for *S. aureus*) [108]. As described elsewhere, qPCR has the capability to quantitatively determine the microbial composition of a chronic infection in less than two hours [109]. The speed of analysis allows clinicians to monitor the microbiota of the wound and the efficiency of the therapy being applied, thus improving patient care and reducing overall costs by supporting more efficient antibiotic selection [110].

Bacterial and fungal tag-encoded FLX amplicon pyrosequencing methods (bTEFAP and fTEFAP, respectively) have been developed to elucidate the microbiota of chronic wounds regardless of the phenotype. [5, 111–113] Similar to qPCR, unique genetic information is used in the bTEFAP and fTEFAP next generation molecular diagnostic technologies to identify micro-organisms [108, 112]. Unlike qPCR, the universal molecular probe sets for either bacterial or fungal small ribosomal subunits allow the identification and quantification of most unknown bacteria or fungi in a sample [114]. Therefore, the analyst does not need to predetermine a select group of target organisms in order to detect and identify it. Though significantly more powerful, the bTEFAP and fTEFAP techniques are more costly and typically take several days to complete. Therefore, should a qPCR panel through the use of universal primers show the presence of significant microbial DNA as a first method of analysis but cannot specifically identify them, the sample can be referred to the bTEFAP and fTEFAP methods of analysis for explication. This illustrates the value of combining molecular platforms to more fully define chronic infections (Table 3).

Table 3. Hospital isolated microorganisms by sequencing

Microorganism	# Patients	# Occurrences	Microorganism	# Patients	# Occurrences
Finegoldia magna	75	82	*Candidatus Peptoniphilus*	11	13
Pseudomonas aeruginosa	74	92	*Clostridium hiranonis*	11	12
Staphylococcus aureus	73	79	*Fusobacterium nucleatum*	11	12
Staphylococcus epidermidis	71	84	*Parvimonas micra*	11	12
Anaerococcus vaginalis	45	48	*Prevotella buccalis*	11	12
Corynebacterium striatum	36	42	*S. piscifermentans*	11	11
Enterococcus faecalis	36	40	*Terrimonas ferruginea*	11	11
Serratia marcescens	34	36	*Burkholderia ambifaria*	10	10
Anaerococcus lactolyticus	33	35	*Corynebacterium jeikeium*	10	13
Propionibacterium acnes	28	28	*Peptoniphilus lacrimalis*	10	12
C. tuberculostearicum	27	32	*Staphylococcus capitis*	10	11
Pelomonas saccharophila	26	26	*Staphylococcus hominis*	10	10
Peptoniphilus indolicus	24	27	*Prevotella melaninogenica*	9	10
Streptococcus agalactiae	23	24	*Acinetobacter baumannii*	9	9
Escherichia coli	19	19	*Staphylococcus caprae*	9	9
Peptoniphilus ivorii	19	19	*Bacteroides fragilis*	8	9
Anaerococcus octavius	17	17	*C. aurimucosum*	8	8
Ralstonia pickettii	17	17	*Porphyromonas levii*	8	9
Streptococcus mitis	17	18	*Prevotella bivia*	8	9
Porphyromonas somerae	16	17	*Acinetobacter junii*	7	7
Anaerococcus prevotii	14	17	*Bacteroides thetaiotaomicron*	7	7
Peptoniphilus harei	13	14	*Candida albicans*	7	8
Anaerococcus hydrogenalis	12	12	*Staphylococcus lugdunensis*	7	7
Corynebacterium xerosis	12	13	*Streptococcus parasanguinis*	7	7
Pseudomonas hibiscicola	12	13	*Streptococcus sanguinis*	7	7
Ruminococcus obeum	12	12	*Streptococcus thermophilus*	7	8
Staphylococcus haemolyticus	12	12	*Veillonella parvula*	7	8
Stenotrophomonas maltophilia	12	12	*Actinomyces europaeus*	6	6

Note: only the top 56 organisms were reported although 584 organisms were identified.

It is now well documented within various scientific disciplines that traditional microbiological culturing methods of analysis are limited to the detection of a very small percentage of naturally occurring bacteria, which includes the bacteria found in chronic infections [115–117]. Traditional cultural methods of analysis are typically accomplished by enriching the sample bacteria in various laboratory media, isolating individual colonies after a period of growth, and then characterizing them biochemically to determine identity. This type of analysis has several drawbacks, in some instances due to the differences between laboratory conditions and those of *in vivo* environments. There are obvious identification biases that form for bacteria that readily reproduce in the laboratory culture media [118, 119]. Although common clinical pathogens (*Staphylococcus*, *Pseudomonas*, *Enterococcus* spp., etc) may only represent a small percentage of the microbiota within a biofilm sample, they will falsely dominate a culture analysis in the lab. Analogously, a major *in vivo* component of the same biofilm may only propagate under very specialized laboratory conditions, and therefore be underrepresented or completely missed by a standard cultural protocol [107].

We instituted molecular diagnostic methodologies in our clinic because many of "surgical site infections" that continually were reported as "no growth" by the microbiology laboratory. Also we observed the same 5-10 bacterial genera (*Staphylococcus*, *Pseudomonas*, *Enterococcus*, *Klebsiella*, *Streptococcus* spp., etc) always reported as the cause of infection in chronic wounds (diabetic foot ulcers, decubitus ulcers, and venous leg ulcers). Yet when we treated based upon these culture results we did not see appreciable improvement in the healing trajectory. In fact, the literature suggests that managing chronic wounds based on culture results does not improve wound healing outcomes [120, 121]. We soon realized from the DNA based results that our training and trust in clinical microbiology culture results had been inherently flawed especially in regard to polymicrobial biofilm infections.

Molecular methods such as PCR and newer sequencing technologies have an immense advantage over cultivation methods for the evaluation of clinical infections. These molecular methods are far more accurate, faster, and able to identify bacteria that cannot be grown routinely or easily in the clinical microbiology laboratory, can evaluate genetic antibiotic resistance factors, detect bacteria which are otherwise viable but not culturable, evaluate bacteria that are in biofilms, and add the indispensable piece of information of quantification. Molecular methods are more complete in that all DNA present is used to identify the infection rather than selecting out the single organism which can be easily propagated in the laboratory to be credited as the cause of the infection. By understanding each of the microbial species present within the infection, a clearer picture can be obtained and a better treatment regimen may be instituted.

Molecular methods have emerged to address the problem of selection bias and inability to grow certain species and especially biofilm phenotype bacteria. The molecular methods have been borrowed from the environmental scientists who, early in their surveys of the environment, realized that they could see many more bacteria than could be grown. Molecular methods revealed a very robust, complex, and diverse microbial reality for chronic infections. Utilizing molecular methods for chronic infections will continue to identify diversity and complexity which will dramatically impact how infectious diseases are managed.

There are several arguments that arise from the echelon which like "cultivation based technology" of culture. This group rightfully point out that direct assessment of antibiotic sensitivities cannot be obtained by molecular methods. Also, there is the possibility that nonviable and/or naked microbial DNA may give a false result of the microorganisms present in the chronic infection. Some worry that the massive amount of information obtained through PCR, sequencing and other modern diagnostic tools cannot be made clinically relevant, but overwhelmingly the concern is that molecular methods are identifying mainly "contaminants" causing inappropriate use or even worse over use of antibiotics.

Cultivation methods actually grow bacteria from the source of chronic infection. However, as noted above, there is clear selection bias and an inability to grow most organisms. Yet even when microorganisms from the chronic infection propagate in the artificial setting of the Petri dish they are growing in planktonic phenotype. Therefore antibiotic sensitivities are limited value. Far more valuable is the molecular determination of mobile genetic information (mecA cassette, ESBL, van, tet, etc.) which yields valuable insight as to actual resistance within the biofilm infection.

Microbial DNA within a host setting is rapidly degraded [122]. Ehrlich demonstrated in a chinchilla ear model that DNA contributed from Bacteria rapidly degrades [123] It is clear that any microbial DNA that can be detected from a chronic infection should be considered as being contributed by a viable microorganism.

There is no question that moderate methods whether molecular microorganism identification, biomarkers, imaging or any of the emerging technologies, will produce overwhelming information to the clinician. However, bioinformatics methods have already been developed to reduce this information into clinically relevant formats. For example, the large number of microorganisms identified in chronic infections can be evaluated at a computer level. First criteria can be set for which species to not treat and just watch. For example, microorganisms that are in very minimally represented can be reported but not included in treatment regimens. Also, microorganisms that are treated the same can be grouped. The final report of these results then only suggests a handful of treatments allowing the information to be clinically useful.

The overwhelming pushback against molecular methods is that it is identifying microorganisms that are not contributing to the chronic infection. We have taken a position that if a microorganism can propagate itself in a host location that is producing a lesion clinically significant enough to warrant culturing that it must be considered clinically relevant. Most of what we have come to consider pathogens have developed from the incomplete knowledge we have obtained from clinical cultures. Therefore, until there is clear scientific evidence that a microorganism cannot contribute to a chronic infection, it cannot be considered a "contaminant."

Also, the emergence of new imaging technologies such as scanning electron microscopy, transmission electron microscopy, confocal laser scanning microscopy, atomic force microscopy, and many other imaging technologies are now defining an amazing array of architectures and three dimensional structures which will radically redefine microbial infections. These imaging technologies are defining a much more complex host pathogen relationship and multiple different strategies which microorganisms can pursue to establish and maintain infections.

Other advanced technology such as metabolomics, proteomics, transcriptomics and metagenomics hold great promise not only for understanding biofilm infection but also for a development of much needed diagnostics. These developing disciplines have already characterized important pathways of microorganisms, the host and their interaction that allow us to better understand biofilm infection and where medicine can intervene.

Metabolomics, the study of the unique molecular products that specific cellular processes produce, is useful for the quantitative analysis of molecular phenotypes. Multiple advancements in mass spectrometry and hyphenated techniques (chromatographic and spectral methods) along with bioinformatics has caused metabolomics to be in the forefront of what some call as the post-genomic era [124].

Proteomics, the large-scale study of proteins, particularly their structures and functions, has also benefited from technological advances. Proteomics holds great promise in defining not only host pathways but also bacteria in chronic infections [125, 126]. Protein centric investigation should be able to identify molecular markers/targets for diagnosis, therapy and prognosis. Multiple chronic infections are being evaluated including a focused proteomic studies on chronic wounds [127].

Transcriptomics is the set of all RNA molecules, including messenger RNA, ribosomal RNA, transfer RNA, and other non-coding RNA produced in a population of cells. Because of the human genome project, remarkable technologies have emerged allowing comparative genomics to merge with host pathogen transcriptomics allowing the clear definition of different pathogenic phenotypes and processes. [128] At the end of the day, these technologies will define a much more robust microbial reality than has even been postulated today.

Treatment

Medical Biofilm Management

Whatever regimen is chosen for the management of a chronic infection, to be successful it must contain the simultaneous use of multiple strategies targeting different aspects of biofilm development and physiology. Any single strategy is doomed to failure because biofilm possesses great complexity and the ability to use its different species members in any way that allows the colony to survive. Therefore, it seems important to use multiple strategies simultaneously to attack the biofilm, and then when the biofilm community adapts to those challenges, it is important to change management strategies. We have coined the phrase "biofilm-based wound care" to convey the principles of multiple concurrent strategies applied dynamically. These strategies include use of physical disruption, anti-biofilm agents, and bacteriocidal agents such as antibiotics, biocides and/or bacteriophages. Managing chronic infections is so very different than the "one and done" methods of treating acute infections. The clinician who takes on the challenge of managing chronic infections must have a clear understanding of all biofilm properties, have multiple diagnostic and therapeutic tools, and most important be persistent.

Attachment is the first committed step of biofilm formation. Therefore, when new tissue is exposed, either through trauma or physical disruption of the Medical Biofilm on the host surface, blocking attachment/reattachment becomes a primary strategy. Multiple different methods have been used including lavage, lactoferrin [129], surfactants, materials [130] and direct electrical current [131, 132]. By interfering with the attachment process for just a brief period of time even impaired host defenses are given an opportunity to clear the planktonic phenotype bacteria before biofilm is established.

If attachment is successful, the bacteria quickly upregulates biofilm behaviors secreting extracellular polymeric substances along with commandeering host products/cells and rendering senescent local eukaryotic cells. The bacteria divide in the protection of this matrix until a sufficient number present to produce quorum sensing signaling, which produces the architectural structure of mature biofilm. It is quorum sensing pathways which create the 3-dimensional structure arranged with different phenotypes in the different regions of the biofilm [133]. These regions can be roughly divided into thirds with the outer third containing a phenotype which is more planktonic and, in fact, it is this region that is responsible for seeding dispersal as well as interaction with the environment [53]. The middle third of the biofilm is spatially and physiologically optimized for horizontal gene transfer, and respond to stresses placed on the biofilm. The innermost region of the biofilm tends to be anoxic and metabolically challenged due to waste products. The phenotype of individual bacterium residing in this area is most consistent with stationary phase and also contains high numbers of persisters [134]. It is clear that if quorum sensing pathways can be disrupted, this formidable organization of the community constituents can be suppressed. This makes the field of quorum sensing inhibitors a very exciting potential for novel treating agents of chronic infections [135–138].

Quorum sensing inhibitors have the unique property of only affecting the biofilm behavior of bacteria. Quorum sensing molecules are specific signaling agents that control the behavior of biofilms by regulating the gene expression of each of the colony bacteria. The original name given to these signaling molecules was quorum sensing molecules because it requires a critical density of individual bacteria (quorum) to trigger biofilm gene expression. In other words, a certain number of these molecules need to be present to cause upregulation or downregulation of suites of genes (operons and regulons) responsible for biofilm formation. It was clear from the beginning that if this communication language could be understood, certain behaviors could be manipulated (inhibited), thus averting the impenetrable defenses of biofilm.

As outlined in the previous section *S. aureus* produces RNA III Activating Peptide (RAP). If RAP accumulates in high enough concentrations near a receptor named Target of RNA III Activating Peptide (TRAP), that receptor is upregulated. This causes phosphorylation of TRAP, which in turn upregulates the agr regulon. The agr regulon is responsible for production of RNA III, but also contains a number of other genes responsible for biofilm behavior. RNA III works directly to upregulate productions of toxins such as hemolysin, enterotoxin B, and several other staphylococcus virulence factors. Therefore, it was hoped that by blocking the upregulation of the agr regulon, *S. aureus* would show less biofilm behavior and not cause the tissue destruction associated with its virulence factors [139–142].

A 7 amino acid peptide that blocks TRAP has been identified. Because it inhibited the effects of RAP and therefore decreased the expression of RNA III, it was given the name RNA III inhibitory peptide (RIP). RNA III inhibitory peptide is a gram-positive quorum sensing inhibitor. RIP blocks RAP and therefore the upregulation of the agr regulon. RIP is a true quorum sensing inhibitor blocking only biofilm behavior and does not kill planktonic bacteria.[143] Animal studies using mice with infected metal devices (biofilm) show that the mice treated with RIP were able to clear their biofilms with the appropriate antibiotics. On the contrary, mice not treated with RIP but only the appropriate antibiotics were not able to clear the biofilm from the metal devices [144, 145]. This is a very exciting step towards specific, well-understood agents that may be useful in a clinical setting.

For several years investigators have been looking to plants for new sources of quorum sensing inhibitors. It has been found that the conformational structure of hamamelitannin is very similar to RIP. In studies hamamelitannin has been shown to be an effective quorum sensing inhibitor [146]. Now, it is known that aspirin works to block the same receptor and has shown some clinical benefits in preventing *Staphylococcus* infections [147, 148]. It seems only logical that plants, with a very limited immune system, would develop these kinds of molecules. This may be the science behind the popular emphasis on phytosterols, tannins and plant derived antioxidants.

The biofilm matrix is one of the most formidable parts of the biofilm fortress. The matrix is comprised of extracellular polymeric substances secreted from the bacteria itself, as well as commandeered host proteins and products. The composition of the biofilm matrix can be mainly DNA, protein, or polymeric sugars. This requires multiple strategies to breach this defense. For host and bacterial derived proteins, topical proteases can be successful.

Treatment is difficult, however, because biofilm complexity is just now fully being defined and more importantly biofilm presents a moving target for therapeutic goals. The synergies which biofilm possesses, allows biofilm to change the composition of the matrix rapidly with the stress of any treatment. What this means functionally is, the biofilm that existed at the initial treatment will be significantly different than the biofilm community later in the treatment. Therefore, rapid, quantitative, and reliable methods must be utilized to frequently reevaluate chronic infections.

Biofilm in general and its developmental stages, are well documented [2, 149–153]. In short, planktonic (single cell) bacteria undergo a radical metamorphosis through attachment, micro-colony formation, and then through robust quorum sensing mechanisms and progress into mature biofilms with an organized matrix and heterogenously differentiated members [154]. Genotypic and phenotypic heterogenicity is a seminal survival strategy for mature biofilms [155–157]. Therefore it follows that if a biofilm's resistance increases as it matures, any procedure that physically disrupts the biofilm, such as sharp debridement, may create a "window of opportunity" by removing the bulk of mature biofilm and forcing the residual bacteria to reconstitute the biofilm. Clinicians may be able to exploit the short duration of time from the debridement procedure until the biofilm has reaccumulated and matured into a climax community as a therapeutic window, allowing the host to gain more opportunity to heal instead of fight infection. This is vital, especially in patients with co-morbidities, as fewer host resources may be available to contribute toward the physiological healing process.

Wound biofilm, or any biofilm adapted to a host niche, exhibits genotypic and phenotypic diversity [23]. This diversity is directly responsible for multiple biofilm defenses against antibiotics such as impeded penetration, decreased growth/metabolic rate, stress response and quorum sensing pathways. It is the simultaneous expression of these defenses in conjunction with the biofilm structure and then combined with the incalculable synergies between multiple microbial species that "can prevent the eradication of polymicrobial infections"[158] caused by biofilm. Impaired penetration caused by the biofilm self-secreted matrix may be much more important relative to biocides than to antibiotics [159–164]. Biocides, by nature, tend to be reactive and thus are consumed by the extracellular polymeric substance. Although antibiotics can react, they tend to be adsorbed, be limited by size or be electrostatically influenced by the matrix [162]. And, it is clear that most antibiotics are not perturbed significantly by the glycocalyx.[164–166] From a genetic resistance standpoint, Anderl et al., [167] demonstrated that beta-lactamase producing cells could protect non-beta-lactamase producing bacterium through accumulation of beta-lactamase within the extracellular polymeric substance. However, the matrix is not impenetrable to most antibiotics and thus can only be considered the first linc of defense.

Another defense is the slow growth/metabolic rate which protects against antibiotics which tend to be more successful against bacteria with highly active metabolism. Indeed, some of the most effective antibiotics target metabolic machinery or cell wall synthesis directly relating to cell metabolism or growth. Immature biofilms, however, are postulated to be highly active as the bacteria engage in reaching a climax community and ultimately a mature stable biofilm. Unless biofilms in chronic wounds are continually forced to remain in an immature state where they are more metabolically active, such as during reformation after debridement, they have the potential to be hundreds of times more resistant than their planktonic phenotype [162, 168].

Indeed the data from our in vivo wound biofilm and clinical biofilm maturity experiments also suggests that early biofilm reaccumulation in the host niche correlates with the pattern seen in our in vitro models. There was increased sensitivity to the antimicrobials tested in one day-old biofilms as compared with mature, three day-old biofilms. Thus, this increased sensitivity to young biofilms may be therapeutically useful. In all the models we can deduce that after 48 hours the biofilm can reach maturity and be very resistant to antibiotic therapy even with higher doses applied topically.

Wound Debridement

Debridement is the cornerstone component in every protocol for the management of every type of chronic wound [169–173]. Wound care specialists universally agree that necrotic tissue and other “harmful substances” negatively impact chronic wound healing [174–176]. In cases such as pyoderma gangrenosum, coagulopathies, and arterial disease, some specialists believe that the risks of debridement may outweigh its benefits. However, the wound care community has reached a consensus that debridement, if it can be done safely, is beneficial to wound healing. The universal acceptance of debridement has led to a number of different tools and techniques. These debridement techniques include sharps/surgical, mechanical/energy transfer (eg. pulse evac, ultrasound, laser), biologic (maggots), enzymatic, and autolytic debridement [177, 178].

Even though the use of debridement is ubiquitous, experts still struggle to understand at a cellular and biochemical level what healing advantages debridement bestows upon the chronic wound. Many theories have sought to explain how wound debridement improves healing, including removing senescent cells [174], balancing the bioburden [179], improving the microcirculation [176], and normalizing the biochemistry [175]. In fact, all of these observations may be correct, but they still do not explain how debridement produces these positive outcomes. We would be just as correct if we said that, clinically, more wounds heal with debridement than without, but we would still fail to understand the scientific explanation for these measurable benefits. Therefore, if we can understand the fundamental cellular and biochemical mechanisms by which debridement confers a competent healing response, we can rationally fine tune our debridement techniques and treatment regimens.

Biofilm-host Interaction

The biochemistry and cellularity seen on chronic wound beds may be explained by a biofilm-host interaction [180]. It has long been known that the wound bed of every chronic wound, regardless of the etiology, is defined by a persistent inflammation. Chronic wounds are stuck in a chronic inflammatory state exhibiting markedly elevated proinflammatory cytokines (interleukin 1, tumor necrosing factor alpha, gamma interferon), elevated matrix metalloproteases (MMP2, MMP8, MMP9), and excessive neutrophils [60, 181–184].

The biofilm-host interaction in cystic fibrosis has been well described. Costerton and Stewart reported that at the cellular and molecular level of cystic fibrosis, biofilm causes intense migration of neutrophils and macrophages to the area of the biofilm through stimulation of the innate immune system [185]. The pathogenic biofilm also produces collateral damage to the host tissue by increasing phagocytic enzymes, such as elastase and matrix metalloproteases. Other chronic infections possess the same molecular and cellular findings such as ulcerative colitis [186, 187], chronic rhinosinusitis [188] and chronic obstructive pulmonary disease [189, 190].

Debriding to Suppress Biofilm

Dentistry is decades ahead of the rest of the medical community in treating Medical Biofilms espousing frequent removal of dental plaque (biofilm) in 1965, 13 years before Costerton's landmark article defining biofilm [7]. Before the discovery of biofilm, dentistry clinically recognized the role of plaque in producing dental disease. Cavities and periodontal disease are treated quite differently than other Medical Biofilms. In dentistry, if maintenance debridement (brushing) fails, the response of the dental community is not to abandon brushing but to become more aggressive with removal [191]. Therefore, brushing is recommended at increased intervals; flossing and water pick are also often added to further suppress oral plaque. Toothpaste, a multivalent cocktail of antibacterial biofilm agents is then applied to continue to suppress and prevent the reaccumulation of biofilm. Toothpaste includes lactoferrin, xylitol, triclosan, fluoride, and other agents. Toothpaste can be combined with professional cleanings, biocide rinses, and other strategies. These are all done concurrently for the purpose of suppressing the oral plaque below the level at which it produces disease. Antibiotics are often used as adjuncts in periodontal disease, but none of us would continue to utilize a dentist that prescribed only antibiotics and recommended against brushing, flossing, or other debridement techniques. It is universally understood among the dental community as well as the patients that they serve that frequent removal of oral plaque is critical to the management of oral biofilm disease.

Some members of the medical community have come to the same conclusions about the necessity of wound debridement. Debridement has been validated by one randomized control trial [192] and has recently been validated by review of data from other studies [193, 194]. This rather lean literature is in contrast to the universal acceptance of debridement among wound care professionals.

Healing Nonhealable Infections

If a patient's wound demonstrates ischemia, osteomyelitis, or other severe impairments, many doctors will recommend amputation. Also, if a wound does not heal after several weeks of debridement, doctors may also choose to amputate. However, amputation is a failed strategy. Several studies have demonstrated very high 5-year mortality rates (74-80%) for diabetic patients who undergo major limb amputations [195, 196]. Armstrong suggests the mortality and the suffering imposed by major limb amputation rivals cancer [197]. Studies suggest age matched norms with diabetes and peripheral vascular disease without amputation have a 5-year mortality rate of 25-30% [196, 197] so amputation more than doubles the patient's mortality rate. But these patients do not just die. Quality of life assessments show that they rate their "misery" index as worse than patients that have cancer. Also, up to 50% of these patients will lose their contralateral limb within 3 years. A diabetic who undergoes a major limb amputation suffers, leaves this world a piece at a time, and dies at a younger age than if his or her limb had not been amputated. Medicine's justification for major limb amputation is that it prevents sepsis and the death of the patient. However, sepsis is a rare complication for chronic infections in general and specifically in wounds [198, 199].

Physical removal of biofilm is the mainstay of biofilm management. Experience and the literature have shown us that debridement is universally beneficial to wounds. Frequent, aggressive debridement is the primary tool for healing chronic wounds because of a number of positive effects it has on the wound bed. Callous must be removed from around the wound because it prevents wound contraction. By removing epithelial cells from the edge of the wound, which are proliferative but not migratory, more normal keratinocytes are exposed to the wound bed [200]. It has also been stated that the wound bed can rarely be "over-debrided" because the removed reactive granulation tissue, fibrosis, and even "normal" host tissues (fat, connective tissue, capillaries, etc.), is always replaced, at least transiently, by non-senescent cells. By removing "senescent" fibroblasts and other types of abnormal wound bed cells, more normal host cells are exposed that can synthesize, proliferate, and migrate in a more normal healing pattern.

The first priority in debridement of a chronic wound is to alter its anatomy to favor the host over the biofilm. This requires opening all undermining and tunnels, removing all devitalized and marginal (poorly perfused) tissue, as well as shaping the wound topography. Many wounds possess vegetative growths (polyps), crevices, or other niches where wound surfaces touch (overlap). These types of biofilm-friendly environments need to be sharply removed, all with the goal of providing a well-perfused wound bed free of any environmental niches that favor biofilm.

Eradication of biofilm is not possible. In oral biofilm, once a patient gets control of cavities or periodontal disease, the strategy is to maintain the oral biofilm at a level that does not cause disease again. Similarly in the wound, once the anatomy is altered to favor the host and the wound begins healing, a maintenance level of debridement is necessary. The goal of this maintenance debridement is to suppress wound biofilm below a level that causes disease (Figure 1).

For any chronic infection it will be important to disrupt the biofilm matrix. It is relatively easy to stress the biofilm producing periodontal disease, wounds and external infections but most chronic infections reside deep within the patient. Even with current therapeutic tools, reaching biofilm on a heart valve, in a lung or other internal infections is just too difficult.

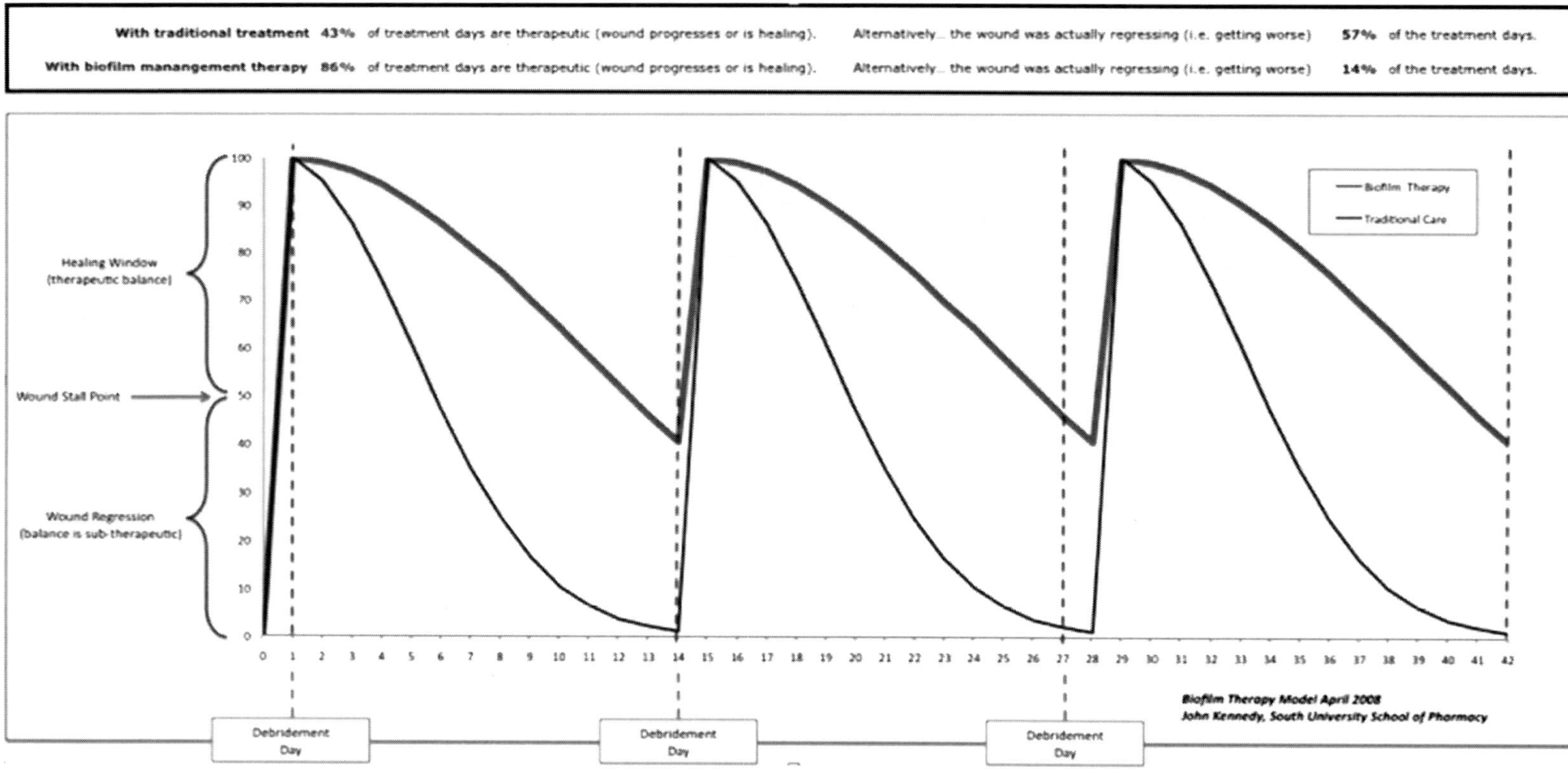

Figure 1. This diagram demonstrates that by providing treatment simultaneously biofilm is suppressed below point by which it can inhibit wound healing. Suppressing biofilm below this point is the goal for managing chronic wounds.

This may require enzymatic, chemical, bacteriophage products or other strategies to degrade the biofilm matrix. However, once the matrix is breeched it is clear that bacterialcidal agents will be important along with the anti-biofilm agents mentioned above.

Bactericidal Agents

Bacteriophage

Bacteriophages are naturally occurring, ubiquitous viruses that specifically infect bacterial cells. It is estimated that 10^{31} phages exist on earth - ten-fold more than their associated hosts. These bacterial viruses come in many shapes and sizes, but the most commonly used therapeutic phages are the tailed bacteriophages. These phages can have a lysogenic or lytic lifecycle. Upon infection, lysogenic phages can incorporate into the host genome and perpetuate their genetic information through host cell division. The lytic phages do not have the ability to incorporate into the host genome, so they can only reproduce their genetic information by replicating within the host cell and subsequently lysing the cell to release the newly assembled virions. These lytic phages have been used extensively in medicine for almost a century, but their use in the United States and the Western World diminished as the use of penicillin commenced [201]. Therapeutic bacteriophages have been administered topically, orally, as an inhalant, and intravenously.

There have been no reports of serious complications associated with the use of phages. Some minor side-effects during phage therapy may be triggered by the *in vivo* lytic activity of phages against the etiologic agent. For example, liver area pain in a patient undergoing phage therapy in Poland was suggested to be related to extensive liberation of endotoxins from the lysed pathogenic bacteria [202], which can be similarly observed during antibiotic therapy. During the 1970s-1990s in the United States, purified phage phi X174 was extensively used to monitor humoral immune function in humans because the virus is detectable by the immune system but does not cause disease. Recipients of this phage, included healthy volunteers, patients with Down's Syndrome [203], Wiskott-Aldrich syndrome [204], HIV infections [205], and other immunodeficiency conditions - without any apparent side effects.

The onset of increasing numbers of antimicrobial resistant infections has brought with it a burgeoning interest in alternatives to traditional antibiotics, including a renewed interest in bacteriophage therapy. Scientists understanding of bacteriophage biology has matured over the last several decades, which gives clinicians hope that newly developed treatment regimens involving bacteriophages can be better designed than ever was possible in the past.

Doolittle et al., [206] demonstrated that bacteriophages can infect and replicate within *E. coli* biofilms, and similar findings were later reported [207]. Studies of phage-biofilm interactions have been performed with other bacterial pathogens, including *Enterobacter agglomerans*, *Listeria monocytogenes*, *S. aureus*, *Yersinia ruckeri*, and *P. aeruginosa*. The activity of bacteriophage polysaccharide depolymerase was examined in 1956 by Adams and Park. [208] Hughes et al., [209, 210] later identified that similar phage polysaccharide depolymerases work to degrade the polysaccharide component of the extracellular matrix of some biofilms. Dispersin B has good potential to become a commercially available bacteriophage product to degrade certain extra polymeric polysaccharides [211]. However, the biofilm matrix is complex and these depolymerases will need to be directed by molecular bacterial identification and/or molecular characterization of the matrix.

These depolymerase enzymes facilitate the penetration of phage into a biofilm and disrupt the biofilm by a combination of (i) extracellular matrix degradation by the depolymerase, and (ii) infection and subsequent bacterial lysis by the phage. Upon lysis of the bacterium, more phages are released that are capable of infecting and lysing neighboring bacteria. Therefore, the use of bacteriophages as an antimicrobial agent would theoretically concentrate the antimicrobial agent (i.e., bacteriophage) within the location of highest bacterial concentration. Bacteriophages can act as self-replicating, self-concentrating antimicrobials.

Bacteriophage therapy is a common form of antimicrobial therapy in countries such as the Republic of Georgia and Poland. In these countries, phage therapy has empirically proven to be an effective and economically sound alternative or adjunct to antibiotic therapy. However, the West is still only considering the re-entrance of phage therapy into its medicine because of the limited number of modern randomized-controlled trials (RCTs) that have been performed using bacteriophages. Unfortunately, until these expensive and time consuming RCTs are performed and prove the utility of bacteriophages as an antimicrobial agent, this alternative to conventional antibiotics will not be widely available.

Biocides

Nonspecific biocides include compounds such as alcohol, peroxide, acetic acid, and iodine. These types of bacteriacidal agents cause much damage to host tissue as to the biofilm. Selective biocides such as silver and cadexomer iodine possess very little host cell toxicity but demonstrate significant killing power to wound biofilm. The best studied selective biocide relative to wound biofilm is ionic silver.

Silver has been used for hundreds of years in many applications, and in its ionic form (Ag^+) it has been shown to be effective against a broad range of microorganisms [212, 213]. Consequently, products containing or coated with ionic silver are being used extensively in healthcare to bolster the antimicrobial effects of the device. Such products now include wound bandages, water systems [214], and silver-coated catheters [215].

Products that control the release of iodine such as cadexomer iodine markedly decreased host toxicity while maintaining the biocidal effect towards biofilm. The same principle can be used for methylene blue and crystal violet (Hydrofera Blue®). It is conceivable that selective biocides (either narrow or broad spectrum) could be developed to target internal biofilms.

Antibiotics

Antibiotics, both topically and systemically, have historically been used in the management of clinically infected wounds (color, dolor, tumor, rubor), but there has been no agreement on what constitutes an infected wound. The classic signs of Celsus are inadequate for diagnosing infected wounds, so additional signs such as delayed healing, increased exudate and alteration in pain (amongst others) have been suggested to be used for diagnosing infection. [54] The difficulty in determining an "infected" chronic wound may be best explained by biofilm.

Although chronic infections are predominantly produced by biofilms, systemic antibiotics still play a role. There is some evidence that the metabolically active environmental edge may be susceptible to a number of antibiotics [216]. Systemic antibiotics have been shown to help decrease drainage in the wound, thereby decreasing the nutrient source to the biofilm. The problem with systemic antibiotics is they cannot eradicate the biofilm. This suggests that if systemic antibiotics are used they will need to be used as adjuncts to topical antiseptic and anti-biofilm strategies. Combining anti-biofilm strategies with traditional systemic strategies in chronic infections can be used to weaken the biofilm until it reaches a level at which the impaired host's immunity can fight off the chronic infection.

As mentioned previously wound biofilm is polymicrobial and has contributions from both aerobic and anaerobic species [217], Gram positive and Gram negative species, as well as contributions from various phenotypes within each species. Therefore, the antibiotic sensitivity of a biofilm is heterogeneous. This diversity has led to several different strategies in managing wound biofilm with systemic antibiotics.

One strategy is to use systemic broad-spectrum antibiotics together with broad spectrum topical antiseptics, which may help to suppress the multiple subpopulations within the wound biofilm. For example, Ertapenem has been demonstrated when administered intravenously to improve outcomes in diabetic foot ulcers [218]. It may work by suppressing multiple species within the community [219].

A second systemic usage of an antibiotic strategy suggests that targeting the treatment towards specific subpopulations may have some promising results. Some suggest that the presence or prevalence of certain species or guilds of bacteria (e.g. methicillin-resistant *Staphylococcus aureus* MRSA) correlates with wound outcomes [220]. Some researchers have suggested that if antibiotics can effectively target keystone organisms, then the bacterial community in the wound ecosystem will collapse [221].

A third antibiotic strategy could be to use intravenous administered antibiotics that are known to be effective against biofilm bacteria such as antibiotics which insert into the cell membrane [222]. For example, daptomycin and linezolid have been shown to affect some biofilm infections [222, 223]. Topical antibiotics have proven to be superior in managing external biofilm infections such as in the external ear canal, wounds and burns. Common myths about topical antibiotics are that they will increase allergies to the antibiotics and cause an explosion of antibiotic resistance. Another myth is because topical antibiotics are so concentrated they can cause systemic toxicity. However, a 30 day supply of topical antibiotics contains less drug then one systemic intravenous dose of the same antibiotic. There is little evidence in the literature that demonstrates increased hypersensitivity and clearly systemic use of antibiotics is the driving force for antibiotic resistant bacteria.

The advantages of topical antibiotics are they can be compounded to specifically address the major populations of microorganisms and a chronic infection. Also these antibiotics can be delivered in concentrations of 500-1000 times MIC which is what is required to target biofilm phenotype bacteria. Also topical antibiotics can be combined with anti-biofilm agents which degrade biofilm defenses making the topical antibiotics more effective. And topical antibiotics, even in such high doses are much less likely to produce adverse events when compared to systemic antibiotics [224]. Now there is clear clinical evidence that the use of topical antibiotics guided by molecular diagnostics can significantly improve the outcomes for the treatment of chronic infections [225].

Whatever strategy is used, it is clear the use of antibiotics alone will not eradicate wound biofilm and as wounds are polymicrobial, antiseptics with a broad spectrum of activity are clearly more effective topically. Therefore, systemic antibiotics for use in infected wounds should be used in conjunction with multiple strategies including debridement and anti-biofilm agents.

Conclusion

Medical microbiology is at a crossroads. It is almost unimaginable to think, given Medicines current technological capabilities, that the most important responsibility of Medical microbiology-the identification of infecting microorganisms-is relegated to a methodology that is over 150 years old. The use of clinical cultures becomes even more surreal when we consider "pure culture’s" history (developed as a work around) along with the unacceptable shortcomings of this method in diagnosing biofilm based infectious disease. It is the failure of clinical cultures along with tenaciously holding on to the "one organism one infection" dogma that has prevented Medicine from fully appreciating biofilms role in chronic infections.

Appreciating the full scope of the devastation produced by Medical Biofilm will take time. Biofilms role in normal health and well-being through healthy commensals is just now being investigated [226]. If our commensal biofilms are sick then we are sick. The vast carnage of chronic infections produced by pathogenic biofilm (as outlined in the introduction) has not been fully elucidated and therefore cannot be fully comprehended. Furthermore, if even a few of the major diseases that behaved like chronic infections, such as atherosclerosis, amyloid formation in the pancreas and brain, chronic inflammation producing cancer, turn out to be related to biofilm then the importance of Medical Biofilm becomes incalculable.

Medical Biofilm, that is all human disease related to biofilm, must be viewed as a single entity much like cancer. It is clear that chronic infections from different host niches will be different yet biofilm principles should translate to each one. Therefore what we've learned from environmental biofilm, industrial biofilm and all the other biofilm disciplines should be applicable to specific chronic infections. With this understanding along with the development and deployment of new technologies we should make rapid progress in managing biofilm-based disease. Early translational work has shown this to be true and if the translation of biofilm principles to human chronic infections holds up then it is not unreasonable to estimate that Medical Biofilm as a disease can be controlled in our lifetime.

References

[1] [Anonymous] (2009) ^ "Research on microbial biofilms (PA-03-047)". NIH, National Heart, Lung, and Blood Institute. 2002-12-20.

[2] Donlan RM, Costerton JW (2002) Biofilms: survival mechanisms of clinically relevant microorganisms. *Clin. Microbiol. Rev.* 15: 167-193.

[3] Henderson B, Wilson M (1998) Commensal communism and the oral cavity. *J. Dent. Res.* 77: 1674-1683.

[4] del Pozo JL, Patel R (2007) The challenge of treating biofilm-associated bacterial infections. *Clin. Pharmacol. Ther.* 82: 204-209.

[5] Wolcott RD, Ehrlich GD (2008) Biofilms and chronic infections. *JAMA* 299: 2682-2684.

[6] Wolcott RD, Rhoads DD, Bennett ME, Wolcott BM, Gogokhia L, Costerton JW, Dowd SE (2010) Chronic wounds and the medical biofilm paradigm. *J. Wound Care* 19: 45-53.

[7] Costerton JW, Geesey GG, Cheng KJ (1978) How bacteria stick. *Sci. Am.* 238: 86-95.

[8] Donlan RM, Costerton JW (2002) Biofilms: Survival mechanisms of clinically relevant microorganisms. *Clin. Microbiol. Rev.* 15: 167-193.

[9] Stoodley P, Sauer K, Davies DG, Costerton JW (2002) Biofilms as complex differentiated communities. *Annu. Rev. Microbiol.* 56: 187-209.

[10] Lam JS, Mutharia LM, Hancock RE, Hoiby N, Lam K, Baek L, Costerton JW (1983) Immunogenicity of *Pseudomonas aeruginosa* outer membrane antigens examined by crossed immunoelectrophoresis. *Infect. Immun.* 42: 88-98.

[11] Lam JS, MacDonald LA, Lam MY, Duchesne LG, Southam GG (1987) Production and characterization of monoclonal antibodies against serotype strains of *Pseudomonas aeruginosa*. *Infect. Immun.* 55: 1051-1057.

[12] Leid JG, Willson CJ, Shirtliff ME, Hassett DJ, Parsek MR, Jeffers AK (2005) The exopolysaccharide alginate protects *Pseudomonas aeruginosa* biofilm bacteria from IFN-gamma-mediated macrophage killing. *J. Immunol.* 175: 7512-7518.

[13] Leid JG, Shirtliff ME, Costerton JW, Stoodley AP (2002) Human leukocytes adhere to, penetrate, and respond to Staphylococcus aureus biofilms. *Infect. Immun.* 70: 6339-6345.

[14] Fux CA, Costerton JW, Stewart PS, Stoodley P (2005) Survival strategies of infectious biofilms. *Trends Microbiol.* 13: 34-40.

[15] Stewart PS, Costerton JW (2001) Antibiotic resistance of bacteria in biofilms. *Lancet* 358: 135-138.

[16] Stewart PS (2006 February) Matrix mysteries hold keys to controlling biofilms. http://www.biofilmsonline.com/cgi-bin/biofilmsonline/00345.html.

[17] Stewart PS, Grab L, Diemer JA (1998) Analysis of biocide transport limitation in an artificial biofilm system. *J. Appl. Microbiol.* 85: 495-500.

[18] Slater MS, Komanapalli CB, Tripathy U, Ravichandran PS, Ungerleider RM (2007) Treatment of endocarditis: a decade of experience. *Ann. Thorac. Surg.* 83: 2074-2079.

[19] Stoodley P, Sauer K, Davies DG, Costerton JW (2002) Biofilms as complex differentiated communities. *Annual Review of Microbiology* 56: 187-209.

[20] Kolenbrander PE, Ganeshkumar N, Cassels FJ, Hughes CV (1993) Coaggregation: specific adherence among human oral plaque bacteria. *FASEB J.* 7: 406-413.

[21] Min KR, Rickard AH (2009) Coaggregation by the freshwater bacterium *Sphingomonas natatoria* alters dual-species biofilm formation. *Appl. Environ. Microbiol.* 75: 3987-3997.

[22] Malic S, Hill KE, Hayes A, Percival SL, Thomas DW, Williams DW (2009) Detection and identification of specific bacteria in wound biofilms using peptide nucleic acid fluorescent in situ hybridization (PNA FISH). *Microbiology* 155: 2603-2611.

[23] Wolcott RD, Gontcharova V, Sun Y, Dowd SE (2009) Evaluation of the bacterial diversity among and within individual venous leg ulcers using bacterial tag-encoded FLX and titanium amplicon pyrosequencing and metagenomic approaches. *BMC Microbiol.* 9: 226.

[24] Lewis K (2010) Persister cells. *Annu. Rev. Microbiol.* 64: 357-372.

[25] Dickschat JS (2010) Quorum sensing and bacterial biofilms. *Nat. Prod. Rep.* 27: 343-369.

[26] Sung JM, Chantler PD, Lloyd DH (2006) Accessory gene regulator locus of *Staphylococcus intermedius*. *Infect. Immun.* 74: 2947-2956.

[27] Balaban N, Goldkorn T, Gov Y, Hirshberg M, Koyfman N, Matthews HR, Nhan RT, Singh B, Uziel O (2001) Regulation of *Staphylococcus aureus* pathogenesis via target of RNAIII-activating Protein (TRAP). *J. Biol. Chem.* 276: 2658-2667.

[28] Smith EE, Buckley DG, Wu Z, Saenphimmachak C, Hoffman LR, D'Argenio DA, Miller SI, Ramsey BW, Speert DP, Moskowitz SM, Burns JL, Kaul R, Olson MV (2006) Genetic adaptation by *Pseudomonas aeruginosa* to the airways of cystic fibrosis patients. *Proc. Natl. Acad. Sci. U. S. A.* 103: 8487-8492.

[29] Rickard AH, Colacino KR, Manton KM, Morton RI, Pulcini E, Pfeil J, Rhoads D, Wolcott RD, James G (2010) Production of cell-cell signalling molecules by bacteria isolated from human chronic wounds. *J. Appl. Microbiol.* 108: 1509-1522.

[30] Katsuyama M, Kobayashi Y, Ichikawa H, Mizuno A, Miyachi Y, Matsunaga K, Kawashima M (2005) A novel method to control the balance of skin microflora Part 2. A study to assess the effect of a cream containing farnesol and xylitol on atopic dry skin. *J. Dermatol. Sci.* 38: 207-213.

[31] Katsuyama M, Ichikawa H, Ogawa S, Ikezawa Z (2005) A novel method to control the balance of skin microflora. Part 1. Attack on biofilm of *Staphylococcus aureus* without antibiotics. *J. Dermatol. Sci.* 38: 197-205.

[32] Abee T, Kovacs AT, Kuipers OP, van d, V (2010) Biofilm formation and dispersal in Gram-positive bacteria. *Curr. Opin. Biotechnol.* S0958-1669(10)00217-X [pii];10.1016/j.copbio.2010.10.016 [doi].

[33] Cao C, Wan Y (2009) Parameters of protection against ultraviolet radiation-induced skin cell damage. *J. Cell Physiol.* 220: 277-284.

[34] Wang R, Khan BA, Cheung GY, Bach TH, Jameson-Lee M, Kong KF, Queck SY, Otto M (2010) *Staphylococcus epidermidis* surfactant peptides promote biofilm maturation and dissemination of biofilm-associated infection in mice. *J. Clin. Invest.* 42520 [pii];10.1172/JCI42520 [doi].

[35] Grande R, Di GM, Bessa LJ, Di CE, Baffoni M, Guarnieri S, Cellini L (2010) Extracellular DNA in *Helicobacter pylori* biofilm: a backstairs rumour. *J. Appl. Microbiol.* 10.1111/j.1365-2672.2010.04911.x [doi].

[36] Ma L, Conover M, Lu H, Parsek MR, Bayles K, Wozniak DJ (2009) Assembly and development of the Pseudomonas aeruginosa biofilm matrix. *PLoS Pathog.* 5: e1000354.

[37] Landini P, Antoniani D, Burgess JG, Nijland R (2010) Molecular mechanisms of compounds affecting bacterial biofilm formation and dispersal. *Appl. Microbiol. Biotechnol.* 86: 813-823.

[38] Hornemann JA, Lysova AA, Codd SL, Seymour JD, Busse SC, Stewart PS, Brown JR (2008) Biopolymer and water dynamics in microbial biofilm extracellular polymeric substance. *Biomacromolecules* 9: 2322-2328.

[39] Dominiak DM, Nielsen JL, Nielsen PH (2010) Extracellular DNA is abundant and important for microcolony strength in mixed microbial biofilms. *Environ. Microbiol.* 10.1111/j.1462-2920.2010.02375.x [doi].

[40] Arciola CR (2010) Host defense against implant infection: the ambivalent role of phagocytosis. *Int. J. Artif. Organs.* D866B1A7-ACA9-48D0-AC25-F5EC3B0B57E9 [pii].

[41] Katsuyama M, Kobayashi Y, Ichikawa H, Mizuno A, Miyachi Y, Matsunaga K, Kawashima M (2005) A novel method to control the balance of skin microflora Part 2. A study to assess the effect of a cream containing farnesol and xylitol on atopic dry skin. *J. Dermatol. Sci.* 38: 207-213.

[42] Katsuyama M, Ichikawa H, Ogawa S, Ikezawa Z (2005) A novel method to control the balance of skin microflora. Part 1. Attack on biofilm of *Staphylococcus aureus* without antibiotics. *J. Dermatol. Sci.* 38: 197-205.

[43] Torres VJ, Stauff DL, Pishchany G, Bezbradica JS, Gordy LE, Iturregui J, Anderson KL, Dunman PM, Joyce S, Skaar EP (2007) A Staphylococcus aureus regulatory system that responds to host heme and modulates virulence. *Cell Host Microbe* 1: 109-119.

[44] Sundaramoorthy R, Fyfe PK, Hunter WN (2008) Structure of *Staphylococcus aureus* EsxA suggests a contribution to virulence by action as a transport chaperone and/or adaptor protein. *J. Mol. Biol.* 383: 603-614.

[45] Burts ML, DeDent AC, Missiakas DM (2008) EsaC substrate for the ESAT-6 secretion pathway and its role in persistent infections of Staphylococcus aureus. *Mol. Microbiol.* 69: 736-746.

[46] Mills E, Baruch K, Charpentier X, Kobi S, Rosenshine I (2008) Real-time analysis of effector translocation by the type III secretion system of enteropathogenic *Escherichia coli*. *Cell Host Microbe* 3: 104-113.

[47] Kim M, Ashida H, Ogawa M, Yoshikawa Y, Mimuro H, Sasakawa C (2010) Bacterial interactions with the host epithelium. *Cell Host Microbe* 8: 20-35.

[48] Sansonetti PJ, Di Santo JP (2007) Debugging how bacteria manipulate the immune response. *Immunity* 26: 149-161.

[49] d'Hauteville H, Khan S, Maskell DJ, Kussak A, Weintraub A, Mathison J, Ulevitch RJ, Wuscher N, Parsot C, Sansonetti PJ (2002) Two msbB genes encoding maximal acylation of lipid A are required for invasive Shigella flexneri to mediate inflammatory rupture and destruction of the intestinal epithelium. *J. Immunol.* 168: 5240-5251.

[50] Jahoor A, Patel R, Bryan A, Do C, Krier J, Watters C, Wahli W, Li G, Williams SC, Rumbaugh KP (2008) Peroxisome Proliferator Activated Receptors Mediate Host Cell Pro-inflammatory Responses to P. aeruginosa Autoinducer. *J. Bacteriol.*

[51] Llewelyn M, Cohen J (2002) Superantigens: microbial agents that corrupt immunity. *Lancet Infect. Dis.* 2: 156-162.

[52] Cornelis GR (2006) The type III secretion injectisome. *Nat. Rev. Microbiol.* 4: 811-825.

[53] Purevdorj-Gage B, Costerton WJ, Stoodley P (2005) Phenotypic differentiation and seeding dispersal in non-mucoid and mucoid Pseudomonas aeruginosa biofilms. *Microbiology* 151: 1569-1576.

[54] Khandaker MH, Xu L, Rahimpour R, Mitchell G, DeVries ME, Pickering JG, Singhal SK, Feldman RD, Kelvin DJ (1998) CXCR1 and CXCR2 are rapidly down-modulated by bacterial endotoxin through a unique agonist-independent, tyrosine kinase-dependent mechanism. *J. Immunol.* 161: 1930-1938.

[55] Schooling SR, Beveridge TJ (2006) Membrane vesicles: an overlooked component of the matrices of biofilms. *J. Bacteriol.* 188: 5945-5957.

[56] Jensen ET, Kharazmi A, Garred P, Kronborg G, Fomsgaard A, Mollnes TE, Hoiby N (1993) Complement activation by Pseudomonas aeruginosa biofilms. *Microb. Pathog.* 15: 377-388.

[57] Bayles KW (2007) The biological role of death and lysis in biofilm development. *Nat. Rev. Microbiol.* 5: 721-726.

[58] Schindler R, Beck W, Deppisch R, Aussieker M, Wilde A, Gohl H, Frei U (2004) Short bacterial DNA fragments: detection in dialysate and induction of cytokines. *J. Am. Soc. Nephrol.* 15: 3207-3214.

[59] Prince LR, Bianchi SM, Vaughan KM, Bewley MA, Marriott HM, Walmsley SR, Taylor GW, Buttle DJ, Sabroe I, Dockrell DH, Whyte MK (2008) Subversion of a lysosomal pathway regulating neutrophil apoptosis by a major bacterial toxin, pyocyanin. *J. Immunol.* 180: 3502-3511.

[60] Diegelmann RF (2003) Excessive neutrophils characterize chronic pressure ulcers. *Wound Repair Regen.* 11: 490-495.

[61] Toth B, Alexander M, Daniel T, Chaudry IH, Hubbard WJ, Schwacha MG (2004) The role of gammadelta T cells in the regulation of neutrophil-mediated tissue damage after thermal injury. *J. Leukoc. Biol.* 76: 545-552.

[62] Aujla SJ, Dubin PJ, Kolls JK (2007) Interleukin-17 in pulmonary host defense. *Exp. Lung Res.* 33: 507-518.

[63] Johansson C, Ingman M, Jo WM (2006) Elevated neutrophil, macrophage and dendritic cell numbers characterize immune cell populations in mice chronically infected with Salmonella. *Microb. Pathog.* 41: 49-58.

[64] Montemurro P, Nishioka H, Dundon WG, de BM, Del GG, Rappuoli R, Montecucco C (2002) The neutrophil-activating protein (HP-NAP) of Helicobacter pylori is a potent stimulant of mast cells. *Eur. J. Immunol.* 32: 671-676.

[65] Jones SA, Wolf M, Qin S, Mackay CR, Baggiolini M (1996) Different functions for the interleukin 8 receptors (IL-8R) of human neutrophil leukocytes: NADPH oxidase and phospholipase D are activated through IL-8R1 but not IL-8R2. *Proc. Natl. Acad. Sci. U. S. A.* 93: 6682-6686.

[66] Schmausser B, Josenhans C, Endrich S, Suerbaum S, Sitaru C, Andrulis M, Brandlein S, Rieckmann P, Muller-Hermelink HK, Eck M (2004) Downregulation of CXCR1 and CXCR2 expression on human neutrophils by *Helicobacter pylori*: a new pathomechanism in *H. pylori* infection? *Infect. Immun.* 72: 6773-6779.

[67] Tikhonov I, Doroshenko T, Chaly Y, Smolnikova V, Pauza CD, Voitenok N (2001) Down-regulation of CXCR1 and CXCR2 expression on human neutrophils upon activation of whole blood by S. aureus is mediated by TNF-alpha. *Clin. Exp. Immunol.* 125: 414-422.

[68] Fadok VA, Bratton DL, Guthrie L, Henson PM (2001) Differential effects of apoptotic versus lysed cells on macrophage production of cytokines: role of proteases. *J. Immunol.* 166: 6847-6854.

[69] Fadok VA, de CA, Daleke DL, Henson PM, Bratton DL (2001) Loss of phospholipid asymmetry and surface exposure of phosphatidylserine is required for phagocytosis of apoptotic cells by macrophages and fibroblasts. *J. Biol. Chem.* 276: 1071-1077.

[70] Fadok VA, Voelker DR, Campbell PA, Cohen JJ, Bratton DL, Henson PM (1992) Exposure of phosphatidylserine on the surface of apoptotic lymphocytes triggers specific recognition and removal by macrophages. *J. Immunol.* 148: 2207-2216.

[71] Baran J, Guzik K, Hryniewicz W, Ernst M, Flad HD, Pryjma J (1996) Apoptosis of monocytes and prolonged survival of granulocytes as a result of phagocytosis of bacteria. *Infect. Immun.* 64: 4242-4248.

[72] Guzik K, Bzowska M, Smagur J, Krupa O, Sieprawska M, Travis J, Potempa J (2007) A new insight into phagocytosis of apoptotic cells: proteolytic enzymes divert the recognition and clearance of polymorphonuclear leukocytes by macrophages. *Cell Death Differ.* 14: 171-182.

[73] Coxon A, Tang T, Mayadas TN (1999) Cytokine-activated endothelial cells delay neutrophil apoptosis in vitro and in vivo. A role for granulocyte/macrophage colony-stimulating factor. *J. Exp. Med.* 190: 923-934.

[74] do VA, Costa-Ramos C, Silva A, Silva DS, Gartner F, dos Santos NM, Silva MT (2007) Systemic macrophage and neutrophil destruction by secondary necrosis induced by a bacterial exotoxin in a Gram-negative septicaemia. *Cell Microbiol.* 9: 988-1003.

[75] Baumann R, Casaulta C, Simon D, Conus S, Yousefi S, Simon HU (2003) Macrophage migration inhibitory factor delays apoptosis in neutrophils by inhibiting the mitochondria-dependent death pathway. *FASEB J.* 17: 2221-2230.

[76] Gamberale R, Giordano M, Trevani AS, Andonegui G, Geffner JR (1998) Modulation of human neutrophil apoptosis by immune complexes. *J. Immunol.* 161: 3666-3674.

[77] Anwar S, Whyte MK (2007) Neutrophil apoptosis in infectious disease. *Exp. Lung Res.* 33: 519-528.

[78] Hartl D, Latzin P, Hordijk P, Marcos V, Rudolph C, Woischnik M, Krauss-Etschmann S, Koller B, Reinhardt D, Roscher AA, Roos D, Griese M (2007) Cleavage of CXCR1 on neutrophils disables bacterial killing in cystic fibrosis lung disease. *Nat. Med.* 13: 1423-1430.

[79] Sabroe I, Whyte MK (2007) Incapacitating the immune system in cystic fibrosis. *Nat. Med.* 13: 1417-1418.

[80] Birrer P, McElvaney NG, Rudeberg A, Sommer CW, Liechti-Gallati S, Kraemer R, Hubbard R, Crystal RG (1994) Protease-antiprotease imbalance in the lungs of children with cystic fibrosis. *Am. J. Respir. Crit. Care Med.* 150: 207-213.

[81] Vandivier RW, Fadok VA, Hoffmann PR, Bratton DL, Penvari C, Brown KK, Brain JD, Accurso FJ, Henson PM (2002) Elastase-mediated phosphatidylserine receptor cleavage impairs apoptotic cell clearance in cystic fibrosis and bronchiectasis. *J. Clin. Invest.* 109: 661-670.

[82] Nwomeh BC, Yager DR, Cohen IK (1998) Physiology of the chronic wound. *Clin. Plast. Surg.* 25: 341-356.

[83] Hartl D, Latzin P, Hordijk P, Marcos V, Rudolph C, Woischnik M, Krauss-Etschmann S, Koller B, Reinhardt D, Roscher AA, Roos D, Griese M (2007) Cleavage of CXCR1 on neutrophils disables bacterial killing in cystic fibrosis lung disease. *Nat. Med.* 13: 1423-1430.

[84] Yang JJ, Kettritz R, Falk RJ, Jennette JC, Gaido ML (1996) Apoptosis of endothelial cells induced by the neutrophil serine proteases proteinase 3 and elastase. *Am. J. Pathol.* 149: 1617-1626.

[85] Venaille TJ, Ryan G, Robinson BW (1998) Epithelial cell damage is induced by neutrophil-derived, not pseudomonas-derived, proteases in cystic fibrosis sputum. *Respir. Med.* 92: 233-240.

[86] Korkmaz B, Moreau T, Gauthier F (2008) Neutrophil elastase, proteinase 3 and cathepsin G: physicochemical properties, activity and physiopathological functions. *Biochimie* 90: 227-242.

[87] Rohde H, Burdelski C, Bartscht K, Hussain M, Buck F, Horstkotte MA, Knobloch JK, Heilmann C, Herrmann M, Mack D (2005) Induction of Staphylococcus epidermidis biofilm formation via proteolytic processing of the accumulation-associated protein by staphylococcal and host proteases. *Mol. Microbiol.* 55: 1883-1895.

[88] Harris IR, Yee KC, Walters CE, Cunliffe WJ, Kearney JN, Wood EJ, Ingham E (1995) Cytokine and protease levels in healing and non-healing chronic venous leg ulcers. *Exp. Dermatol.* 4: 342-349.

[89] Sosroseno W, Herminajeng E (2002) The role of macrophages in the induction of murine immune response to Actinobacillus actinomycetemcomitans. *J. Med. Microbiol.* 51: 581-588.

[90] Wolcott RD, Rhoads DD (2008) A study of biofilm-based wound management in subjects with critical limb ischaemia. *J. Wound Care* 17: 145-2, 154.

[91] Lam JS, MacDonald LA, Lam MY, Duchesne LG, Southam GG (1987) Production and characterization of monoclonal antibodies against serotype strains of Pseudomonas aeruginosa. *Infect. Immun.* 55: 1051-1057.

[92] Meluleni GJ, Grout M, Evans DJ, Pier GB (1995) Mucoid Pseudomonas aeruginosa growing in a biofilm in vitro are killed by opsonic antibodies to the mucoid exopolysaccharide capsule but not by antibodies produced during chronic lung infection in cystic fibrosis patients. *J. Immunol.* 155: 2029-2038.

[93] Kufer TA, Sansonetti PJ (2007) Sensing of bacteria: NOD a lonely job. *Curr. Opin. Microbiol.* 10: 62-69.

[94] Garrett WS, Lord GM, Punit S, Lugo-Villarino G, Mazmanian SK, Ito S, Glickman JN, Glimcher LH (2007) Communicable ulcerative colitis induced by T-bet deficiency in the innate immune system. *Cell* 131: 33-45.

[95] Dekimpe V, Deziel E (2009) Revisiting the quorum-sensing hierarchy in Pseudomonas aeruginosa: the transcriptional regulator RhlR regulates LasR-specific factors. *Microbiology* 155: 712-723.

[96] Deptula A, Gospodarek E (2010) Reduced expression of virulence factors in multidrug-resistant Pseudomonas aeruginosa strains. *Arch. Microbiol.* 192: 79-84.

[97] Le BR, Nguyen S, Nowak E, Kipnis E, Pierre M, Ader F, Courcol R, Guery BP, Faure K (2008) Quorum-sensing activity and related virulence factor expression in clinically pathogenic isolates of Pseudomonas aeruginosa. *Clin. Microbiol. Infect.* 14: 337-343.

[98] Costerton JW, Stewart PS, Greenberg EP (1999) Bacterial biofilms: a common cause of persistent infections. *Science* 284: 1318-1322.

[99] Kim M, Ashida H, Ogawa M, Yoshikawa Y, Mimuro H, Sasakawa C (2010) Bacterial interactions with the host epithelium. *Cell Host Microbe* 8: 20-35.

[100] Schaber JA, Triffo WJ, Suh SJ, Oliver JW, Hastert MC, Griswold JA, Auer M, Hamood AN, Rumbaugh KP (2007) Pseudomonas aeruginosa forms biofilms in acute infection independent of cell-to-cell signaling. *Infect. Immun.* 75: 3715-3721.

[101] (2007) Robert Koch The Nobel Prize in Physiology or Medicine 1905. The Official Web Site of the Nobel Foundation.

[102] Hager, T (2006) The Demon Under the Microscope. New York, NY: Harmony Books.

[103] Rhoads DD, Cox SB, Rees EJ, Sun Y, Wolcott RD (2012) Clinical identification of bacteria in human chronic wound infections: culturing vs. 16S ribosomal DNA sequencing. *BMC Infect. Dis.* 12: 321.

[104] Mullis K, Faloona F, Scharf S, Saiki R, Horn G, Erlich H (1986) Specific enzymatic amplification of DNA in vitro: the polymerase chain reaction. *Cold Spring Harb. Symp. Quant. Biol.* 51 Pt 1: 263-273.

[105] Cuiv O, Aguirre de CD, Jones M, Klaassens ES, Worthley DL, Whitehall VL, Kang S, McSweeney CS, Leggett BA, Morrison M (2010) The Effects from DNA Extraction Methods on the Evaluation of Microbial Diversity Associated with Human Colonic Tissue. *Microb. Ecol.* 10.1007/s00248-010-9771-x [doi].

[106] Zemanick ET, Wagner BD, Sagel SD, Stevens MJ, Accurso FJ, Harris JK (2010) Reliability of quantitative real-time PCR for bacterial detection in cystic fibrosis airway specimens. *PLoS One* 5: e15101.

[107] Wolcott RD, Dowd SE (2008) A rapid molecular method for characterising bacterial bioburden in chronic wounds. *J. Wound Care* 17: 513-516.

[108] Leake JL, Dowd SE, Wolcott RD, Zischkau AM (2009) Identification of yeast in chronic wounds using new pathogen-detection technologies. *J. Wound Care* 18: 103-4, 106, 108.

[109] Melendez JH, Frankel YM, An AT, Williams L, Price LB, Wang NY, Lazarus GS, Zenilman JM (2010) Real-time PCR assays compared to culture-based approaches for identification of aerobic bacteria in chronic wounds. *Clin. Microbiol. Infect.* 16: 1762-1769.

[110] Wolcott RD, Cox SB, Dowd SE (2010) Healing and healing rates of chronic wounds in the age of molecular pathogen diagnostics. *J. Wound Care* 19: 272-1.

[111] Dowd SE, Sun Y, Secor PR, Rhoads DD, Wolcott BM, James GA, Wolcott RD (2008) Survey of bacterial diversity in chronic wounds using pyrosequencing, DGGE, and full ribosome shotgun sequencing. *BMC Microbiol.* 8: 43. 1471-2180-8-43.

[112] Dowd SE, Wolcott RD, Sun Y, McKeehan T, Smith E, Rhoads D (2008) Polymicrobial nature of chronic diabetic foot ulcer biofilm infections determined using bacterial tag encoded FLX amplicon pyrosequencing (bTEFAP). *PLoS One* 3: e3326.

[113] Sun Y, Dowd SE, Smith E, Rhoads DD, Wolcott RD (2008) In vitro multispecies Lubbock chronic wound biofilm model. *Wound Repair Regen.* 16: 805-813.

[114] Holt RA, Jones SJ (2008) The new paradigm of flow cell sequencing. *Genome Res.* 18: 839-846.

[115] Dowd SE, Sun Y, Wolcott RD, Domingo A, Carroll JA (2008) Bacterial tag-encoded FLX amplicon pyrosequencing (bTEFAP) for microbiome studies: bacterial diversity in the ileum of newly weaned Salmonella-infected pigs. *Foodborne Pathog. Dis.* 5: 459-472.

[116] Dowd SE, Wolcott RD, Sun Y, McKeehan T, Smith E, Rhoads D (2008) Polymicrobial nature of chronic diabetic foot ulcer biofilm infections determined using bacterial tag

encoded FLX amplicon pyrosequencing (bTEFAP). *PLoS One* 3: e3326. 10.1371/journal.pone.0003326 [doi].

[117] Dowd SE, Sun Y, Secor PR, Rhoads DD, Wolcott BM, James GA, Wolcott RD (2008) Survey of bacterial diversity in chronic wounds using Pyrosequencing, DGGE, and full ribosome shotgun sequencing. *BMC Microbiol.* 8: 43.

[118] Bonham PA (2009) Swab cultures for diagnosing wound infections: a literature review and clinical guideline. *J. Wound Ostomy Continence Nurs.* 36: 389-395.

[119] Brook I (2007) Treatment of anaerobic infection. *Expert Rev. Anti Infect. Ther.* 5: 991-1006.

[120] Moore K, Huddleston E, Stacey MC, Harding KG (2007) Venous leg ulcers - the search for a prognostic indicator. *Int. Wound J.* 4: 163-172.

[121] Gardner SE, Frantz RA (2008) Wound bioburden and infection-related complications in diabetic foot ulcers. *Biol. Res. Nurs.* 10: 44-53.

[122] Ehrlich GD (1996) The Need for PCR-based Panel Testing for Syndromic Infectious Diseases. *Mol. Diagn.* 1: 83-87.

[123] Ehrlich GD, Veeh R, Wang X, Costerton JW, Hayes JD, Hu FZ, Daigle BJ, Ehrlich MD, Post JC (2002) Mucosal biofilm formation on middle-ear mucosa in the chinchilla model of otitis media. *JAMA* 287: 1710-1715.

[124] Weckwerth W (2010) Metabolomics: an integral technique in systems biology. *Bioanalysis* 2: 829-836.

[125] Windle HJ, Brown PA, Kelleher DP (2010) Proteomics of bacterial pathogenicity: therapeutic implications. *Proteomics Clin. Appl.* 4: 215-227.

[126] Hare NJ, Cordwell SJ (2010) Proteomics of bacterial pathogens: *Pseudomonas aeruginosa* infections in cystic fibrosis - a case study. *Proteomics Clin. Appl.* 4: 228-248.

[127] Broadbent J, Walsh T, Upton Z (2010) Proteomics in chronic wound research: potentials in healing and health. *Proteomics Clin. Appl.* 4: 204-214.

[128] Sturdevant DE, Virtaneva K, Martens C, Bozinov D, Ogundare O, Castro N, Kanakabandi K, Beare PA, Omsland A, Carlson JH, Kennedy AD, Heinzen RA, Celli J, Greenberg DE, DeLeo FR, Porcella SF (2010) Host-microbe interaction systems biology: lifecycle transcriptomics and comparative genomics. *Future Microbiol.* 5: 205-219.

[129] Ammons MC, Ward LS, Fisher ST, Wolcott RD, James GA (2009) In vitro susceptibility of established biofilms composed of a clinical wound isolate of *Pseudomonas aeruginosa* treated with lactoferrin and xylitol. *Int. J. Antimicrob. Agents* 33: 230-236.

[130] Tang X, Flint SH, Bennett RJ, Brooks JD, Morton RH (2009) Biofilm growth of individual and dual strains of Klebsiella oxytoca from the dairy industry on ultrafiltration membranes. *J. Ind. Microbiol. Biotechnol.* 36: 1491-1497.

[131] Costerton JW, Ellis B, Lam K, Johnson F, Khoury AE (1994) Mechanism of electrical enhancement of efficacy of antibiotics in killing biofilm bacteria. *Antimicrob. Agents Chemother.* 38: 2803-2809.

[132] Del Pozo JL, Rouse MS, Euba G, Kang CI, Mandrekar JN, Steckelberg JM, Patel R (2009) The electricidal effect is active in an experimental model of Staphylococcus epidermidis chronic foreign body osteomyelitis. *Antimicrob. Agents Chemother.* 53: 4064-4068.

[133] Stoodley P, Sauer K, Davies DG, Costerton JW (2002) Biofilms as complex differentiated communities. *Annu. Rev. Microbiol.* 56: 187-209.

[134] Lewis K (2007) Persister cells, dormancy and infectious disease. *Nat. Rev. Microbiol.* 5; 48-56.

[135] Macedo AJ, Abraham WR (2009) Can infectious biofilm be controlled by blocking bacterial communication? *Med. Chem.* 5: 517-528.

[136] Costerton JW, Stewart PS, Greenberg EP (1999) Bacterial biofilms: A common cause of persistent infections. *Science* 284: 1318-1322.

[137] Donlan RM, Costerton JW (2002) Biofilms: survival mechanisms of clinically relevant microorganisms. *Clin. Microbiol. Rev.* 15: 167-193.

[138] Fux CA, Costerton JW, Stewart PS, Stoodley P (2005) Survival strategies of infectious biofilms. *Trends Microbiol.* 13: 34-40.

[139] Balaban N, Stoodley P, Fux CA, Wilson S, Costerton JW, Dell'Acqua G (2005) Prevention of staphylococcal biofilm-associated infections by the quorum sensing inhibitor RIP. *Clin. Orthop. Relat. Res.* 48-54. 00003086-200508000-00009 [pii].

[140] Balaban N, Cirioni O, Giacometti A, Ghiselli R, Braunstein JB, Silvestri C, Mocchegiani F, Saba V, Scalise G (2007) Treatment of Staphylococcus aureus biofilm infection by the quorum-sensing inhibitor RIP. *Antimicrob. Agents Chemother.* 51: 2226-2229.

[141] Cirioni O, Ghiselli R, Minardi D, Orlando F, Mocchegiani F, Silvestri C, Muzzonigro G, Saba V, Scalise G, Balaban N, Giacometti A (2007) RNAIII-inhibiting peptide affects biofilm formation in a rat model of staphylococcal ureteral stent infection. *Antimicrob. Agents Chemother.* 51: 4518-4520.

[142] Giacometti A, Cirioni O, Ghiselli R, Dell'Acqua G, Orlando F, D'Amato G, Mocchegiani F, Silvestri C, Del Prete MS, Rocchi M, Balaban N, Saba V, Scalise G (2005) RNAIII-inhibiting peptide improves efficacy of clinically used antibiotics in a murine model of staphylococcal sepsis. *Peptides* 26: 169-175.

[143] Wolcott R (2008) Clinical Wound Healing Using Signal Inhibitors. In: Balaban N, editors. Control of Biofilm Infections by Signal Manipulation. Springer. pp. 157-170.

[144] Giacometti A, Cirioni O, Ghiselli R, Dell'Acqua G, Orlando F, D'Amato G, Mocchegiani F, Silvestri C, Del Prete MS, Rocchi M, Balaban N, Saba V, Scalise G (2005) RNAIII-inhibiting peptide improves efficacy of clinically used antibiotics in a murine model of staphylococcal sepsis. *Peptides* 26: 169-175..

[145] Lopez-Leban F, Kiran MD, Wolcott R, Balaban N (2010) Molecular mechanisms of RIP, an effective inhibitor of chronic infections. *Int. J. Artif. Organs* 33: 582-589.

[146] Brackman G, Cos P, Maes L, Nelis HJ, Coenye T (2011) Quorum sensing inhibitors increase the susceptibility of bacterial biofilms to antibiotics in vitro and in vivo. *Antimicrob. Agents Chemother.* 55: 2655-2661.

[147] Demirag MK, Esen S, Zivalioglu M, Leblebicioglu H, Keceligil HT (2007) The effect of aspirin on adherence of slime-producing, coagulase-negative staphylococci to vascular grafts. *Ann. Vasc. Surg.* 21: 464-467.

[148] Sedlacek M, Gemery JM, Cheung AL, Bayer AS, Remillard BD (2007) Aspirin treatment is associated with a significantly decreased risk of Staphylococcus aureus bacteremia in hemodialysis patients with tunneled catheters. *Am. J. Kidney Dis.* 49: 401-408.

[149] Costerton JW, Geesey GG, Cheng KJ (1978) How bacteria stick. *Sci. Am.* 238: 86-95.

[150] Costerton JW (1999) Introduction to biofilm. *Int. J. Antimicrob. Agents* 11: 217-221.

[151] Davies DG, Parsek MR, Pearson JP, Iglewski BH, Costerton JW, Greenberg EP (1998) The involvement of cell-to-cell signals in the development of a bacterial biofilm. *Science* 280: 295-298.

[152] Hall-Stoodley L, Costerton JW, Stoodley P (2004) Bacterial biofilms: from the natural environment to infectious diseases. *Nat. Rev. Microbiol.* 2: 95-108.

[153] Stoodley P, Sauer K, Davies DG, Costerton JW (2002) Biofilms as complex differentiated communities. *Annu. Rev. Microbiol.* 56: 187-209.

[154] Stewart PS, Franklin MJ (2008) Physiological heterogeneity in biofilms. *Nat. Rev. Microbiol.* 6: 199-210.

[155] Fux CA, Costerton JW, Stewart PS, Stoodley P (2005) Survival strategies of infectious biofilms. *Trends Microbiol.* 13: 34-40.

[156] Allegrucci M, Hu FZ, Shen K, Hayes J, Ehrlich GD, Post JC, Sauer K (2006) Phenotypic characterization of Streptococcus pneumoniae biofilm development. *J. Bacteriol.* 188: 2325-2335. 188/7/2325 [pii];10.1128/JB.188.7.2325-2335.2006 [doi].

[157] Ehrlich GD, Hu FZ, Shen K, Stoodley P, Post JC (2005) Bacterial plurality as a general mechanism driving persistence in chronic infections. *Clin. Orthop. Relat. Res.* 20-24.

[158] Fux CA, Costerton JW, Stewart PS, Stoodley P (2005) Survival strategies of infectious biofilms. *Trends in Microbiology* 13: 34-40.

[159] Rose H, Baldwin A, Dowson CG, Mahenthiralingam E (2009) Biocide susceptibility of the Burkholderia cepacia complex. *J. Antimicrob. Chemother.* 63: 502-510.

[160] Costerton JW, Stewart PS (2001) Battling biofilms. *Sci. Am.* 285: 74-81.

[161] Stewart PS, Costerton JW (2001) Antibiotic resistance of bacteria in biofilms. *Lancet* 358: 135-138.

[162] Stewart PS (2002) Mechanisms of antibiotic resistance in bacterial biofilms. *Int. J. Med. Microbiol.* 292: 107-113.

[163] Stewart PS, Rayner J, Roe F, Rees WM (2001) Biofilm penetration and disinfection efficacy of alkaline hypochlorite and chlorosulfamates. *J. Appl. Microbiol.* 91: 525-532.

[164] Stewart PS, Davison WM, Steenbergen JN (2009) Daptomycin rapidly penetrates a Staphylococcus epidermidis biofilm. *Antimicrob. Agents Chemother.* 53: 3505-3507.

[165] Mah TF, O'Toole GA (2001) Mechanisms of biofilm resistance to antimicrobial agents. *Trends Microbiol.* 9: 34-39.

[166] Stewart PS (2003) Diffusion in biofilms. *J. Bacteriol.* 185: 1485-1491.

[167] Anderl JN, Franklin MJ, Stewart PS (2000) Role of antibiotic penetration limitation in Klebsiella pneumoniae biofilm resistance to ampicillin and ciprofloxacin. *Antimicrob. Agents Chemother.* 44: 1818-1824.

[168] Mah TF, O'Toole GA (2001) Mechanisms of biofilm resistance to antimicrobial agents. *Trends Microbiol.* 9: 34-39.

[169] Scanlon E, Karlsmark T, Leaper DJ, Carter K, Poulsen PB, Hart-Hansen K, Hahn TW (2005) Cost-effective faster wound healing with a sustained silver-releasing foam dressing in delayed healing leg ulcers--a health-economic analysis. *Int. Wound J.* 2: 150-160.

[170] Niezgoda JA, Van Gils CC, Frykberg RG, Hodde JP (2005) Randomized clinical trial comparing OASIS Wound Matrix to Regranex Gel for diabetic ulcers. *Adv. Skin Wound Care* 18: 258-266.

[171] Brem H, Sheehan P, Rosenberg HJ, Schneider JS, Boulton AJ (2006) Evidence-based protocol for diabetic foot ulcers. *Plast. Reconstr. Surg.* 117: 193S-209S.

[172] Brem H, Lyder C (2004) Protocol for the successful treatment of pressure ulcers. *Am. J. Surg.* 188: 9-17.

[173] Brem H, Kirsner RS, Falanga V (2004) Protocol for the successful treatment of venous ulcers. *Am. J. Surg.* 188: 1-8.

[174] Panuncialman J, Falanga V (2007) The science of wound bed preparation. *Clin. Plast. Surg.* 34: 621-632.

[175] Brem H, Stojadinovic O, Diegelmann RF, Entero H, Lee B, Pastar I, Golinko M, Rosenberg H, Tomic-Canic M (2007) Molecular markers in patients with chronic wounds to guide surgical debridement. *Mol. Med.* 13: 30-39.

[176] Falanga V, Saap LJ, Ozonoff A (2006) Wound bed score and its correlation with healing of chronic wounds. *Dermatol. Ther.* 19: 383-390.

[177] Wolcott RD, Rumbaugh KP, James G, Schultz G, Phillips P, Yang Q, Watters C, Stewart PS, Dowd SE (2010) Biofilm maturity studies indicate sharp debridement opens a time- dependent therapeutic window. *J. Wound Care* 19: 320-328.

[178] Wolcott RD, Kennedy JP, Dowd SE (2009) Regular debridement is the main tool for maintaining a healthy wound bed in most chronic wounds. *J. Wound Care* 18: 54-56.

[179] Sibbald RG, Woo K, Ayello EA (2006) Increased bacterial burden and infection: the story of NERDS and STONES. *Adv. Skin Wound Care* 19: 447-461.

[180] Wolcott RD, Rhoads DD, Dowd SE (2008) Biofilms and chronic wound inflammation. *J. Wound Care* 17: 333-341.

[181] Armstrong DG, Jude EB (2002) The role of matrix metalloproteinases in wound healing. *J. Am. Podiatr. Med. Assoc.* 92: 12-18.

[182] Nwomeh BC, Liang HX, Cohen IK, Yager DR (1999) MMP-8 is the predominant collagenase in healing wounds and nonhealing ulcers. *J. Surg. Res.* 81: 189-195.

[183] Trengove NJ, Stacey MC, MacAuley S, Bennett N, Gibson J, Burslem F, Murphy G, Schultz G (1999) Analysis of the acute and chronic wound environments: the role of proteases and their inhibitors. *Wound Repair Regen.* 7: 442-452.

[184] Yager DR, Nwomeh BC (1999) The proteolytic environment of chronic wounds. *Wound Repair Regen.* 7: 433-441.

[185] Costerton JW, Stewart PS, Greenberg EP (1999) Bacterial biofilms: a common cause of persistent infections. *Science* 284: 1318-1322.

[186] Sventoraityte J, Zvirbliene A, Kiudelis G, Zalinkevicius R, Zvirbliene A, Praskevicius A, Kupcinskas L, Tamosiunas V (2008) Immune system alterations in patients with inflammatory bowel disease during remission. *Medicina* (Kaunas) 44: 27-33.

[187] Tanaka T, Okanobu H, Yoshimi S, Murakami E, Kogame A, Imagawa H, Numata Y, Kuga Y, Moriya T, Ohya T, Kajiyama G (2008) In patients with ulcerative colitis, adsorptive depletion of granulocytes and monocytes impacts mucosal level of neutrophils and clinically is most effective in steroid naive patients. *Dig. Liver Dis.*

[188] Healy DP (2007) Macrolide immunomodulation of chronic respiratory diseases. *Curr. Infect. Dis. Rep.* 9: 7-13.

[189] O'Reilly PJ, Gaggar A, Blalock JE (2008) Interfering with extracellular matrix degradation to blunt inflammation. *Curr. Opin. Pharmacol.*

[190] Mroz RM, Noparlik J, Chyczewska E, Braszko JJ, Holownia A (2007) Molecular basis of chronic inflammation in lung diseases: new therapeutic approach. *J. Physiol. Pharmacol.* 58 Suppl 5: 453-460.

[191] Toto PD, Rapp GW, Goljan KR (1965) The effect of manual and mechanical toothbrushing on oral hygiene, carries conduciveness and ph of the oral cavity. *J. Oral Ther. Pharmacol.* 58: 612-617.

[192] Steed DL, Donohoe D, Webster MW, Lindsley L (1996) Effect of extensive debridement and treatment on the healing of diabetic foot ulcers. Diabetic Ulcer Study Group. *J. Am. Coll. Surg.* 183: 61-64.

[193] Williams D, Enoch S, Miller D, Harris K, Price P, Harding KG (2005) Effect of sharp debridement using curette on recalcitrant nonhealing venous leg ulcers: a concurrently controlled, prospective cohort study. *Wound Repair Regen.* 13: 131-137.

[194] Wolcott RD, Rhoads DD (2008) A study of biofilm-based wound management in subjects with critical limb ischaemia. *Journal of Wound Care* 17: 145-155.

[195] Pohjolainen T, Alaranta H (1999) Epidemiology of lower limb amputees in Southern Finland in 1995 and trends since 1984. *Prosthet. Orthot. Int.* 23: 88-92.

[196] Faglia E, Favales F, Morabito A (2001) New ulceration, new major amputation, and survival rates in diabetic subjects hospitalized for foot ulceration from 1990 to 1993: a 6.5-year follow-up. *Diabetes Care* 24: 78-83.

[197] Armstrong DG, Wrobel J, Robbins JM (2007) Guest Editorial: are diabetes-related wounds and amputations worse than cancer? *Int. Wound J.* 4: 286-287.

[198] Glover JL, Weingarten MS, Buchbinder DS, Poucher RL, Deitrick GA, III, Fylling CP (1997) A 4-year outcome-based retrospective study of wound healing and limb salvage in patients with chronic wounds. *Adv. Wound Care* 10: 33-38.

[199] Chang BB, Darling RC, III, Paty PS, Lloyd WE, Shah DM, Leather RP (1996) Expeditious management of ischemic invasive foot infections. *Cardiovasc. Surg.* 4: 792-795.

[200] Stojadinovic O, Pastar I, Vukelic S, Mahoney MG, Brennan D, Krzyzanowska A, Golinko M, Brem H, Tomic-Canic M (2008) Deregulation of keratinocyte differentiation and activation: A hallmark of venous ulcers. *J. Cell Mol. Med.*

[201] Hanlon GW (2007) Bacteriophages: an appraisal of their role in the treatment of bacterial infections. *Int. J. Antimicrob. Agents* 30: 118-128.

[202] Slopek S, Weber-Dabrowska B, Dabrowski M, Kucharewicz-Krukowska A (1987) Results of bacteriophage treatment of suppurative bacterial infections in the years 1981-1986. *Arch. Immunol. Ther. Exp.* (Warsz) 35: 569-583.

[203] Lopez V, Ochs HD, Thuline HC, Davis SD, Wedgwood RJ (1975) Defective antibody response to bacteriophage phichi 174 in Down syndrome. *J. Pediatr.* 86: 207-211.

[204] Ochs HD, Lum LG, Johnson FL, Schiffman G, Wedgwood RJ, Storb R (1982) Bone marrow transplantation in the Wiskott-Aldrich syndrome. Complete hematological and immunological reconstitution. *Transplantation* 34: 284-288.

[205] Fogelman I, Davey V, Ochs HD, Elashoff M, Feinberg MB, Mican J, Siegel JP, Sneller M, Lane HC (2000) Evaluation of CD4+ T cell function In vivo in HIV-infected patients as measured by bacteriophage phiX174 immunization. *J. Infect. Dis.* 182: 435-441.

[206] Doolittle MM, Cooney JJ, Caldwell DE (1995) Lytic infection of *Escherichia coli* biofilms by bacteriophage T4. *Can. J. Microbiol.* 41: 12-18.

[207] Corbin BD, McLean RJ, Aron GM (2001) Bacteriophage T4 multiplication in a glucose-limited Escherichia coli biofilm. *Can. J. Microbiol.* 47: 680-684.
[208] ADAMS MH, PARK BH (1956) An enzyme produced by a phage-host cell system. II. The properties of the polysaccharide depolymerase. *Virology* 2: 719-736.
[209] Hughes KA, Sutherland IW, Jones MV (1998) Biofilm susceptibility to bacteriophage attack: the role of phage-borne polysaccharide depolymerase. *Microbiology* 144 (Pt 11): 3039-3047.
[210] Hughes KA, Sutherland IW, Clark J, Jones MV (1998) Bacteriophage and associated polysaccharide depolymerases--novel tools for study of bacterial biofilms. *J. Appl. Microbiol.* 85: 583-590.
[211] Chaignon P, Sadovskaya I, Ragunah C, Ramasubbu N, Kaplan JB, Jabbouri S (2007) Susceptibility of staphylococcal biofilms to enzymatic treatments depends on their chemical composition. *Appl. Microbiol. Biotechnol.*
[212] Lansdown AB, Sampson B, Laupattarakasem P, Vuttivirojana A (1997) Silver aids healing in the sterile skin wound: experimental studies in the laboratory rat. *Br. J. Dermatol.* 137: 728-735.
[213] Russell AD, Hugo WB (1994) Antimicrobial activity and action of silver. *Prog. Med. Chem.* 31: 351-370.
[214] Kool JL, Bergmire-Sweat D, Butler JC, Brown EW, Peabody DJ, Massi DS, Carpenter JC, Pruckler JM, Benson RF, Fields BS (1999) Hospital characteristics associated with colonization of water systems by Legionella and risk of nosocomial legionnaires' disease: a cohort study of 15 hospitals. *Infect. Control Hosp. Epidemiol.* 20: 798-805.
[215] Sampath LA, Chowdhury N, Caraos L, Modak SM (1995) Infection resistance of surface modified catheters with either short-lived or prolonged activity. *J. Hosp. Infect.* 30: 201-210.
[216] Stewart PS (2002) Mechanisms of antibiotic resistance in bacterial biofilms. *Int. J. Med. Microbiol.* 292: 107-113.
[217] James G (2007) 50 wound study.
[218] Rogers LC, Bevilacqua NJ, Armstrong DG (2006) Ertapenem for diabetic foot infections. *Drugs Today* (Barc) 42: 695-701.
[219] Lipsky BA, Armstrong DG, Citron DM, Tice AD, Morgenstern DE, Abramson MA (2005) Ertapenem versus piperacillin/tazobactam for diabetic foot infections (SIDESTEP): prospective, randomised, controlled, double-blinded, multicentre trial. *Lancet* 366: 1695-1703.
[220] Rao N, Lipsky BA (2007) Optimising Antimicrobial Therapy in Diabetic Foot Infections. *Drugs* 67: 195-214.
[221] De Leo GA, Levin S (1997) The Multifaceted Aspects of Ecosystem Integrity. *Conservation Ecology* [online] 1:3.
[222] Rybak MJ, Bailey EM, Lamp KC, Kaatz GW (1992) Pharmacokinetics and bactericidal rates of daptomycin and vancomycin in intravenous drug abusers being treated for gram-positive endocarditis and bacteremia. *Antimicrob. Agents Chemother.* 36: 1109-1114.
[223] Curtin J, Cormican M, Fleming G, Keelehan J, Colleran E (2003) Linezolid compared with eperezolid, vancomycin, and gentamicin in an in vitro model of antimicrobial lock therapy for *Staphylococcus epidermidis* central venous catheter-related biofilm infections. *Antimicrob. Agents Chemother.* 47: 3145-3148.

[224] Dohar J, Giles W, Roland P, Bikhazi N, Carroll S, Moe R, Reese B, Dupre S, Wall M, Stroman D, McLean C, Crenshaw K (2006) Topical ciprofloxacin/dexamethasone superior to oral amoxicillin/clavulanic acid in acute otitis media with otorrhea through tympanostomy tubes. *Pediatrics* 118: e561-e569.

[225] Dowd SE, Wolcott RD, Kennedy J, Jones C, Cox SB (2011) Molecular diagnostics and personalised medicine in wound care: assessment of outcomes. *J. Wound Care* 20: 232, 234-232, 239.

[226] Finegold SM, Dowd SE, Gontcharova V, Liu C, Henley KE, Wolcott RD, Youn E, Summanen PH, Granpeesheh D, Dixon D, Liu M, Molitoris DR, Green JA, III (2010) Pyrosequencing study of fecal microflora of autistic and control children. *Anaerobe* 16: 444-453.

In: Biofilms in Bioengineering
Editors: Manuel Simões and Filipe Mergulhão
ISBN: 978-1-62948-161-6

Chapter X

Monofunctional to Multifunctional Coatings: The Changing Paradigm to Control Biofilm Formation on Biomedical Implants

***Yi Wang*[1,2]*, Matthew Libera*[2]*, Henk J. Busscher*[1] *and Henny C. van der Mei*[1,*]**

[1]Department of Biomedical Engineering, W. J. Kolff Institute,
University Medical Center Groningen,
University of Groningen, The Netherlands
[2]Department of Chemical Engineering and Materials Science,
Stevens Institute of Technology, Hoboken, NJ, US

Abstract

Infection is the number one cause of failure of biomaterials implants and devices, despite decades of research into the development of anti-adhesive coatings. In this chapter we postulate that whereas anti-adhesive coatings may be of value for particular applications, such as urinary or intravenous catheters, contact lens cases, and voice prostheses, the paradigm underlying the ideal biomaterial for biofilm control on totally internal, permanent implants needs to change if we want to effectively reduce the occurrence of biomaterials-associated infections. Rather than focusing on the design of monofunctional coatings, multifunctional coatings need to be developed. These must include both anti-adhesive and anti-microbial functionalities, promote a proper immune response, and stimulate tissue integration while simultaneously reducing the threat of bacterial colonization of an implant surface.

[*] E-mail: h.c.van.der.mei@umcg.nl.

Keywords: Anti-adhesive coatings; Biofilms; Biomaterials-associated infections; Monofunctional coatings; Multifunctional coatings; polyethylene glycol

Introduction

Biomaterials-associated infection (BAI) of medical implants and devices is a serious complication that often occurs after an otherwise technically successful surgical procedure or after the prolonged use of contact lenses, catheters, or other percutaneous devices. BAI of totally internal implants nearly always results in the surgical removal and replacement of the implant, because antibiotics are not able to effectively kill bacteria in their biofilm mode of growth that is typical on an implant surface. Similarly, the removal of an infected, temporary percutaneous device, such as a central intravenous catheter, poses a clinical dilemma. Removal is often synonymous to interruption of a desired therapy, and this can often have disastrous consequences in cases such as chemotherapies. On the other hand, not controlling the infection can also evolve into a life-threatening situation.

Over the past decades our knowledge of how bacteria adhere to surfaces and develop into biofilms has grown rapidly. Substratum surface properties like charge [1-3], hydrophobicity [4-6], and roughness [7-9] are now known to be major determinants of bacterial adhesion. Positively charged surfaces usually stimulate bacterial adhesion, because most bacterial strains carry a net negative cell surface charge [10]. However, at the same time, substrata carrying a positive surface charge through immobilization of, for example, quaternary ammonium compounds are known to become bactericidal upon contact [11, 12]. Acid-base interactions form the basis for the hydrophobicity of bacterial cell and substratum surfaces [13]. They also play a role in bacterial adhesion and, depending on the environmental conditions, can increase or decrease bacterial adhesion. Often hydrophobic surfaces attract equal numbers of adhering bacteria as do hydrophilic surfaces, but they appear free of biofilm after several hours or days under fluctuating environmental shear conditions [14]. Similarly, surface roughness has an effect on bacterial adhesion, but this depends on the scale of the roughness. Some studies have demonstrated that roughness on a scale much smaller than bacterial dimensions does not influence adhesion [15, 16], but other studies have shown that even nanometer-scale roughness may impact bacterial adhesion [16-18]. Interestingly, bacteria do not adhere preferentially on surface features such as scratches, but they appear to have better opportunities to grow into a biofilm once they adhere to surface irregularities of bacterial or larger dimensions due to the fact that they are protected against environmental shear forces [19].

The body of literature on bacterial adhesion mechanisms has stimulated the development of anti-adhesive coatings for biomaterials. Whereas anti-adhesiveness may be a sufficient single functionality for a coating when applied on such devices as urinary or intravenous catheters, contact lens cases, and voice prostheses, an anti-adhesive functionality is not sufficient for totally internal, permanent implant coatings like vascular grafts, surgical meshes, and hip or knee prostheses. Tissue integration has been described to offer the best protection of such implants [20, 21]. However, many microbial strains known to cause BAI, adhere to the same serum- or plasma-coated surface receptor sites as do mammalian cells [22], and they can thus colonize unmodified implant surfaces. When a coating is applied to such a surface, however, the anti-adhesive functionality is typically not limited to microorganisms but also to mammalian cells. Hence, many anti-adhesive coatings, while successfully repelling bacteria also repel the desirable tissue cells.

Since anti-adhesiveness is never absolute and may involve at best a 2 or 3-log units reduction in the number of adherent bacteria, implant surfaces can still become contaminated during surgery and hospitalization or become colonized by blood-born organisms from infections elsewhere in the body when not fully tissue integrated. Furthermore, the natural defenses offered by macrophages are often frustrated in the presence of a biomaterial, and macrophages are thus less able to eradicate the infecting organisms. Consequently, in addition to anti-adhesive functionalities, antimicrobial functionalities are desirable attributes for an implant coating, as well [23]. Antimicrobial functionalities can be introduced either by immobilizing antimicrobials on the surface yielding contact killing or by the application of antibiotic-release coatings.

Significantly, depending on the application, approximately 95% or more of all patients who receive biomaterial implants do not suffer adverse, long-term effects due to microbial interactions with the implant. However, the consequences when BAI does occur are severe. There can be substantial patient discomfort, complications that can lead to significant morbidity or mortality, significant stress on physicians and medical facilities, and a huge financial burden on the health care system. Clinically speaking, the major difference between revision patients receiving a secondary implant after BAI and primary implant patients receiving their first implant is that in revision surgery the tissue is compromised by infecting organisms [24]. Clearing the infection from surrounding tissue requires high local antibiotic concentrations, which can often be achieved by antibiotic-releasing beads, spacers, and sponges [25]. Thus, while antibiotic-release implant coatings can benefit primary orthopedic implants [26], such coatings are particular important for implants used in revision surgeries, and revision patients after BAI may require coatings with different functionalities than those used in a primary implant.

In this chapter we first briefly review past developments of biomaterials coatings and discuss their advantages and disadvantages. We argue that monofunctional coatings are recommended only for certain applications, and the community must look beyond these for greater success in applications where an implant cannot be easily removed and where mammalian cells and microbial cells compete for surface colonization. Effective, infection-resisting, permanent implants will require multifunctional surfaces that: (i) are not only anti-adhesive and antimicrobial but also stimulate tissue integration; (ii) do not hamper proper immune responses to microbial presence, and (iii) in revision surgery after BAI, clear infecting organisms from the tissue surrounding the implant.

Monofunctional Coatings

Table 1 summarizes some of the many different monofunctional coatings and surfaces that have been designed over the past decades. There are clearly no coatings that can be expected to perform well in each and every application. The general concern with many of these coatings, especially those showing anti-adhesive functionality, is tissue integration.

Table 1. Different monofunctional coatings that influence bacterial adhesion and biofilm formation together with some of their advantages and disadvantages

Types of coatings	Basis	Advantages (+) and disadvantages (-)	Ref
Anti-adhesive	Hydrophobic coatings	(+) reduced biofilm formation under fluctuating shear, depending on the strain involved (-) no tissue integration	[27-29]
	Hydrophilic coatings	(+) reduced adhesion, depending on the strain involved (+) good mammalian cell interaction	[30-32]
	Polymer brush coatings	(+) very low bacterial adhesion (-) biofilms do develop ultimately, though weakly adhering (-) no tissue integration	[33-38]
	Nano-patterned surfaces	(+) reduced biofilm formation (+) easy to clean	[39-41]
Tissue integrating	Fibronectin-based coatings	(+) good adhesion and spreading of tissue cells (-) many bacterial strains use the same receptor-sites for their adhesion as tissue cells	[42-44]
Immune-friendly	Polymer brush coatings	(+) allow macrophage mobility on a surface (-) no tissue integration	[45-48]
Antimicrobial	Positively-charged surfaces	(+) kill adhering organisms upon contact (-) layer of dead bacteria is left for subsequent adhesion	[49, 50]
	TiO_2 photocatalytic surfaces	(+) kill adhering bacteria after activation (+) can be re-activated depending on the application	[51-54]
Antibiotic-releasing	Antibiotic-loaded cements	(+) high burst release of antibiotics in surrounding tissue (-) long-term, low concentration tail-release may induce antibiotic resistance	[55]
	Antibiotic-loaded cements for beads, spacers and sponges	(+) high burst release of antibiotics in surrounding tissue (+) low tail release and short stay in the human body	[56]
	Antibiotic-releasing coatings	(+) high burst release of antibiotics in surrounding tissue (-) no tail-release due to limited reservoir volume	[50, 57, 58]

Multifunctional Coatings: A Shift in Paradigm

Baier in 1982 [59] suggested that scientists and engineers should take lessons from nature when designing infection-resisting biomaterials coatings. Unfortunately, his approach was not widely adopted at that time, possibly because Baier defined the ideal biomaterials surface predominantly in terms of the so-called critical surface tension. This concept has since been recognized as too simple to cover all functionalities of the natural endothelium, designed by nature to withstand colonization by infecting organisms. Following Baier and taking lessons from nature, we postulate that the ideal biomaterials coating should possess multiple functionalities. The design of such multifunctional surfaces clearly requires a change in paradigm, abandoning the common concept of monofunctionality.

In the next section of this chapter, we briefly describe several examples of bi-functional coatings. Then, once the concept of multifunctional coatings is adopted, the number of possibilities on how to differentially control surface interactions with mammalian cells and with infecting pathogens becomes huge.

Selected Examples

The Anti-adhesiveness of PEG-based Coatings

Poly(ethylene glycol) (PEG)-based coatings are highly hydrated gel-like structures that constitute the most anti-adhesive surfaces hitherto known [60]. Often alternatively described as PEO for poly(ethylene oxide), PEG chains attached to a flat surface can exist in two different conformations, the so called "mushroom structure" at low chain grafting densities and the "brush structure" at higher chain grafting densities, where the chains are forced to stretch in the medium (see Figure 1). Both structures are expected to create a barrier between the microorganism and the surface thereby preventing adhesion.

Since PEG-based coatings have superior anti-adhesive properties and, importantly, also allow for enhanced macrophage mobility, they are increasingly being used as a starting point for adding other functionalities [61-64]. Among these are antimicrobial character and/or tissue-integrating factors. In the design of a multifunctional PEG-based coating, the difference in dimensions between mammalian cells (average range from about 10 to 50 μm depending upon their degree of spreading) and bacteria (on average having a diameter of around 1 μm) needs to be taken into account. Antimicrobial or tissue-integrating functionalities must be applied in surface densities that do not eliminate the anti-adhesiveness of a multifunctional coating against the smaller bacteria.

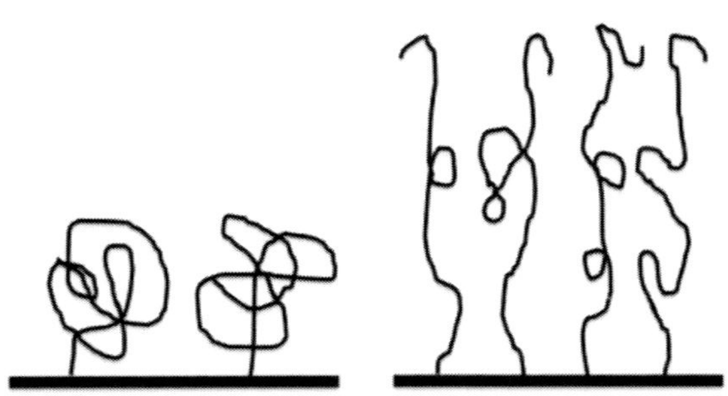

Figure 1. Schematic presentation of a monolayer PEG coating: mushroom conformation (left) and brush conformation (right).

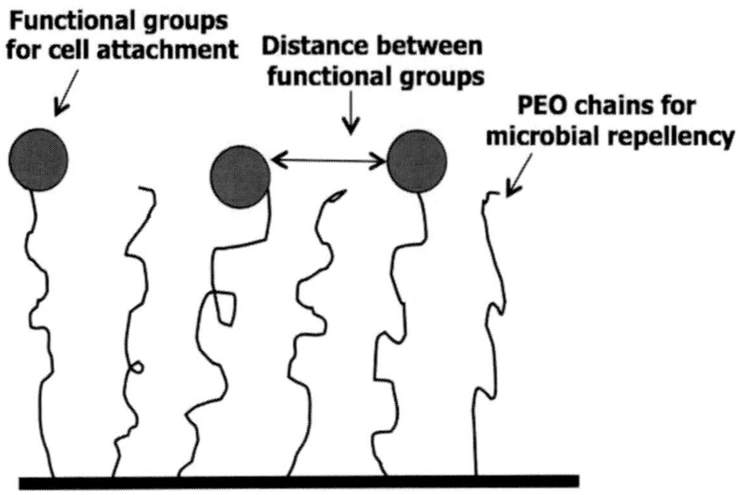

Figure 2. Schematic presentation of modified PEO polymer chains combining microbial repellency with cell-adhesive functional groups. The distance between the functional groups for cell attachment critically determines whether microbial repellency will be maintained or not.

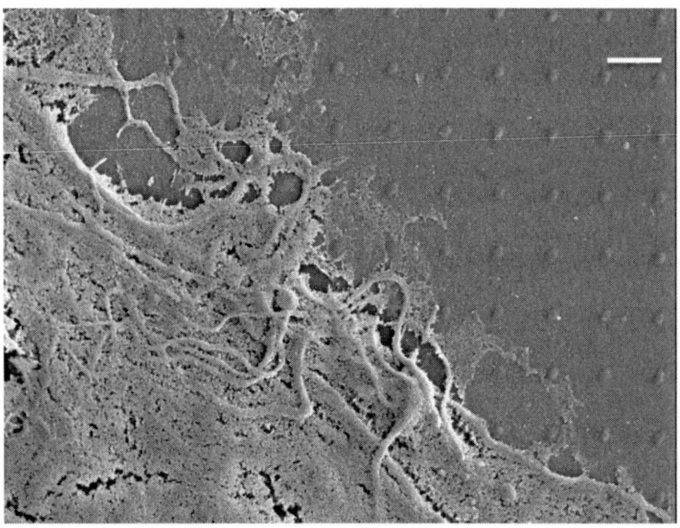

Figure 3. Scanning electron microscope image of a mammalian cell spreading on a PEG-microgel patterned surface with an inter-gel spacing of 1.5 μm. Dotted structures are PEG-microgels, while extensions from the cell circumference can be seen to grow over the microgels in search of the adhesive surface sites. The bar represents 1.5 μm.

PEG-based Anti-adhesive Coating with an Antimicrobial Functionality

PEG-based coatings on silicone rubber strongly reduce bacterial adhesion, but, despite 2-log units reduction in initial adhesion, significant numbers of biofilms can nevertheless form [65]. Conjugation of lysozyme to PEG-molecules and coating of silicone rubber surfaces with 68/32% PEG/lysozyme has yielded a bi-functional polymer brush, possessing both anti-adhesive activity due to the polymer brush combined with the antibacterial activity of lysozyme [66].

PEG-based Anti-adhesive Coating with RGD-containing Peptides

The anti-adhesiveness of PEG-based coatings extends to mammalian cells. In order to prepare bi-functional coatings which prevent microbial adhesion while supporting tissue cell growth, a biologically inert poly(L-lysine)-graft-poly(ethylene glycol) (PLL-g-PEG) copolymer has been equipped with the arginine-glycine-aspartic acid (RGD) peptide sequence [67]. This RGD peptide is known as one of the major recognition sites of integrin receptors through which cells connect to their extracellular matrix [68]. Reduced bacterial adhesion on bi-functional PLL-g-PEG/PEG-RGD-modified surfaces has been demonstrated separately from their ability to support tissue cell growth [69, 70]. Significantly, when the "race for the surface" between staphylococci and osteoblasts was studied on such bi-functional coatings in a co-culture experiment [71], mammalian cells appeared to be much more at an advantage than on common biomaterials surfaces.

Differentially Adhesive Surfaces

PEG-based submicron-sized, non-adhesive microgels patterned on an otherwise cell-adhesive surface have recently been described as another example of a bi-functional, anti-adhesive and tissue-integrating coating. Such a structure creates a surface which is largely cell-adhesive but contains non-adhesive features, which, when spaced at distances comparable to bacterial dimensions, yield dramatically reduced staphylococcal deposition rates [72]. This has been attributed to the fact that staphylococci have relatively rigid cell walls that cannot easily conform to the substrate under conditions of modulated adhesiveness. In contrast, mammalian cells have flexible cell membranes and adhere to extracellular structures via transmembrane integrins. Mammalian cells can consequently adhere to surfaces with modulated cell adhesiveness by adhering to adhesive patches in between non-adhesive structures (Figure 3).

Conclusion

A great variety of coatings have been described in the literature and claim to be promising for the prevention of biomaterials-associated infection. Most are based on reducing bacterial adhesion and the subsequent formation of surface-attached biofilms. While such coatings are useful in certain applications, there has been a general lack of clinically favorable outcomes when using these coatings on internal biomedical devices such as implanted prostheses. Applications such as these, which are meant to be both totally internal and permanent, require more cleverly designed coatings possessing multiple functionalities. Importantly, this approach represents a significant paradigm change from one where coatings are designed to reduce infection by minimizing all biological interaction to one where interactions with desirable tissue cells are amplified while bacteria are simultaneously killed or blocked from the surface. Such an approach will require a far better understanding of how multiple cell types interact with synthetic surfaces together with a far greater level of creativity about how to design surfaces with differential cell-interactive properties.

References

[1] Jones, C. R., Adams, M. R., Zhdan, P. A. & Chamberlain, A. H. (1999). The role of surface physicochemical properties in determining the distribution of the autochthonous microflora in mineral water bottles. *J. Appl. Microbiol. 86*, 917-927.

[2] Legeay, G., Poncin-Epaillard, F. & Arciola, C. R. (2006). New surfaces with hydrophilic/hydrophobic characteristics in relation to (no)bioadhesion. *Int. J. Artif. Organs. 29*, 453-461.

[3] Gottenbos, B., Van der Mei, H. C., Klatter, F., Grijpma, D. W., Feijen, J., Nieuwenhuis, P. & Busscher, H. J. (2003) Positively charged biomaterials exert antimicrobial effects on gram-negative bacilli in rats. *Biomaterials. 24*, 2707-2710.

[4] Andersen, T. E., Kingshott, P., Palarasah, Y., Benter, M., Alei, M. & Kolmos, H. J. (2010). A flow chamber assay for quantitative evaluation of bacterial surface colonization used to investigate the influence of temperature and surface hydrophilicity on the biofilm forming capacity of uropathogenic *Escherichia coli. J. Microbiol. Methods. 81*, 135-140.

[5] Schwarz, F., Sculean, A., Wieland, M., Horn, N., Nuesry, E., Bube, C. & Becker, J. (2007). Effects of hydrophilicity and microtopography of titanium implant surfaces on initial supragingival plaque biofilm formation. A pilot study. *Mund Kiefer Gesichtschir. 11*, 333-338.

[6] Almaguer-Flores, A., Olivares-Navarrete, R., Wieland, M., Ximenez-Fyvie, L. A., Schwartz, Z. & Boyan, B. D. (2012). Influence of topography and hydrophilicity on initial oral biofilm formation on microstructured titanium surfaces in vitro. *Clin. Oral Implants Res. 23*, 301-307.

[7] Wu, Y., Zitelli, J. P., TenHuisen, K. S., Yu, X. & Libera, M. R. (2011). Differential response of Staphylococci and osteoblasts to varying titanium surface roughness. *Biomaterials. 32*, 951-960.

[8] Tang, L., Pillai, S., Revsbech, N. P., Schramm, A., Bischoff, C. & Meyer, R. L. (2011). Biofilm retention on surfaces with variable roughness and hydrophobicity. *Biofouling. 27*,111-121.

[9] Charman, K. M., Fernandez, P., Loewy, Z. & Middleton, A. M. (2009). Attachment of *Streptococcus oralis* on acrylic substrates of varying roughness. *Lett. Appl. Microbiol. 48*, 472-477.

[10] Jucker, B. A., Harms, H. & Zehnder, A. J. (1996). Adhesion of the positively charged bacterium *Stenotrophomonas (Xanthomonas) maltophilia* 70401 to glass and Teflon. *J. Bacteriol. 178*, 5472-5479.

[11] Tiller, J. C., Liao, C. J., Lewis, K. & Klibanov, A. M. (2001). Designing surfaces that kill bacteria on contact. *Proc. Natl. Acad. Sci. U. S. A. 98*, 5981-5985.

[12] Majumdar, P., Lee, E., Patel, N., Ward, K., Stafslien, S. J., Daniels, J., Chrisholm, B. J., Boudjouk, P., Callo, M. J., Callow, J. A. & Thompson, S. E. (2008). Combinatorial materials research applied to the development of new surface coatings IX: an investigation of novel antifouling/fouling-release coatings containing quaternary ammonium salt groups. *Biofouling. 24*, 185-200.

[13] Van Oss, C. J. (1995) Hydrophobicity of biosurfaces -- Origin, quantitative determination and interaction energies. *Colloids and Surfaces B: Biointerfaces. 5*, 91-110.

[14] Bos, R., Van der Mei, H. C. & Busscher, H. J. (1999). Physico-chemistry of initial microbial adhesive interactions--its mechanisms and methods for study. *FEMS Microbiol. Rev. 23*, 179-230.

[15] Li, B. & Logan, B. E. (2004). Bacterial adhesion to glass and metal-oxide surfaces. *Colloids Surf. B Biointerfaces. 36*, 81-90.

[16] Mitik-Dineva, N., Wang, J., Mocanasu, R. C., Stoddart, P. R., Crawford, R. J. & Ivanova, E. P. (2008). Impact of nano-topography on bacterial attachment. *Biotechnol. J. 3*, 536-544.

[17] Mitik-Dineva, N., Wang, J., Truong, V. K., Stoddart, P. R., Malherbe, F., Crawford, R. J. & Ivanova, E. P. (2009). Differences in colonisation of five marine bacteria on two types of glass surfaces. *Biofouling. 25*, 621-631.

[18] Emerson, R. J. 4th, Bergstrom, T. S., Liu, Y., Soto, E. R., Brown, C. A., McGimpsey, W. G. & Camesano, T. A. (2006). Microscale correlation between surface chemistry, texture, and the adhesive strength of *Staphylococcus epidermidis*. *Langmuir. 22*, 11311-11321.

[19] Charman, K. M., Fernandez, P., Loewy, Z. & Middleton, A. M. (2009). Attachment of *Streptococcus oralis* on acrylic substrates of varying roughness. *Lett. Appl. Microbiol. 48*, 472-477.

[20] Gristina, A. G. (1987). Biomaterial-centered infection: microbial adhesion versus tissue integration. *Science. 237*, 1588-1595.

[21] Deacon, J. M., Pagliaro, A. J., Zelicof, S. B. & Horowitz, H. W. (1996). Prophylactic use of antibiotics for procedures after total joint replacement. *J. Bone Joint Surg. Am. 78*, 1755-1770.

[22] Patti, J. M., Allen, B. L., McGavin, M. J. & Hook, M. (1994). MSCRAMM-mediated adherence of microorganisms to host tissues. *Annu. Rev. Microbiol. 48,* 585-617.

[23] Boelens, J. J., Dankert, J., Murk, J. L., Weening, J. J., Van der Poll, T., Dingemans, K. P., Koole, L., Laman, J. D. & Zaat, S. A. J. (2000). Biomaterial-associated persistence of *Staphylococcus epidermidis* in pericatheter macrophages. *J. Infect. Dis. 181*, 1337-1349.

[24] Broekhuizen, C. A. N., Schultz, M. J., Van der Wal, A. C., Boszhard, L., De Boer, L., Vandenbroucke-Grauls, C. M. J. E. & Zaat, S. A. J. (2008). Tissue around catheters is a niche for bacteria associated with medical device infection. *Crit. Care Med. 36*, 2395-2402.

[25] Neut, D., Van de Belt, H., Van Horn, J. R., Van der Mei, H. C. & Busscher, H. J. (2003). Residual gentamicin-release from antibiotic-loaded polymethylmethacrylate beads after 5 years of implantation. *Biomaterials. 24*, 1829-1831.

[26] Engesaeter, L. B., Lie, S. A., Espehaug, B., Furnes, O., Vollset, S. E. & Havelin, L. I. (2003). Antibiotic prophylaxis in total hip arthroplasty: effects of antibiotic prophylaxis systemically and in bone cement on the revision rate of 22,170 primary hip replacements followed 0-14 years in the Norwegian Arthroplasty Register. *Acta Orthop. Scand. 74*, 644-651.

[27] Legeay, G., Poncin-Epaillard, F. & Arciola, C. R. (2006). New surfaces with hydrophilic/hydrophobic characteristics in relation to (no)bioadhesion. *Int. J. Artif. Org. 29*, 453-461.

[28] Harris, L. G., Mead, L., Muller-Oberlander, E. & Richards, R. G. (2006). Bacteria and cell cytocompatibility studies on coated medical grade titanium surfaces. *J. Biomed. Mater. Res A. 78*, 50-58.

[29] Karakecili, A. G. & Gumusderelioglu, M. (2002). Comparison of bacterial and tissue cell initial adhesion on hydrophilic/hydrophobic biomaterials. *J. Biomater Sci. Polym. Ed. 13*, 185-196.

[30] Tenke, P., Riedl, C. R., Jones, G. L., Williams, G. J., Stickler, D. & Nagy, E. (2004). Bacterial biofilm formation on urologic devices and heparin coating as preventive strategy. *Int. J. Antimicrob. Agents. 23*, S67-74.

[31] An, Y. H., Bradley, J., Powers, D. L. & Friedman, R. J. (1997). The prevention of prosthetic infection using a cross-linked albumin coating in a rabbit model. *J. Bone Joint Surg. Br. 79*, 816-819.

[32] Wilson, M. & Harvey, W. (1989). Prevention of bacterial adhesion to denture acrylic. *J. Dent. 17*, 166-170.

[33] Pechey, A., Elwood, C. N., Wignall, G. R., Dalsin, J. L., Lee, B. P., Vanjecek, M., Welch, I., Ko, R., Razvi, H. & Cadieux, P. A. (2009). Anti-adhesive coating and clearance of device associated uropathogenic *Escherichia coli* cystitis. *J. Urol. 182*, 1628-1636.

[34] Zoulalian, V., Zurcher, S., Tosatti, S., Textor, M., Monge, S. & Robin, J. J. (2010). Self-assembly of poly(ethylene glycol)-poly(alkyl phosphonate) terpolymers on titanium oxide surfaces: synthesis, interface characterization, investigation of nonfouling properties, and long-term stability. *Langmuir. 26*, 74-82.

[35] Paulsson, M., Kober, M., Freij-Larsson, C., Stollenwerk, M., Wesslen, B. & Ljungh, A. (1993). Adhesion of staphylococci to chemically modified and native polymers, and the influence of preadsorbed fibronectin, vitronectin and fibrinogen. *Biomaterials.14*, 845-853.

[36] Sankar, S. & Rajalakshmi, T. (2007). Application of poly ethylene glycol hydrogel to overcome latex urinary catheter related problems. *Biofactors. 30*, 217-225.

[37] Muller, R., Ruhl, S., Hiller, K. A., Schmalz, G. & Schweikl, H. (2008). Adhesion of eukaryotic cells and *Staphylococcus aureus* to silicon model surfaces. *J. Biomed. Mater. Res. A. 84*, 817-827.

[38] Saldarriaga Fernandez, I. C., Van der Mei, H. C., Metzger, S., Grainger, D. W., Engelsman, A. F., Nejadnik, M. R., & Busscher, H. J. (2010). In vitro and in vivo comparisons of staphylococcal biofilm formation on a cross-linked poly(ethylene glycol)-based polymer coating. *Acta Biomater. 6*, 1119-1124.

[39] Hannig, M., Kriener, L., Hoth-Hannig, W., Becker-Willinger, C. & Schmidt, H. (2007). Influence of nanocomposite surface coating on biofilm formation in situ. *J. Nanosci. Nanotechnol. 7*, 4642-4648.

[40] Diaz, C., Schilardi, P. L., Salvarezza, R. C. & De Mele, M. F. (2007). Nano/microscale order affects the early stages of biofilm formation on metal surfaces. *Langmuir. 23*, 11206-11210.

[41] Battin, T. J., Kammer, F. V., Weilhartner, A., Ottofuelling, S. & Hofmann, T. Nanostructured TiO2: transport behavior and effects on aquatic microbial communities under environmental conditions. (2009). *Environ. Sci. Technol. 43*, 8098-8104.

[42] Tweden, K. S., Harasaki, H., Jones, M., Blevitt, J. M., Craig, W. S., Pierschbacher, M. & Helmus M. N. (1995). Accelerated healing of cardiovascular textiles promoted by an RGD peptide. *J. Heart Valve Dis. 4*, S90-97.

[43] Clubb, F. J., Jr., Clapper, D. L., Deferrari, D. A., Hu, S. P., Seare, W. J., Jr., Capek, P. P., Armstrong, J., McGee, M. G., Billings, L. A., Fuqua, J. M. & Parins, S. M. (1999). Surface texturing and coating of biomaterial implants: effects on tissue integration and fibrosis. *Asaio J. 45*, 281-287.

[44] Petrie, T. A., Raynor, J. E., Reyes, C. D., Burns, K. L., Collard, D. M. & Garcia, A. J. (2008). The effect of integrin-specific bioactive coatings on tissue healing and implant osseointegration. *Biomaterials. 29*, 2849-2857.

[45] Hemmerlein, J. B., Trerotola, S. O., Kraus, M. A., Mendonca, M. S. & Desmond, L. A. (1997). In vitro cytotoxicity of silver-impregnated collagen cuffs designed to decrease infection in tunneled catheters. *Radiology. 204*, 363-367.

[46] Atiyeh, B. S., Costagliola, M., Hayek, S. N. & Dibo, S. A. (2007). Effect of silver on burn wound infection control and healing: review of the literature. *Burns. 33*, 139-148.

[47] Widgerow, A. D. (2010). Nanocrystalline silver, gelatinases and the clinical implications. *Burns. 36*, 965-974.

[48] Bright, K. R., Gerba, C. P. & Rusin, P. A. (2002). Rapid reduction of *Staphylococcus aureus* populations on stainless steel surfaces by zeolite ceramic coatings containing silver and zinc ions. *J. Hosp. Infect. 52*, 307-309.

[49] Cokeliler, D., Goktas, H., Tosun, P. D. & Mutlu, S. (2010). Infection free titanium alloys by stabile thiol based nanocoating. *J. Nanosci. Nanotechnol. 10*, 2583-2589.

[50] Norowski, P. A., Jr. & Bumgardner, J. D. (2009). Biomaterial and antibiotic strategies for peri-implantitis: a review. *J. Biomed. Mater. Res. B Appl. Biomater. 88*, 530-543.

[51] Hochbaum, A. I. & Aizenberg, J. (2010). Bacteria pattern spontaneously on periodic nanostructure arrays. *Nano Lett. 10*, 3717-3721.

[52] Kappell, G. M., Grover, J. P. & Chrzanowski, T. H. (2009). Micro-scale surface-patterning influences biofilm formation. *Elec. J. Biotechnol. 12*, 1-8.

[53] Krsko, P., McCann, T. E., Thach, T. T., Laabs, T. L., Geller, H. M. & Libera, M. R. (2009). Length-scale mediated adhesion and directed growth of neural cells by surface-patterned poly(ethylene glycol) hydrogels. *Biomaterials. 30*, 721-729.

[54] Chung, K. K., Schumacher, J. F., Sampson, E. M., Burne, R. A., Antonelli, P. J. & Brennan, A. B. (2007). Impact of engineered surface microtopography on biofilm formation of *Staphylococcus aureus. Biointerphases. 2*, 89-94.

[55] Neut, D., Dijkstra, R. J., Thompson, J. I., Van der Mei, H. C. & Busscher, H. J. (2011). Antibacterial efficacy of a new gentamicin-coating for cementless prostheses compared to gentamicin-loaded bone cement. *J. Orthop. Res. 29*, 1654-1661.

[56] Diefenbeck, M., Muckley, T. & Hofmann, G. O. (2006). Prophylaxis and treatment of implant-related infections by local application of antibiotics. *Injury. 37*, 95-104.

[57] Kwok, C. S., Horbett, T. A., Ratner, B. D. (1999). Design of infection-resistant antibiotic-releasing polymers. II. Controlled release of antibiotics through a plasma-deposited thin film barrier. *J. Control Release. 62*, 301-311.

[58] Davidoff, S. N., Call, B. P., Hogrebe, P. C., Grainger, D. W. & Brooks, A. E. (2010). A robust method to coat allograft bone with a drug-releasing polymer shell - biomed 2010. *Biomed. Sci. Instrum. 46*, 184-189.

[59] Baier, R. E. (1982). Conditioning surfaces to suit the biomedical environment: recent progress. *J. Biomech. Eng. 104*, 257-271.

[60] Schulte, V. A., Diez, M., Hu, Y., Moller, M. & Lensen, M. C. (2010). Combined influence of substrate stiffness and surface topography on the antiadhesive properties of Acr-sP(EO-stat-PO) hydrogels. *Biomacromolecules. 11*, 3375-3383.

[61] Zhang, Y., Kohler, N. & Zhang, M. (2002). Surface modification of superparamagnetic magnetite nanoparticles and their intracellular uptake. *Biomaterials. 23*, 1553-1561.

[62] Jang, J. Y., Lee, D. Y., Park, S. J. & Byun, Y. (2004). Immune reactions of lymphocytes and macrophages against PEG-grafted pancreatic islets. *Biomaterials. 25*, 3663-3669.

[63] Schiffelers, R., Storm, G. & Bakker-Woudenberg, I. (2001). Liposome-encapsulated aminoglycosides in pre-clinical and clinical studies. *J. Antimicrob. Chemother. 48*, 333-344.

[64] Park, D., Wu, W. & Wang, Y. (2011). A functionalizable reverse thermal gel based on a polyurethane/PEG block copolymer. *Biomaterials. 32*, 777-786.

[65] Nejadnik, M. R., Van der Mei, H. C., Norde, W. & Busscher, H. J. (2008). Bacterial adhesion and growth on a polymer brush-coating. *Biomaterials. 29*, 4117-4121.

[66] Muszanska, A. K., Busscher, H. J., Herrmann, A., Van der Mei, H. C. & Norde W. Pluronic-lysozyme conjugates as anti-adhesive and antibacterial bifunctional polymers for surface coating. *Biomaterials. 32*, 6333-6341.

[67] VandeVondele, S., Voros, J. & Hubbell, J. A. (2003). RGD-grafted poly-L-lysine-graft-(polyethylene glycol) copolymers block non-specific protein adsorption while promoting cell adhesion. *Biotechnol. Bioeng. 82*, 784-790.

[68] LeBaron, R. G. & Athanasiou, K. A. (2000). Extracellular matrix cell adhesion peptides: functional applications in orthopedic materials. *Tissue Eng. 6*, 85-103.

[69] Maddikeri, R. R., Tosatti, S., Schuler, M., Chessari, S., Textor, M., Richards, R. G. & Harris, L. D. (2008). Reduced medical infection related bacterial strains adhesion on bioactive RGD modified titanium surfaces: a first step toward cell selective surfaces. *J. Biomed. Mater. Res. A. 84*, 425-435.

[70] Shi, Z., Neoh, K. G., Kang, E. T., Poh, C. & Wang, W. (2008). Bacterial adhesion and osteoblast function on titanium with surface-grafted chitosan and immobilized RGD peptide. *J. Biomed. Mater. Res. A. 86*, 865-872.

[71] Subbiahdoss, G., Kuijer, R., Grijpma, D. W., Van der Mei, H. C. & Busscher, H. J. (2009). Microbial biofilm growth vs. tissue integration: "the race for the surface" experimentally studied. *Acta Biomater. 5*, 1399-1404.

[72] Wang, Y., Subbiahdoss, G., Swartjes, J., Van der Mei, H. C., Busscher, H. J. & Libera, M. (2011). Length-scale mediated differential adhesion of mammalian cells and microbes. *Adv. Funct. Mater. 21,* 3916-3923.

In: Biofilms in Bioengineering
Editors: Manuel Simões and Filipe Mergulhão
ISBN: 978-1-62948-161-6

Chapter XI

Antibiofilm Strategies

Ankita Pagedar Singh Ph. D.[1,*] ***and Jitender Singh Ph. D.***[2,†]
[1]A&R Patel Institute of Integrated Study & Research in Biotechnology and Allied Sciences (ARIBAS), New Vallabh Vidyanagar, Gujarat, India
[2]Product and Process Development Group,
National Dairy Development Board, Gujarat, India

Abstract

The cell-to-cell and cell-to-surface interactions play a pivotal role in the initiation and establishment of biofilms, which pose a multitude of problems in almost every aspect of our day to day lives. The success of any antibiofilm strategy lies in its efficacy to reduce initial adhesion, establishment and enhance disintegration of the biofilms. In healthcare industry, application of antibiotics alone or in combination with biocide is the first line of action to treat biofilms on animate and inanimate surfaces. However, other industries rely on economically viable chemical interventions in combination with physical methods. Nevertheless, the associated environmental issues of residual toxicity and the emergence of antimicrobial resistance have highlighted the need for alternative approaches. In this book chapter, various antibiofilm strategies, both established and emerging, comprising physical methods such as: photodynamic antimicrobial chemotherapy, bioelectric effect and sonication; biological methods using bacteria, bacteriophages, biological molecules (enzymes, antibodies, fatty acids, and maggot excretion/secretion); and chemical methods using: antimicrobial surfaces, inhibitors of modified nucleotide synthesis, quorum sensing inhibitors, phytochemicals, uncouplers of oxidative phosphorylation, and biocides are discussed. Unfortunately, a single key antibiofilm solution is not applicable and/or equally effective across the various microbial groups and industries. Probably, antibiofilm strategies will continue to evolve with a better understanding of the underlying mechanisms of biofilm formation.

* Email: ankitapagedar@gmail.com.
† E-mail: jitender9@gmail.com.

Keywords: Bacteriophages; Biocides; Biofilm control, Modified nucleotide; Quorum sensing; Uncoupler

Introduction

The concept and effect of surface adhered microbes evolved during mid 1920s to the late 1930s, [1] and decades later the term "biofilm" came into existence [2]. The biofilm involves microbial consortium enclosed in matrix of extracellular polymeric substances (EPS) that provide structural/social integrity, protection against antimicrobials and environmental stress. On one hand, the biofilm is most successful growth mode exhibited by more than 99% of the organisms in natural habitat [3], while on the other it poses multitude of problems in various industries. According to National Institute of Health, 80% of all the infections are biofilm related [4]. Biofilms in food industry are responsible for obstructions in heat exchange processes, inadequate cleaning regimes, food spoilage and foodborne outbreaks. In water and sewage treatment systems, regular replacement of the filtration membrane due to its clogging by microbial biofilms is the major financial burden. In paper industry, the biofilm produced microbial slime imparts poor quality to the paper. In shipping and marine industry, the biofouling mediated corrosion is responsible for reduced life of ship hulls. Degradation of petroleum products and clogging of oil pipelines are biofilm mediated concerns in fuel and energy industry. In short, there is hardly any aspect of human society where biofilms do not make their presence felt.

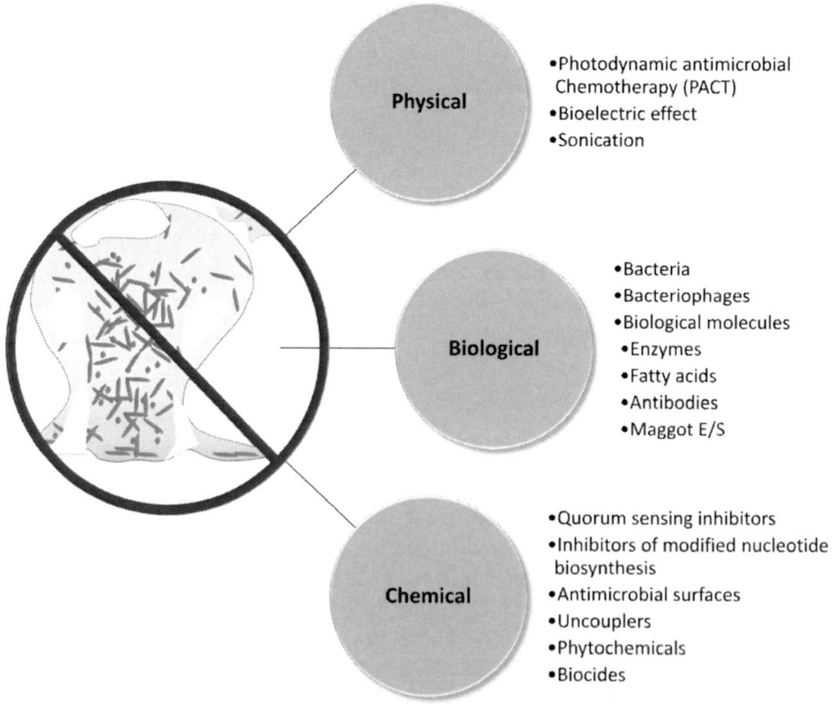

Figure 1. Antibiofilm strategies: physical, chemical and biological methods.

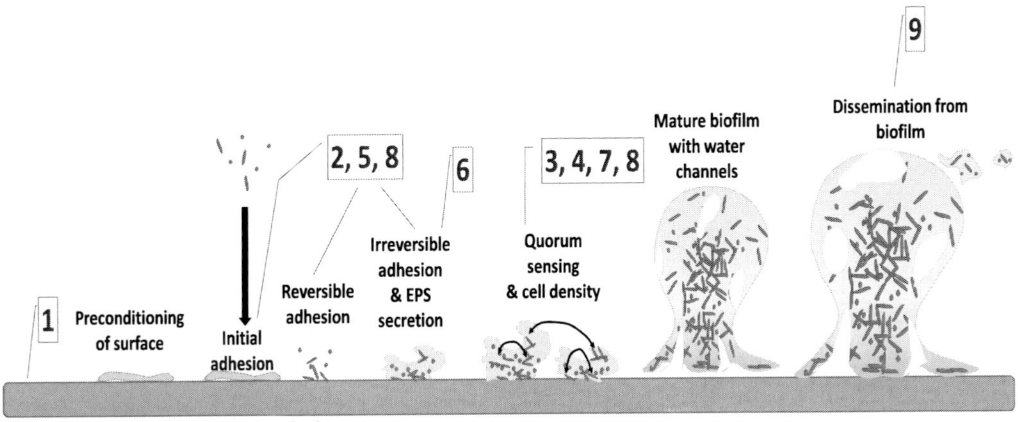

1) Modified surface (A, L, N, O).
2) Antimicrobial substance (A, E, L, M, O).
3) Altered cell components and membrane (B, E, F, G, J, N, O).
4) Enhanced cellular metabolism to increase susceptibility to biocides and antibiotics (C, D).
5) Prevention of adherence (I, O).
6) EPS and matrix degradation (F, G, K, M).
7) Inhibition of quorum sensing (D, H, J).
8) Competitive exclusion (E, I).
9) Enhanced biofilm disintegration and detachment (C, E).

Figure 2. Target sites of antibiofilm strategies during the process of biofilm formation. A - Antimicrobial surface (prevents adhesion and exudes antimicrobial substance); B - Photodynamic antimicrobial chemotherapy (PACT) (reactive radicals attacking macromolecules, cell components & membrane); C - Bioelectric effect (electro-osmotic fluid/ions, detachment & disintegration of biofilms, enhance cellular metabolism); D - Sonication (microstreaming disrupts autoinducer gradient, enhance cellular metabolism); E - Microorganisms (antimicrobial metabolite, free radicals, competitive exclusion, enhanced dispersion); F - Bacteriophage (cell lysis, EPS degradation); G - Enzymes (degradation of EPS and other matrix components); H - Quorum sensing inhibitors (inhibit autoinducer production, their receptors & effector proteins); I - Immunological (reduce microbial interaction; enhance opsonization of matrix & cells); J - Fatty acids (cell lysis, disruption of quorum sensing, nutrient uptake & enzymatic activity); K - Inhibitors of modified nucleotide biosynthesis (inhibit EPS production); L - Antimicrobial surface (prevent initial adhesion, release antimicrobial substance); M - Maggot E/S (degrade EPS; production of antimicrobial substance); N - Uncouplers of oxidative phosphorylation (inhibition of ATP synthesis, enhance membrane permeability, oxidative stress); O - Biocides (antimicrobial substance, modify surface properties to prevent adherence, disintegration of biofilm.

The control of biofilms is very arduous such that even wonder drugs like antibiotics are rendered ineffective. Overall, an ideal antibiofilm strategy should involve approaches or combinations thereof to prevent the initial adhesion step, disintegrate, and disinfect the preformed biofilm. Throughout this chapter, approaches exhibiting any one or more of these qualities are discussed as antibiofilm. Practically, there is nothing like universal antibiofilm approach that can be applied simultaneously on animate and inanimate objects. The selection criteria of single or combinatorial approaches are cost, site of application, delivery at site, dosing, frequency of application, specificity, contact time, biofilm stage and heterogeneity, residual effect, nature of biofilm (static or dynamic), safety guidelines (if any), etc. The biofilm control measures can be broadly categorized into physical, chemical and biological (Figure 1). Healthcare industries generally rely on the application of biochemical methods for biofilm mitigation. These methods involve application of enzymes, bacteriophages, quorum sense inhibitors, surface modification and novel compounds and metabolites whereas non healthcare industries majorly rely more on physico-chemical methods such as application of biocides, sonication, detergents, uncouplers and surface modification, although other methods or combinations are also being explored. The present chapter entails the established and emerging strategies of biofilm control in various industries along with their pros and cons.

Antibiofilm Strategies

Physical Methods

Photodynamic Antimicrobial Chemotherapy (PACT)

The antimicrobial effect of photodynamic therapy is derived from photosensitization/irradiation of certain chemicals such as phenothiazinium, phenazine, acridine, cyanine, porphyrin, phthalocyanine, psoralen, perylenequinonoid which release microbial cell specific reactive radicals [5]. The photosensitization/irradiation is done with light of a specific wavelength which corresponds to the absorption maxima of the phosensitizers/chemical used. The light source may vary from lasers to tungsten filament lamps. Different combinations of light source parameters like illumination time and power density are used to achieve the desired antimicrobial/antibiofilm effect. Optimal antimicrobial effect is observed when the absorption maxima of sensitizer at longer wavelengths is capable to generate reactive radicals and penetrate the biofilm matrix or biofilm invaded tissue farther without collateral damage of surrounding tissue [6].

The antimicrobial effect of PACT depends on preferential localization of photosensitizer in the bacterial cell components and macromolecules such as cell wall/membrane, unsaturated lipids, amino acids, peptides, viral proteins, respiratory chain, cytoplasmic or viral enzymes, and nucleic acids (typically guanosine). Photosensitizer, toluidine blue O (TBO), affects cell membrane permeability while methylene blue causes nucleic acid strand breakage in *E. coli* [7]. Hematoporphyrin has been reported to be effective against a number of organisms such as *Staphylococcus aureus, Bacillus subtilis* and *Streptococcus faecalis* in planktonic phase [8]. Photosensitizers such as aluminium disulphonated phthalocyanine against *S. sanguis* [9], photoporphyrin IX against *Pseudomonas aeruginosa* [10], merocyanine 540 [11] and TBO [12] against biofilm and planktonic *S. aureus* have been proven effective. The use of δ-aminolaevulinic acid as photosensitizer (20mM with 240 J/cm^2) against *P. aeruginosa* biofilms resulted in complete disinfection although regrowth of biofilms (after 24 h) was observed when the PACT treated biofilm was allowed to grow in a sterile reactor [10]. Single PACT treatment might not be sufficient to prevent regrowth of biofilms but double PACT treatment showed promising results to prevent biofilm regrowth. Erythrosin irradiated with white light (500–650 nm) has been reported to be effective against an aged biofilm of *S. mutans* [13], while TBO was effective against a younger biofilm [14]. Such stage specific eradication of biofilm might be useful in selecting a specific photosensitizer as per severity/stage of biofilm mediated infections.

The biofilm matrix may limit the penetrability of the photosensitizers or the impinging light. Hence, in comparison to suspension, inactivation of *S. aureus* biofilm has been reported to require higher dose of either photosensitizer (40 μM threshold concentration of TBO) or light (25 to 200 J/cm^2) [15]. However, the applications of metal chelators have proven to complement poor penetrability of photosensitizer by rendering the biofilms more permeable [16]. Recently, investigators have used polymer-based nanoparticles such as the Food and Drug Administration (FDA) approved poly(lactic-co-glycolic acid) (PLGA) for photosensitizer delivery and release systems (especially biodegradable, biocompatible and biodistributable) [17]. The excited state of the photosensitizer is quenched when it is within the PLGA and phototoxicity is regained when nanoparticles are incubated with target cells [18]. Pagonis et al used methylene blue loaded PLGA nanoparticles against planktonic and root canal biofilms formed by *Enterococcus faecalis* and observed approximately 44% and 58% reduction of bacterial viability, respectively [17]. Synergism was observed against *E. faecalis* biofilm (1 log reduction) for the combination of nanoparticles (50 mg ml^{-1} equivalent to methylene blue) and impinging light (red light at 665 nm; 30 J/cm^2).

PACT has been approved for clinical treatment in the United States, the European Union, Canada, Russia, and Japan [19]. The therapy has been widely used against dental biofilm [13] and disinfection of whole blood samples, but further studies are required for its effective implementation to cure tissue invaded by biofilms to a deeper level, catheter tubes and implants. Overall there are various advantages associated with PACT viz. rapid application, non-mutagenic, suitable for localized application which prevents spread of multiresistant pathogens, readily available and inexpensive. Such advantage of PACT makes photosensitizers attractive for treatment against biofilms. However, PACT can only be applied on easily accessible surfaces such as dental biofilms, and localized wound/organ infection but not for systemic infections.

PACT as an antibiofilm therapy is still in its initial days and future research addressing the underlying mechanisms and standardization of procedures (in terms of required dose of photosensitizers, impinging light duration, intensity), side effects like host toxicity, triggering of the inflammatory immune response along with the discovery/synthesis of broad spectrum photosensitizers or those specific to organ/cell, immune response and microorganisms would be helpful in its wider application for biofilm mitigation.

Bioelectric Effect

Bioelectric effect is a synergistic effect of antimicrobials in combination with electric current (alternating or direct), magnetic and electromagnetic fields against biofilms [20]. While introducing the concept, Costerton et al., [20] observed that lower concentration of antibiotics is required to remove biofilm when used in combination with direct current (DC) electric fields of about 1.5 to 20 V/cm (current densities of about 15 X 10^{-6} to 2.1 X 10^{-3} A/cm^2).

Block current application on the biofilm generates the flow of electro-osmotic fluid and hydrated ions, resulting in enhanced detachment of the adhered cells [21]. However, such circulation might be inhibited due to dense EPS matrix in already established biofilms [22]. The use of electric current along with antibiotics (tobramycin, gentamicin, cycloheximide) has resulted in synergistic effect for eradication of biofilm formed by various pathogens such as *P. aeruginosa, S. epidermis, Klebsiella pneumoniae, S. gordonii,* and *Candida albicans* [23, 24, 25]. The antibiotics are less effective against biofilm inhabitants due to their reduced metabolic rate and poor penetrability of antibiotics. However, in conjunction with the electric current, not only the penetrability of antibiotics is enhanced but the metabolic rate of inhabitants is also increased due to better access to nutrients and hydrated ions rendering them vulnerable to antibiotics. The bioelectric effect has potential to be applied against biofilm infections in indwelling implants but certain radicals generated due to direct current may have broad spectrum toxic effects on host tissue. Moreover, to cure a deep seated lesion, high intensity current would be required which might be conducted throughout the body [26].

Initial studies were focused on deciphering the antibiofilm effect of various electrical parameters such as intensity and duration of electric field, current density, and the direction of flow of the current. In this regard, Hong et al., [27] while investigating the adhesion of *P. aeruginosa* PAO1 on the electrode itself, reported that anodic current inactivates the organism while the cathodic current increases detachment. The antibacterial and antibiofilm mechanism of bioelectric current is speculated to be due to the transport of ions between the electrodes, modification of the pH, and the production of additional biocides by electrolysis [28, 25, 29]. However, negating these possible mechanisms, Caubet et al., [30] demonstrated bioelectric effect using radio frequency current (RFC; 10 MHz, low effective intensity 6 mA/cm^2) and antibiotics (gentamicin or oxytetracycline) against *E. coli* biofilms. The resulting magnetic fields were non-ionizing, non-electroporating and did not produce a major heating effect. The antibiofilm effect of RFC is exerted by increased O_2 tension leading to the enhanced metabolic activity of biofilm inhabitants [31] and vibration of charged particles and/or polar subsystems of EPS resulting in its weakening, thus providing better access to the antibiotics throughout biofilm matrix. To optimize the RFC frequencies against EPS matrix, dielectric spectroscopic analysis of EPS over the range 0 to 10 MHz would be necessary [30].

The low intensity alternating current (AC) is also being explored for its potential bioelectric effect. The use of non-conductive electrodes to generate low intensity AC reduces the heating effect as observed in the case of direct current. Using the same approach, Giladi et al., [26] evaluated the bioelectric effect of AC (in absence/presence of chloramphenicol and in a temperature controlled manner) on planktonic cells of *P. aeruginosa* and *S. aureus*. They observed that cells exhibited defective phospholipid synthesis [32] and cell division [33] due to disturbed enzyme-substrate equilibrium of membrane associated enzymes. Such mechanisms might exert an antibiofilm effect by increasing biofilm matrix instability and reducing nutrient availability to slow dividing cells within the matrix.

Changing electric fields generate electromagnetic effects. Low frequency electromagnetic fields (50 Hz; 1 miliTelsa of magnetic flux intensity) influence synthesis of heat shock proteins [34], and the cell-to-cell and cell-to-surface interactions required for initial adhesion of cells to different surfaces [35]. An exposure to low frequency electromagnetic fields during biofilm formation and established formed biofilm of *Helicobacter pylori* resulted in decreased cell clustering and adhesion, altered morphology, and reduced potential of cells to attain viable persistent morphology [36]. Although the above mentioned reports demonstrated antibiofilm effects using low frequency electromagnetic waves, static magnetic field has been reported to aggravate the antibiotic resistance in *E. coli* [37]. The implementation of an electromagnetic effect as antibiofilm strategy has certain limitations like its reduced efficacy in gel like biofilm structure [38] and scarcity of *in vivo* studies which would require the generation of magnetic fields in a controlled manner [39].

Although an ever increasing number of evidence supports the efficacy of bioelectric current against various biofilm systems, contrary reports do exist. There was lack of bioelectric effect in combination with vancomycin on *S. epidermidis* [40] and piperacillin on *P. aeruginosa* [41]. Moreover, the bioelectric effect is not a generalizable phenomenon across microorganisms and antimicrobial agents [42]. The standardization of bioelectric effect *in vitro* is subjected to the type of electrode, current, media, and antimicrobial (anodic, cathodic or uncharged) used. Moreover, single isolated colonies are usually used to study the bioelectric effect and may not affect the *in vivo* biofilm (including mature structure, nutrients, and protein cover) in the exact same way. Further studies are required to elucidate the mechanism(s) of bioelectric effect of various antimicrobial agents against mixed species biofilms. Such insights and in depth understanding of bioelectric effect would help in its effective *in situ* application.

Sonication

Sonication, the bombardment of high energy ultrasonic pulses, is a common method used to lyse and dislodge microbial cells and deposited debris. Sonication as an antibiofilm strategy has the advantage of being non-toxic, eco-friendly, and with wide range of application. Its antimicrobial and antibiofilm effects are speculated to be due to the creation of microstreamings which disrupt autoinducer gradients, enhance cellular metabolism and thus rendering microbial cells more susceptible to antibiotics [43]. However, the efficient application of sonication is dependent on various factors such as the heterogeneity of biofilm matrix, type of microbial cell, intensity of ultrasound, frequency, duration of treatment and distance from ultrasound transducer [44]. In medical implants such as prosthetic devices sonication is the method of choice for removal of biofilm from retrieved implants and has been shown to be more effective for detachment of the cells than conventional methods such as swabbing. In order to remove biofilm from orthopedic implants, sonication for 1 to 5 min has been found to be quite effective [44]. The sonication and vortexing sampling technique to dislodge biofilm from removed spinal implant has been proven to be better than peri-implant tissue culture [45]. However, in comparison to direct sonication and vortexing, ultrasonic bath has been proved to be superior in dislodging bacteria from Dacron® graft cylinders (vascular prostheses) [46].

In water treatment and purification systems, biofouling of the membranes leads to huge economical and productivity losses. Sonication for 10 min cleans microfiltration-membrane associated biofilms in activated sludge wastewater. Though the flux recovery is satisfactory after sonication, the membrane pore blockage due to the smaller particles generated by sonication is highly plausible [47]. Thus, sonication treatment may be more effective in removal of cake-like biofilm on membranes, which can be rendered porous or permeable before sonication [48], than clearing of plugged membrane pores. In food industry, sonication is well established to remove biofilms from common surfaces such as metal, ceramic and plastic [49]. In boilers, sonication removes mineral deposits [50] in addition to *P. fluorescens* biofilms inside the tubes [51]. A portable ultrasonic apparatus producing sound waves at frequency of 40 kHz has proven effective to dislodge biofilm on food processing equipments. In a lab scale model of a meat industry biofilm, the performance of the equipment was four times better than the conventional swabbing method [52]. Closed surfaces in the food industry are amongst one of the most tedious areas to clean. A curved device, specifically designed to clean unaccessible areas, could only achieve 33% and 66% removal of *E. coli* and *S. aureus* biofilm (40kHz/10s), respectively. However, in combination with chelating agents like ethylene diamine tetra acetic acid (EDTA) or ethylene glycol tetra acetic acid EGTA (0.05mM) complete removal of *E. coli* biofilm could be achieved [53]. These findings highlight that synergism with other antibiofilm strategies might be helpful to broaden the spectrum of sonication as an effective antibiofilm strategy. On similar lines, synergistic effect of trypsin (30 min) or protease (15 min) with ultrasonic waves (40kHz/10 sec) was observed against *E. coli* biofilms [53]. Such strategy can be effectively applied in both food and indwelling implants. An interesting factor that should be kept in mind is that the low energy acoustic waves inhibit biofilm formation but strangely high energy acoustic waves have been reported to enhance bacterial adhesion [54]. The sharp transition to opposite effect on biofilm development that occurs due to different acoustic energy dose/frequency levels, indicates that using a piezoelement in indwelling implants and catheters is not advisable. Although sonication has been extensively used to dislodge biofilms in retrieved implants, its application as antibiofilm strategy is still very limited. To widen its scope of application, further studies addressing various issues such as inclusion in the system of treatment, safety and combinatorial approaches with other therapies would be helpful.

Biological Methods

Bacteria

This strategy deploys non-pathogenic microorganisms to inhibit, kill and/or displace biofilms formed by pathogens. Such effect can be achieved by means of various mechanisms such as production of antibiofilm/antimicrobial metabolites, free radicals, and competitive exclusion. Bacteria as antibiofilm agents can be used on variety of surfaces, both animate (gut and dentures) and inanimate (prostheses, metallic surfaces and concrete)**.**Probiotics are living microorganisms which when ingested in sufficient number provide wide range of health benefits including protection against foodborne diseases. Commonly, probiotics are consumed by incorporation in the food products and as pharmaceutical formulations. Probiotic *Lactobacillus reuteri* RC-14 and *L. iners*, effectively displaced the three days old biofilm of *Gardnerella vaginalis*, which is responsible for bacterial vaginosis [55]. In voice prostheses, a silicon rubber device used in speech rehabilitation of patients subjected to laryngectomy, biofilm infestation results in leakage of food and liquids and increased resistance to air flow making it necessary to replace the device in three to four months. Surfactants produced by probiotic bacteria *Lactococcus lactis* 53 and *S. thermophilus* A resulted in 96 and 83% reduction in prostheses biofilm count, respectively. Similarly, the fungal count showed 85 and 74% reduction on a *in vitro* model [56]. On similar lines, surfactant produced by *L. acidophilus* effectively reduced adhesion of *S. aureus* and *S. epidermidis* in a dose dependent manner. Additionally, it reduced the overall biofilm formation potential and enhanced dispersion of the preformed biofilm [57].

Development of dental caries is one of the most vivid examples of biofilm mediated health problems. Hence the application of probiotic as antibiofilm strategy is being extensively explored in oral health. Generally, the probiotic is expected to reduce the number of caries forming bacteria such as *S. mutans* and others. In this context, the administration of the milk containing probiotic produced by *L. rhamnosus* GG for seven months resulted in significant decrease in the risk of caries. In comparison to the control, the probiotic administered group showed lesser caries and lower counts of *S. mutans* [58]. The same probiotic culture incorporated in cheese lowered the risk factor for dental caries by the reduction in number of *S. mutans* and salivary yeast in a short term study [59]. Probiotic fortified chewing gum has also been shown to be quite effective in reduction of in the number of the salivary *S. mutans* [60]. In a recent study, probiotic cultures, *L. rhamnosus* GG, *L. palntarum* 299v and, *L. reuteri* strains were compared for their antibiofilm effect on *S.*

mutans biofilms. The findings of the study indicated that the latter two were more efficient in *in vitro* inhibition of *S. mutans* biofilms on glass surface [61].

In food industry, the biofilm formed by pathogens in food processing equipments poses a serious health threat. Hence, the possibility of application of a food origin organism to control pathogenic biofilms is being explored vigorously. *Enterococcus durans* and *L. lactis* ssp. *lactic* inhibited *in vitro* bioflm formation by *Listeria monocytogenes* of food origin on stainless steel surfaces. Both cultures led to growth inhibition of >5 log cfu of *Listeria monocytogenes*/cm^2, by competitive exclusion [62]. Introducing the concept of food grade adjunct for microbial biofilm control, Kim et al., [63] reported 94% inhibition of *E. coli* O157:H7 biofilms on polyvinyl chloride surfaces due to released exopolysaccharide (rEPS) of *L. acidophilus* A4. The addition of rEPS led to down regulation of genes associated with curli production and chemotaxis leading to lesser biofilm formation [63]. However, prior to use of such adjunct culture, their interactions with the starter cultures and effects on the product quality need to be taken into consideration.

In sewage industry, biocorrosion of sewer concrete pipes, wastewater collection systems, and treatment plants is a major concern. When the concrete surface is exposed to sewer waters, sulphate reducing bacteria convert the hydrogen sulphide and sulphur into sulphuric acid which reacts with lime [64] forming compounds like gypsum and ettringite [65]. The latter results in pit formation and cracking [66], thus compromising integrity and strength of the sewer system. A low cost and effective strategy to displace/inhibit sulphate reducing bacterial biofilms (*Desulfovibrio, Desulfotomaculum, Desulfomonas* and *Acidithiobacillus thiooxidans*), is the use of a biobarrier formed by a certain microbial biofilm that may reduce oxygen availability at the surface, release antibacterial substances, and/or produce biofilm matrix degrading enzymes [67, 68]. An aerobic and iron oxidizing bacterium *Leptothrix discophora* SP-6 is usually found on peripheral biofilms formed by sulfate reducing *Desulfosporosinus orientis* on steel and aluminium surfaces [69] resulting in corrosive synergy. Gramicidin-S producing *Bacillus brevis* 18-3 biofilms was 20-fold more efficient in combating the synergistic corrosion attack by *L. discophora* and *D. orientis* in comparison to non-Gramicidin-S producing *Paenibacillus polymyxa* [70].

Bacteria as antibiofilm agents are non-toxic, easily available, and efficiently deliverable. However, there is a need to develop industry specific intervention taking into account the safety, delivery and efficacy of selected microorganism(s).

Bacteriophages

Bacteriophages, the viruses that infect bacteria, bind to specific receptor(s) of the host and cause cell lysis during the release of its progeny (lytic or lysogenic cycle). The specificity of host cell receptors determines the level of interaction of a virus at genus, species, or strain level. Till date most of the phage based strategies have been developed for treatment of biofilm inflicted chronic wounds and infections.

Phages as biofilm control agents have apparent benefits over chemical agents by being inexpensive, specific to a host or host range, easy to isolate, eco-friendly and with minimum side effects. Although, the phage interactions with a biofilm are not fully understood, the general basis of bacteriophage application deploys its virtue of propagation on bacterial pathogen, EPS degrading enzyme (depolymerase) [71] or easier penetrability through the intact biofilm matrix [72].

A recombinant T7 phage, was designed to express a polymer degrading enzyme to simultaneously attack biofilm forming *E. coli* cells interspersed in inner and outer layers of a biofilm, resulting in more than 99.9% disinfection [73]. A recent study has shown that application of depolymerase producing lytic bacteriophage (KPO1K2) lowered *Klebsiella pneumoniae* count in mature biofilm by inducing structural changes in the biofilm matrix and restricting microcolony size. Synergistic effect was observed with ciprofloxacin and KPO1K2 in reducing the *K. pneumonia* biofilm count but the same effect was not obtained with *K. pneumoniae* B5055 [74]. However, this treatment decreased the number of resistant variants within B5055 biofilms, their potential to form biofilms or produce EPS and increased their susceptibility to the mouse model immune system [75]. Lytic phages have also been reported to effectively reduce (approximately 90%) *Proteus mirabilis* and *E. coli* biofilm count in urinary tract indwelling devices [76]. A recent study showed that in conjugative plasmid bearing *E. coli*, early biofilm formation was completely inhibited by the addition of single stranded DNA bacteriophage f1, although mature biofilm remain unaffected [77]. Fu et al., [78] concluded that although a single phage could considerably reduce the P. aeruginosa biofilm count on indwelling medical devices such as catheters, a phage cocktail is a more effective option to reduce the emergence of phage resistant isolates. A specific phage cocktail has been shown to effectively remove the dual species biofilm of *Staphylococcus lentus and P. fluorescens* with the former being the major partner. Additionally, P. fluorescens specific phiIBB-PF7A phage alone was able to partially dismantle the dual species biofilm matrix and even washed-off the non target host [79]. Such biofilm sloughing action of phage can be used to render the multispecies biofilms more permeable and susceptible to other conventional antibiofilm therapies like antibiotics or disinfectants. A phage cocktail comprising P. aeruginosa phages 14/1 (Myoviridae) and PNM (Podoviridae) and S. aureus phage ISP (Myoviridae) is currently under clinical trial for treatment of burn wounds [80]. A commercial phage cocktail PhageBioDerm™ is also available for treatment of tropic ulcers, infected superficial wounds, bedsores, and burns [81].

In reference to food industry, Sharma et al., [82] reported synergistic effect of bacteriophage and alkaline cleaner against *E. coli* O157:H7 biofilms on a stainless steel coupon. However, the application of phages in food industry has its own problems since they are the major cause of starter failure in dairy industry. In a recent study, application of L. monocytogenes specific bacteriophage P100 led to reduction in biofilm count to a varying extent in 13 serotypes of the organism [83]. Phages are also being explored for mitigating clogging and membrane fouling [84]. Gino et al., [85] reported synergistic effect of glycolic acid and phages against ground water well clogging by *Sphaerotilus natans* biofilm. On similar lines, specific lytic phages diminished biofouling caused by *P. aeruginosa*, *Acinetobacter johnsonii* and *B. subtilis* (single or multispecies biofilms) on ultrafiltration membrane by more than 40 percent.

The application of bacteriophages to control biofilm is still in its infancy and requires further research to address issues pertaining to development of phage resistance and delivery to the site. In a clinical background, the phage is delivered either by impregnation or surface application on the indwelling device or by respiratory aerosole specifically in case the of *P. aeruginosa* biofilms in cystic fibrosis [86]. The efficiency of a phage based approach is affected by various parameters and conditions such as infection strategy (dynamic or static) and multiplicity of infection, host growth phase, temperature and media [79, 87]. Conclusively, the effective implementation of phage(s) as an antibiofouling strategy in different industries would require specific interventions regarding application, delivery and efficacy.

Biological Molecules

Enzymes

EPS, the building block of a biofilm matrix, comprises polysaccharides, proteins, nucleic acids, lipids, mineral ions and cellular debris that together form an interconnected three dimensional polymer network [88]. The shielding effect of the biofilm matrix reduces the efficiency of anti-biofilm agents. Hence, enzymes targeting EPS components are potential tools for biofilm control strategies especially where the limited penetrability of conventional chemical and physical methods is major hindrance. On the basis of the composition of the biofilm matrix, enzyme formulation targeting polysaccharides, nucleic acids and proteins has been explored.

Polysaccharide hydrolyzing enzymes in combination with oxidoreductase have been reported to be effective in disinfection of biofilms on polypropylene and stainless steel surfaces [89]. Alginate lyase and cellulase have been reported to adversely affect mass and count of *P. aeruginosa* biofilm by compromising the structural integrity of the matrix [90]. In addition to bacteriophages, their enzymes (native or engineered) are also being used as a biofilm control tools. Recombinant endolysin of *S. aureus* phage phi 11 could eliminate biofilms with high efficiency comparable to positive control lysostaphin [91]. Indeed, bacteriophages have been engineered to express biofilm degrading enzyme during infection. Using this two way strategy of enzymatic bacteriophage, the biofilm count was reduced by 2-log which was better than that by phage alone [73]. On a recent study, SAL-2, a cell wall degrading enzyme from *S. aureus* bacteriophage (SAP-2) displayed broad spectrum activity within *Staphylococcus* genus including methicillin resistant strains at concentration of 1 μg ml^{-1} [92]. The extracellular enzyme dispersin B, from *Aggregatibacter actinomycetemcomitans*, could efficiently control the biofilms of *S. epidermidis* by hydrolyzing β(1→6)-linked *N*-acetyl-d-glucosamine components of the biofilm matrix [93]. Catheter surfaces pre-coated with dispersin B exhibited strong antibiofilm properties. However, the performance of the enzyme formulation was affected by contact time, concentration, target biofilm count/stage and interfering material.

Proteolytic enzymes such as proteinase K, trypsin, subtilisin have been used for the biofilm inhibition and dispersion over a range of contact surfaces. Addition of proteinases K in growth media (1mg ml^{-1}) inhibited the non-typeable biofilm formation by *H. influenzae* without affecting the planktonic cells [94]. Similarly, as an adjunct in formulations, enzymes have been reported to effectively remove *S. aureus* biofilms [95]. Another proteolytic enzyme, subtilisin, is reported to be effective in controlling the initial adhesion as well as detachment of mature biofilm of *Pseudoalteromonas sp.* D41 in comparison to other commercial enzyme preparations such as Amano Protease A, Papai and Umamizyme [96]. Poele and van der Graaf, [97] reported that application of protease enzyme to control biofouling of ultra-filtration membranes on a waste water plant resulted in a higher clean water efflux than when the alkaline cleaning protocol was used. The combination of enzymes with other biofilm control agents such as detergents, surfactants, dispersants and antiprecipitants is reported to be very effective in controlling immature biofilms on reverse osmosis membrane. The adjunct components of the formulation help the enzyme to penetrate into the matrix and ensure stability against toxic compounds [98]. Application of a proteinase based commercial formulation effectively reduced more than 85% of *P. fluorescens* biofilm count [99]. The application of proteolytic and/or glycolytic enzymes in combination with ultrasonic wave for 10s resulted in 61-96% reduction in *E. coli* biofilms developed on stainless steel coupons in milk [53].

DNA is one of the major components of biofilm matrix and plays various roles such as induction of biofilm formation, maintenance of structural integrity and stability of matrix [100]. Hence, application of DNase has also been explored as a biofilm control strategy. In one such study, DNase application at a concentration of 100 μg ml^{-1} has been reported to cause significant reduction in biofilm formation by *S. pneumonia* without affecting planktonic cells [101]. The enzyme effectively inhibited biofilm formation by *P. aeruginosa* during initial stages, although preformed biofilms were not much affected [102]. Enzyme application at 6 and 12 h after inoculation resulted in diminished biofilm formation by *E. faecalis*, however, only marginal affect was observed at 24 h [103]. The findings can also be interpreted as that at initial biofilm development stage, DNA in biofilm matrix might act as major stabilizing factor, hence DNase can be used to prevent biofilm formation rather than biofilm removal. Cell wall hydrolyzing enzymes, are also being explored as biofilm control agents. Lysozyme hydrolyses 1,4-beta linkages in the peptidoglycan of the bacterial cell wall. At a concentration of 6.25 μg ml^{-1}, lysozyme significantly inhibited biofilm formation by *C. albicans* on denture acrylic surfaces while a higher concentration of 100 μg ml^{-1} effectively killed 45% of the sessile cells [104]. Lysozyme-polyethylene glycol coating (immobilized enzyme) on stainless steel surfaces significantly inhibited bacterial adhesion by *Listeria ivanovii* and *Micrococcus luteus* in comparison to the non-immobilized enzyme control [105]. In order to ensure delivery at site, an ultrafine lysozyme liposome was synthesized and applied for biofilm control. The liposome immobilized enzyme resulted in 86.5% reduction in lipid biomass in comparison to free enzyme which could reduce only 62.4% [100].

Although the enzyme based biofilm control strategies have been reported two decades ago, most of the formulations are patented, rendering the strategy expensive. Various factors affecting their efficacy are the EPS heterogeneity, pH, temperature, requirement of catalyst, etc, which may hamper the efficacy of enzyme formulation. Enzyme immobilization, modification and protein engineering may be useful in making the enzyme based antibiofilm strategies more effective [106]. Further studies involving extensive field of application of enzyme based formulations and mathematical models to evaluate the simulation studies may be helpful in propelling these strategies as reliable antibiofilm alternatives to chemical agents.

Antibodies

The immune system exerts a response against any foreign element that enters in the body. However, these elements which also include pathogens, have devised various evasive strategies, including biofilm formation, leading to chronic infections. In healthcare industry, the traditional approach of biofilm mitigation relies on bactericidal virtue of various antimicrobial agents such as biocides, enzymes and antibiotics. In addition, interventions reducing the initial adhesion and/or surface associated microbial growth may help the immune system to effectively combat pathogens in planktonic stage. Hence, antibodies targeting the extracellular macromolecules, surface binding proteins involved in cell-surface and cell-cell interactions are being explored as antibiofilm strategies that may be administered either through passive immunization [107], or topically applied on biofilm infected wounds [108] and/or coated on the medical devices.

In case of dental caries or periodontal infections, instead of systemic passive immunization, local application of pre-formed immunoreactive serum or antibodies on surfaces of teeth and gingival tissue and/or mouth rinse can be effective antibiofilm strategies [107]. Antibodies can block adhesion, modify the virulence factors and aid in clearance of microorganisms by the immune system [109]. Different vehicles for passive immunization like milk from immunized cows [110], transgenic plants [111] and chimeric recombinant microbial vectors [112] have shown encouraging results in prevention of dental caries and periodontal infections in animal models. However, lack of human studies, altered antigenicity of putative pathogens within oral biofilms, and ecologically driven multi-microbial oral infections question the long-lasting effectiveness and wide application of passive immunization.

Application of pooled human immunoglobulins (25mg ml^{-1}) has been shown to be effective in reducing adhered cells of *P. aeruginosa* from 1.5 x 10^5 cfu/contact lens to 4.4 x 10^3 cfu/contact lens [113]. In order to evaluate the efficacy of non-specific antibodies in the pooled immunoglobulins, the initial adhesion of preopsonized *P. aeruginosa* was reduced by 53 and 28% in presence and absence of *P. aeruginosa* specific antibodies. In the same study, surface growth rate was shown to reduce from 54% to 39% when specific antibodies were removed from the antibody pool used for opsonization [114]. The results apparently indicated that even non-specific antibody pool can be used as prophylactic measure against wound infections by opportunistic pathogens. However, the limited supply and low avidity of monoclonal antibodies might be a discouraging factor for its application directly on wounds and medical devices. Avidity of antibodies can be enhanced by using small sized, multivalent single-chain antibodies which lack Fc portion [115]. The absence of Fc portion would efficiently reduce immunogenicity of the antibodies for illicitation of host inflammatory response responsible for tissue damage. Medical devices exuding such antibodies in controlled-release manner can be effectively used to control device related biofilm infections [116].

Monoclonal or engineered antibodies directed against such specific molecules and/or proteins involved in microbial adhesion, accumulation, colonization and surface adhered growth have shown encouraging results as antibiofilm agents. *S. epidermidis* surface protein C (SesC) is expressed at higher levels in biofilm cells than in planktonic cells. Anti-SesC antibodies have effectively reduced the fibrinogen binding activity and count in established biofilms in rat model [117]. Application of a mixture of monoclonal antibodies and $F(ab)^2$ against accumulation associated protein (aap) of the organism has led to up to 87% reduction on its accumulation on biomedical devices [118]. Polyclonal antibody against Opr86, an outer membrane protein of *P. aeruginosa* PAO1 effectively reduced biofilm formation across non-related strains. The cells treated with the antibody were unable to attach to the contact surface [119]. Serine-rich repeat proteins which are present in various Gram-positive bacteria may be a potential target for antibodies. A recent study has shown that Pneumococcal serine rich repeated protein (PsrP), plays crucial role in bacterial adhesion and formation of biofilm like structure in the lungs, and this effect was shown to be neutralized by anti-PsrP antibodies [120]. Application of anti-*C. albicans* antibodies (IgY) was effective in reduction of *Candida* spp. adherence to human oral epithelial cells in a dose dependent manner [121].

Contrary to encouraging results pertaining to use of antibodies as antibiofilm agents, few studies have indicated that antibodies specific for glucuronoxylomannan, an important component of cryptococcal biofilm matrix, immobilized *Cryptococcus neoformans* cells. The cryptococcal biofilms on such coated medical devices were not only enhanced in comparison to control [122] but the biofilms also exhibited reduced susceptibility to antifungal agents [123]. The antibody based biofilm control strategy needs to go a long way before its effective application as a therapeutic prophylactic agent. The major constraint is the limited availability of antibodies at a high concentration required to achieve the desired effects and intensity of immune response. In addition, efficient delivery at the site, non-specific interactions of the antibodies and effects on other body fluids require further investigation.

Fatty Acids

Fatty acids (FA) are integral component of cell structures like membranes in the form of phospholipids and/or glycolipids. These are usually conjugated with phosphate, protein, glycerol or carbohydrate groups. Free fatty acids (FFA) are enzymatically released from lipids and might act as antimicrobial, cytotoxic, and signaling molecules [124]. Most abundant FFA in human skin are lauric acid (C12:0), myristic acid (C14:0), palmitic acid (C16:0), sapienic acid (C16:1n-10) and cis-8-octadecenoic acid (C18:1n-10) [125]. The antibacterial activity of FFA is probably mediated by resultant acidic pH on host surface, disruption of quorum sensing [126], cell lysis [127], uncoupling of oxidative phosphorylation and electron transport chain [128], disruption of enzymatic activity and nutrient uptake [129] and generation of toxic products [130]. Antibacterial activity of FFA is reported to be governed by the extent of unsaturation [131] especially cis-unsaturation but the same might contribute to their instability [132]. FFA are generally regarded as safe [133] and microbial resistance against FFA is less likely to develop as compared to that against conventional antibiotics [134]. In medical industry, FFA have been successfully used to prevent colonization of opportunistic pathogens like *S. aureus* on skin and nose [135], and stomach ulcers caused by *H. pylori* [136]. The triterpene ursolic acid (10 μg ml^{-1}, added at the time of inoculation or to mature 24 h old biofilm) inhibited biofilm formation by *E. coli, P. aeruginosa* and *Vibrio harveyi* [137]. Asiatic acid and corosolic acid enhanced the susceptibility of *P. aeruginosa* to antibiotics [138]. Baicalein rendered the *P. aeruginosa* biofilms porous and at a concentration of 2 μg ml^{-1} exhibited synergistic effects with ampicillin [139]. However, the synergistic activity of such fatty acids with antibiotics has been reported to be limited to younger biofilms only [138]. One of the major virulence factors of *C. albicans* is its ability to switch between yeast, pseudohyphae and hyphae mode of growth [140] which is known to play pivotal role in causing oropharyngeal candidiasis in immunocompromised individuals [141]. Undecylenic acid, a fatty acid, is known to prevent the morphological transformation of *C. albicans* [142]. Clement et al., [143] demonstrated reduced hyphae formation by *Candida* spp. in presence of whey-derived fatty acids such as lauric acid, myristoleic acid, linoleic acid and arachidonic acid. While investigating the effect of marine polyunsaturated fatty acids on biofilm formation by *C. albicans* and *C. dubliniensis*, Thibane et al., [144] found that fatty acids such as stearidonic acid (18:4 n-3), eicosapentaenoic acid (20:5 n-3), docosapentaenoic acid (22:5 n-3) and docosahexaenoic acid (22:6 n-3) exhibited an inhibitory effect on biofilm biomass of both yeasts. The antibiofilm effect was suggested to be mediated by inhibition of mitochondrial metabolism.

On one hand, fatty acids have been explored for their antibiofilm effects but on the other, the role of fatty acids in enhancing biofilm formation and virulence of biofilm inhabitants has been a major concern for clinicians. Some fatty acids, categorized as diffusible signal factors, facilitate movement of *Stenotrophomonas maltophila* [145] and mediate quorum sensing *of P. aeruginosa*. The fatty acids produced by *S. maltophila* promote increased stress tolerance and antimicrobial resistances in *P. aeruginosa* [146]. Interestingly, a quorum sensing fatty acid, 12-methyl-tetradecanoic acid, produced by *P. aeruginosa* inhibited swarming motility and reduced *S. maltophila* biofilm formation by 31% at 10 μg ml^{-1} [147]. Moreover, 2-cis-decenoic acid has been shown to enhance dispersion and virulence of *P. fluorescens* biofilms [126]. Certain demerits such as non-specific binding of FFA to proteins or other compounds [136], lack of concerned intellectual property rights and their unsavory taste may limit a wider application of FFA as antibiofilm agents [148]. However, these demerits can be overcome by further studies in the area of effective delivery, synergism with other antibiofilm strategies and development chemically engineered FFA. Moreover, clinical trials and *in vivo* studies would be helpful to validate the *in vitro* findings pertaining to antibiofilm effects of fatty acids.

Maggot Excretion/Secretions

In late 1920s William S. Baer, an orthopaedic surgeon at Johns Hopkins Hospital (Baltimore), used maggot debridement therapy (maggots of the green bottle blowfly; *Lucilia sericata*) to treat children suffering from osteomyelitis and gas gangrenous wounds [149]. Later with discovery of antibiotics, maggot therapy almost disappeared. Decades later in 1980s, concerns regarding antibiotic resistances motivated clinicians to look for alternatives like maggot therapy to cure severely infected wounds [150]. Nowadays, benefit of maggot therapy in management of biofilm inflicted chronic wounds is gaining momentum among care providers [151] and is even approved by FDA [152]. In general, the antibiofilm effect of maggot excretion/secretion (E/S) is exerted by production of antibacterial molecules such as p-hydroxybenzoic acid; p-hydroxyphenylacetic acid; and octahydro-dipyrrolo[1,2-a;1′,2′-d] pyrazine-5,10-dione, also known as the cyclic dimer of proline or proline diketopiperazine or cyclo[Pro,Pro] [153]. Additionally, the presence of proteolytic and chymotrypsin like enzymes in maggot E/S mediates degradation of biofilm matrix [154]. Some of these antibacterial substances have been sequenced and found to be homologous to lysozymes [155]. Recent studies have demonstrated antibiofilm effect of maggot E/S on biofilms of *S. epidermidis* [156], *S. aureus* and *P. aeruginosa* [157]. Not only E/S prevented biofilm formation but it also degraded mature biofilms of *S. epidermidis* by protease or glucosaminidase like enzymatic activity [156]. As low as 0.2 μg of E/S was effective for prevention of biofilm formation by *S. aureus*, although higher dose (2 μg) was required for removal of the established biofilm [157]. Though the effectiveness of E/S is well evident against many pathogens, the Gram-positive bacteria are shown to be more susceptible to the therapy [157]. Generally, E/S concentration of 80 to 160 μg ml^{-1} devoid of excess protein resulted in optimal antibiofilm effect [157, 158]. While evaluating the effect of E/S on the biofilms of *P. aeruginosa* on different biomaterials, Cazander et al., [158] observed that even after one month storage at room temperature E/S could prevent, inhibit and breakdown the preformed biofilm. Further studies reported that sterile E/S was able to inhibit and degrade established biofilm of *S. aureus* and *S. epidermidis* on different contact surfaces [159].

However, no direct bactericidal effect of live maggot/ES was observed on the planktonic bacterial growth [160]. Recently in an attempt to investigate the output of secreted protein from maggot, lucifensin, a defensin of maggot, was shown to display antibacterial activity against Gram-positive bacteria, although it was not abundantly secreted [161]. Undoubtedly, maggot E/S is a very interesting and effective antibiofilm therapy especially to counter emergence of resistance. Further investigation to better understand the role of various affecting factors such as rapid changes in pH, requirement of cofactors, synergism of modulatory molecules with other antibiofilm therapies and negative allosteric effects of some enzymes [158] is required for the effective implementation for debridement and disinfection of biofilm inflicted chronic wounds. Moreover, related safety issues such as enhancement of biofilm formation during initial stages which was the basis of recent recall [maggot therapy dressing LeFlap™ (Batch#MM-090406 with CCII-0901)] also need to be addressed.

Chemical Methods

Antimicrobial Surface

Practically, almost every material supports microbial adhesion although the extent may vary from surface to surface [162] In the process of biofilm formation, the hydrophobicities of substrate and cell play a pivotal role [163]. Initial adherence of microorganisms on the surfaces can be prevented either by incorporating the antimicrobial substance/s into the surface polymer itself or coating the surface. In food industry, approved biocide and/or in/organic compound with broad spectrum activity can be incorporated or coated onto the surfaces [164]. Such antimicrobial surfaces can be then suitably used as processing/cutting surfaces to reduce the risk of biofilm formation and dissemination. In food industry, the packaging material impregnated with antimicrobial substances can enhance the shelf life of foods by reducing the surface associated biofilms. The release of antimicrobial substance (eucalyptus oil, bacteriocins, immobilized enzymes, silver ions, covalently-linked peptides and organic acids) [165] from such packaging surfaces might be based on slow-releasing systems, release-on-command systems and non-leaching systems. However, incorporation of active substances can alter the packaging polymer chemistry (oxygen permeability, tensile strength, release rates and brittleness) [166]. Not only antimicrobial coating but preconditioning of the surfaces with surfactants has also been shown effective in preventing biofilm formation [167]. Splendiani et al., [167] evaluated effect of 22 surfactants on inhibition of adhesion and biofilm formation by *Burkholderia* spp. The mechanism of biofilm inhibition by surfactants was shown to be mediated by increase in cell surface charge and reduction in flagella development. Another study documented that preconditioning the surface with nonionic and anionic surfactants efficiently inhibited adhesion of *P. aeruginosa* on stainless steel and glass by 90% [168]. Some materials can be modified to release biocidal components (monomeric oxazolidinones or hydantoins) only when exposed to chlorine or bromide [169]. In other words, the cleaning procedure involving halogen might stimulate the surface to release antimicrobial substance to disinfect and remove the adhered biofilms. These coatings may be applied to glass, plastic, metals, fibers, food wrappers, pipes, tiles, etc. Some surfaces like (-)usnic acid-loaded polymers exhibit antimicrobial effect only after the *S. aureus* cells have adhered, i.e., the antimicrobial compound does not affect rate of attachment but ultimately kills the sessile cells [170]. In textile industries, surface functionalized polymers like polypropylene, rubber, silicone, polyester and polyurethane (raw material used for fabric production) linked with reactive groups have been used to reduce the biofilm formation [171]. The antimicrobial surfaces are attractive target when it comes to wound dressings and swabs. Incorporation of large number of positively charged nitrogen atoms in the main polymer may be used in treatment of microbial infection at sites like skin, mouth and the gastrointestinal tract [172]. Reduced rate of infection has been observed with usage of silicone rubber implants coated with quaternary ammonium compounds [173]. The susceptibility of hydrophilic fiber like silicone to biofilm formation can be reduced by incorporating triclosan or butyl paraben in the silicone elastomer which can be then used to fabricate prosthetics [174]. However, the antimicrobial substances associated with indwelling implants might cause allergic reactions in the host. Such allergic reactions have been reduced by using biocompatible acid precursors (examples include glycolide, lactide, p-dioxanone, glycyl glycolate and lactyl lactate), coated on indwelling implants. The hydrolysis of these precursors reduced microbial adhesion on the implant surfaces [175]. Conclusively, antimicrobial surfaces with controlled release of active substances bear huge potential to be implemented as an antibiofilm strategy. Further studies, pertaining to improved mechanism of the release of antimicrobial compounds, stability of the contact surfaces and minimal side effects would further popularize their application.

Inhibitors of Modified Nucleotide Biosynthesis

EPS is the major component of the biofilm matrix and cyclic di-guanosine monophosphate (c-di-GMP) is an allosteric activator of EPS biosynthesis. It also stimulates other adhesion factors involved in biofilm formation by regulation of gene expression and allosteric activation of protein activity, and protein stabilization [176]. C-di-GMP, also known as secondary signal molecule, is involved in controlling the transition of an organism from planktonic to biofilm mode of growth [177]. Probably inhibition of c-di-GMP leads to its degradation in order to recycle the nucleotide triphosphate for DNA or RNA production [178]. Investigation pertaining to the novel inhibitors of biofilm targeting nucleotide and/or modified nucleotide biosynthesis is relatively a new arena of antiobiofilm research [179, 180]. These compounds at specific concentrations might interfere with nucleotide synthesis resulting in inhibition of synthesis of extracellular DNA [181]. Extracellular DNA is known to play various roles in induction of biofilm formation and its stability, thus, its unavailability is bound to compromise the structural integrity of biofilms. The presence of nucleotide inhibitors such as heavy metal [182] and increased levels of nitric oxide [183] degrade c-di-GMP resulting in enhanced biofilm dissemination [184]. The genes responsible for synthesis of c-di-GMP are conserved in eubacteria and their role in biofilm formation has been shown to be more prominent in Gram-negative bacteria [185]. Moreover the absence of these genes in animal kingdom [186] makes c-di-GMP inhibitors an effective antibiofilm strategy without any side effects. As the c-di-GMP synthesizing enzymes (diguanylate cyclases: DGCs) are structurally and functionally well characterized, the screening of DGC inhibitors becomes easier [187]. The inhibitors of DGCs like YdeH in *E. coli*, PleD in *Caulobacter crescentus* and WspR in *P. aeruginosa* [188, 189] can be used to control biofilm of respective organisms. Based on mass and fluorescent spectrometric assays, substituted purine and pyrimidine nucleotides like MANT [2'(3')-O-(N-methylanthraniloyl)] and TNP [2',3'-O-(2,4,6-trinitrophenyl)] were shown to inhibit biofilm formation by exerting inhibitory effect against YdeH. [177]. Sulfathiazole, an established antimicrobial, is known to reduce biofilm formation by inhibition of c-di-GMP biosynthesis [180]. It interferes with tetrahydrofolate biosynthesis which reduces the cellular pool of thymidine required for microbial DNA synthesis. In addition to abovementioned modified nucleotides, incorporation of a modified uracil, 5-fluorouracil, or mutated uracils in the microbial genes with predominance of uracil nucleotide is known to inhibit DNA replication, microbial adhesion and quorum sensing essential for biofilm formation [179]. Conclusively, biofilm control by nucleotide inhibitors, is a promising strategy that can be directly used to prevent biofilm or as a drug enhancer strategy. Such molecules might not kill the microorganisms themselves but can make them susceptible to other conventional strategies or to the host immune response. However, further studies elucidating nucleotide cyclase structure and function may be helpful in evolving broad spectrum nucleotide synthesis inhibitors.

Quorum Sensing Inhibitors (QSI)

Quorum sensing (QS) is bacterial communication at intra and/or inter species or inter kingdom level, intended to exert an optimal response to establish effective thriving strategies and conditions. This communication process is cell density dependent and leads to the expression of specific genes resulting in production of quorum molecules (autoinducer) such as acyl homoserine (AHL) in Gram negative, autoinducer peptides in Gram positive and autoinducer-II based global signaling system in all bacteria [190]. Quorum sensing is responsible for various metabolic functions in microorganisms including production of virulence factors and biofilm formation. Hence, these molecules are being seen as potential targets for biofilm control. In general, the antibiofilm effect of QSI is mediated through inhibition of autoinducer production, their receptors and downstream effector proteins, although the exact mechanism of biofilm inhibition by most of the quorum inhibitors is yet to be resolved. Amongst various pathogens, *P. aeruginosa* and *S. aureus* are the most studied biofilm models, hence the quorum sensing inhibition strategies to control their biofilms will be discussed in the following section.

P. aeruginosa, a colossal biofilm former, is a leading cause of nosocomial and chronic lung infections such as cystic fibrosis [191]. Amongst various QS systems, expression of the *las* and *rhl* systems has been established to control biofilm formation [192]. A synthetic halogenated furanone molecule derived from australian marine macroalga (*Delisea pulchra*) inhibited the *las* quorum sensing circuit by penetration of microcolonies in the biofilm leading to subsequent instability of matrix [193]. On the contrary, furonones are also known to exert positive effect on biofilm formation at sub-inhibitory concentration [194]. Structure based virtual screening of recognized drugs such as, salicylic acid, nifuroxazide, and chlorzoxazone, showed significant inhibition of QS regulated gene expression and related phenotypes in *P. aeruginosa* [195]. Application of garlic extract and 4-nitro-pyridine-N-oxide (NPO) as QSI has been reported to enhance antibiofilm effect of tobramycin [196]. Additionally, NPO decelerates the process of bacterial adhesion on surface by neutralizing the charge heterogeneities both on bacterial and contact surface, thus providing physiochemical and physiological basis for biofilm control [197]. Although, emergence of antibiotic resistance has been the propelling force behind research efforts directed toward alternate antibiofilm strategies, studies have shown that QSI can be effectively used as potential enhancers of established drugs to inhibit biofilm [198].

Staphylococci mediate virulence through biofilm formation in indwelling devices such as catheters rendering antibiotic therapy less effective. The organism has two quorum sensing systems; *agr* (accessory gene regulator) which modulates biofilm formation by RNAIII upregulation and bacterial detachment, and LuxS which controls the process of polysaccharide intercellular adhesin (PIA) secretion [194]. RNAIII inhibiting peptides (RIP), YSPXTNF, is one of the most studied quorum inhibitors in staphylococci. The peptide along with its synthetic analogue has shown significant reduction of methicillin resistant *S. aureus* (MRSA) biofilm *in vivo* [199], and *in vitro* conditions on mammalian cells and plastic surfaces [200]. Besides suppressing the development of resistant strains, the stability and non-toxicity of the peptide makes it a potential therapeutic molecule. A non-peptide analogue of RIP, 2', 5-di-O-galloyl-d-hamamelose (hamamelitannin) effectively prevented biofilm formation by MRSA and *S. epidermidis* in a rat graft model [201]. A broad spectrum pharmaceutical molecule, ambuic acid, has been shown to inhibit *agr* mediated virulence expression in *S. aureus* [202]. In an interesting observation, auto-inducer of Gram negative bacteria, acyl homoserine and their derivatives, have been shown to inhibit the expression of *agr* gene in *S. aureus* [203].

In water purification systems, reduction in the reverse osmosis (RO) membrane potential due to biofilm formation is one of the major areas of concern. In wastewater treatment plants, QS mediated biofouling in membrane bioreactor leads to higher trans-membrane pressure resulting in reduced efficiency. Use of 2-(5H)-furanone has been shown to effectively inhibit a wide range of AHLs and subsequent biofouling by *Aeromonas hydrophila* [204]. Additionally, application of vanillin on RO membranes, led to significant reduction in biofilm surface coverage, thickness and total biomass through its QSI like activity although the preformed biofilm remained unaffected [205]. Porcine kidney acylase I, has been reported to prevent membrane biofouling by cleaving amide bond of AHL quorum molecules [206].

Biofilm formation by spoilage and pathogenic bacteria is a major concern for food industry, which might compromise food safety and pose a health hazard. Several studies have reported auto-inducer molecules in various food products, thus providing impetus for use of QSI molecules in food products as well [207]. Milk spoilage by *Pseudomonas* spp. predominated microflora was effectively inhibited by the furanone derivative resulting in extended shelf life [208]. Nevertheless, there are number of issues such as cost, effectiveness, usage guidelines/concentration of QSI and impact on food attributes, which need to be addressed prior to its application. Development of modified and economic QSI molecules may offer effective antibiofilm strategy to combat biofouling in industries like petroleum, pulp and paper, and marine transport.

As some of the studies have shown that autoinducers of one organism may enhance the biofilm formation by the other, effective strategies against multispecies biofilm formation and role of external environmental factors, need to be taken in to consideration. Conclusively, application of QSI as antibiofilm strategy is relatively new and various issues such as cost, effective delivery, development of resistant strains and degradability of QSI compounds still needs to be addressed. Moreover, industry specific approaches will be required for its effective implementation.

Phytochemicals

In search of novel antimicrobials, researchers have always re-investigated ancient natural therapies which include usage of plant extracts, barks, leaves, dried flowers, etc in edible or paste form to cure wounds and diseases. Such natural products derived from medicinal plants which have been used since ages could prove potential source of active pharmaceutical ingredients with antimicrobial properties. Various substances such as cinnamon bark oil, papua-mace extracts and clove bud oil and their constituents (cinnamic aldehyde and eugenol) have been evaluated for their antibacterial activities [209]. Tichy and Novak [210] investigated antibiofilm properties of 27 medicinal plant extracts against oral streptococci. The most active extracts included those from *Chelidonium majus* (Papaveraceae), *Ginkgo biloba* (Ginkgoaceae), *Juniperus virginiana* (Cupressaceae), *Pinus virginiana* (Pinaceae), and *Thuja plicata* (Cupressaceae). Garlic, *Allium sativum* (Liliaceae), an important ingredient of south eastern cuisine, contains antimicrobial component called allicin which has more pronounced antimicrobial effect against Gram negative (minimum inhibitory concentration, MIC = 1.1–17.4mg ml^{-1}) than Gram positive pathogens (MIC = 35.7–142.7mg ml^{-1}) [211]. Such broad spectrum antimicrobial phytochemicals may possess the potential to be developed as antibiofilm agents against single/mixed species biofilms. *Harungana madagascariensis* (Hypericaceae) is a native African plant and antimicrobial effect of its soxhlet solvent extract has been suggested to be due to its bioadhesive property [212], which might favor its usage at low concentrations compared to conventional antimicrobials. Extended time of contact would further aid in deeper penetration of the phytochemical extract into the biofilm matrix to exert the desired antimicrobial effect. *Azadirachta indica* (Neem), a Meliaceae family tree, has been used in India since many years for the treatment of several diseases. The aqueous extract of Neem leaves has been reported to decrease adhesion and cell surface hydrophobicity resulting in reduced biofilm formation and colonization by *C. albicans* [213].

Cranberry, *Vaccinium macrocarpon* (Ericaceae), has antibiofilm effect against *H. pylori* in gastric mucosa and oral biofilm forming pathogens like *Porphyromonas gingivalis* and *S. mutans* [214]. It is also a proven remedy for urinary tract infection caused by adhesion of *E. coli* [215]. Bacterial cells treated with cranberry preparation prior to biofilm formation resulted in thinner biofilms with reduced bacterial density. A high molecular weight preparation of cranberry reduced fructosyltransferase and glucosyltransferase activity resulted in enhanced desorption of *S. sobrinus* biofilms [216]. A reduced biofilm formation by *P. gingivalis* was observed at cranberry concentration of 62.5 μg ml^{-1}, but strangely, a higher concentration (250 μg ml^{-1}) did not result in enhanced desorption or antibacterial effect [217]. Polyphenols, especially proanthocyanidins and flavonols have been reported to be active compounds present in cranberry juices responsible for inhibition of *S. mutans* biofilms [214]. *Camellia sinensis* and *Andrographis paniculata* have been reported to exert antibiofilm effect by inhibition of glucosyltransferase activity and glucan-binding lectin activity of oral pathogens [218]. Conversely, the ethanol extract of *Helichrysum italicum* (Compositae) exerts antibiofilm effect by reduction in cell surface hydrophobicity [219].

There is accumulating evidence regarding the potential of phytochemicals which can be effectively harnessed to prevent and cure biofilm mediated pathogenesis. Unfortunately, only 1% of approximately 500000 plant species occurring worldwide has been investigated for discovering novel bioactive compounds [220]. Further *in vivo* studies with purified active compounds in reference to dose optimization and associated side effects (if any) may be helpful in establishing phytochemicals as robust antibiofilm agents.

Uncouplers of Oxidative Phosphorylation

Oxidative phosphorylation is the generation of ATP by a proton motive force, i.e., it provides energy for the energy consuming reactions. Such reactions are considered as coupled reactions. Certain compounds like chlorocarbonyl cyanide phenyl-hydrazone (CCCP), potassium cyanide (KCN), 2,4-dinitrophenol (DNP), and pentachlorophenol (PCP) are reported to uncouple the oxidative phosphorylation from ATP production. This uncoupling reaction increases the H^+ permeability of the cytoplasmic membrane and results in cellular oxidative stress [221]. The inhibition of energy generating functions renders the microbial cells lesser competent to form cell aggregates [222] and may also result in cell autolysis [223]. Moreover, synthesis of universal signaling molecule autoinducer-2 and its precursor from methionine require ATP, in absence of which inter and intra species interactions is inhibited [224]. It has also been shown that in presence of an uncoupler the synthesis of EPS was reduced by 75% [225]. Existing evidence suggests that uncouplers have potential to be used against biofilm formation and to reduce their stability; however, most of the uncouplers are toxic in nature. Uncouplers like DNP can penetrate the biofilms within 4 min and at concentration of 20-40 μM, they significantly affect proton flux in biofilms of *P. aeruginosa* (PAO1) and *Nitrosomonas europaea* (Winogradsky) [226]. The quick penetrability of such uncouplers can be used to improve biofilm disinfection/removal formulations to achieve higher efficacies with shorter contact times. Uncouplers inhibit efflux pump activities which armor the biofilm inhabitants with resistance against broad spectrum of related and unrelated antimicrobial agents [191]. DNP has also been shown to reduce the development of adaptive and cross resistances within *P. aeruginosa* biofilm rendering it more prone to disintegration [227]. In addition, adhesion of *P.aeruginosa* cells exposed to DNP for 1 and 2 h was reduced by 44.1% and 68%. On a nylon membrane and a smoother glass slide, reductions of 73.9% and 82.8% were obtained [224] whereas KCN and CCCP led to reduced aggregation by the organism [228].

Another category of oxidative phosphorylation uncouplers includes weak acids which are non-toxic in nature. The protonated forms of these acids are considerably lipid soluble and can easily diffuse into the cell. In addition, free fatty acids act as uncouplers by inhibiting ATP synthase [229]. A salicylic acid derivative [3,3′,4′,5-tetrachlorosalicylanilide (TCS)] at a concentration of 4 mg l^{-1} resulted in complete disintegration of aerobic sludge biofilms [225]. However, the uncoupling effect of organic acids is neutralized by extreme acidophiles like *Acidithiobacillus ferrooxidans, At. thiooxidans* and *At. caldus* frequently found in acid mine drainage biofilms. Analysis of acidophile genome has revealed putative proton efflux systems and genes responsible for degradation of organic acids [230].

Although the broad spectrum effects of uncoulpers make them attractive strategic ingredients against microbial biofilms, the toxic nature of many chemical uncouplers limits their wide application. Chemical synthesis of non-toxic uncouplers or their analogs may be helpful in widening the scope of their application.

Biocides

Biocides are one of the foremost and established antibiofilm agents finding application in almost every industry where biofouling is a concern. Biocides are defined as any chemical compound which may prevent/control surface adhered bacterial growth and/or remove/disinfect bacteria from the area of application. In addition to disinfection, biocides need to be safe, easy to use and rinse off with no toxic residues after application. The efficiency of biocides is affected by the presence of organic substances and other factors like pH, temperature, water hardness, interfering compounds, in-use concentration and contact time [231].

Table 1. Categories of functional groups of biocides

Category	Biocide	Active constituent	Target site/ mechanism	Application
Alcohols	Ethanol Isopropanol Xylitol	Ethanol Isopropanol Alcohol sugar	Membrane integrity, protein denaturation, metabolism interference	Medical, pharmaceutical, cosmetics
Aldehyde	Glutaraldehyde Formaldehyde Noxythiolin Tauroline Hexamine Dantoin o-Phthalaldehyde	 HCHO HCHO HCHO HCHO	Cell membrane, cross-linking of proteins, Inhibition of RNA DNA protein synthesis	Fixative, sterilant in medical, food industries
Chelating agents	Sodium dodecyl sulfate Ethylene dihydro tetraoxide		Disrupts cell wall and outer membrane	As an adjunct to other antibiofilm strategies
Biguanide	Chlorhexidine-digluconate Alexidine Vantocil	Chlorophenol Ethylhexyl Polyhexamethy-lene biguanides	Membrane permeability, cytosol coagulation, lipid phase separation	Medical, domestic, recreational water,
Halogens	Chloramine-T Isobetadinea Sodium hypochlorite Iodophor Cadexomer iodine™ Methacryloxylethyl cetyl dimethyl ammonium chloride Chlorine dioxide Sodium dichloroisocyanurate Chloramine-T	Sodium-tosyl-chloramide Povidone-iodine Hypochlorite Chloride Iodine Iodine Chloride Chloride Chloride Chloride	Modified nucleotides, inhibition of DNA synthesis, disruption of oxidative phosphorylation,	Medical, domestic, industrial, water treatment, dental adhesive, fresh produce

Category	Biocide	Active constituent	Target site/ mechanism	Application
Peroxygens	Hydrogen peroxide Peracetic acid Ethylene oxide	Reactive oxygen radical Peracetic acid oxide	Oxidation of macromolecules, disruption of sulfur, sulfhydryl bonds	Medical, personal care products, industrial, low temperature liquid sterilant, vapour sterilant
Phenol	Dettol Triclosan Hexachlorophene Chloroxylenol Chlorocresol	Chloroxylenol Bis-phenol Bis-phenol Halophenol	Release of K+ ions, uncoupling of oxidative phosphorylation and proton motive force, coagulation of cytosol	Medical, domestic, surface preservation, dental hygiene, Deodorants
QACs	Cediuma™ Cetrimide Cetylpyridium chloride Benzalkonium chloride	Benzalkonium chloride Cetrimide	Loss of cytoplasmic membrane integrity	Medical, domestic, industrial, pharmaceuticals, cosmetics, surface preservation
Diamidines	Propamidine Dibromopropamidine		Reduced oxygen uptake, leakage of amino acids	Medical, surface preservation
Metallic salts	Silver sulphadiazine Silver acetate Silver nitrate Cupron™	Ionic silver Ionic silver Ionic silver Copper	Attack thiol groups of amino acids, breakage of hydrogen bonds, release of K+ ions,	Bandages, water systems, bioactive fiber, glass fiber, face masks

As biocides comprise diverse active compounds and mechanisms of action, it becomes very difficult to classify biocides by single criterion. However, arbitrarily they are classified on the basis of the reactive group, chemical structure and reactivity, mode of action or target site and chemical radical yielded. In planktonic cell stage, the first interaction of the applied antimicrobial agent seems to be with the outer cell wall. But in biofilm mode of growth, the matrix enclosed cells are well protected, thus diminishing the effective concentration of the biocide [191]. Moreover, the heterogeneity of the cell wall composition of various microbial groups, such as Gram positive, negative, mycobacteria, fungi and spore, might be one of the reasons for the variation of a particular biocide efficacy across different microbial groups [232].

There is plethora of published reports documenting the antibiofilm effect of various biocides on single and/or mixed species biofilms. With time and need, efforts are being invested to evaluate the antibiofilm efficacy of novel chemical molecules alone or in combination with modified biological molecules. Discussion of such patents and reports at length is beyond the scope of this chapter. However, frequently used biocides are listed in Table 1 and few commonly used biocides will be discussed here.

Glutaraldehyde (1,5-pentanedial) is of common application as disinfectant in various industries such as, water systems, paper-pulp industry, oilfield operation, leather tanning industry, poultry industry, and cosmetic field. A recent study has shown that its application weakens the matrix of *P. aeruginosa* biofilms [233]. It has been found to be significantly effective as antibiofilm agent in synergism with other approaches to biofilm control. A combination with biodegradable chelator ethylenediaminedisuccinate, resulted in reduced required dose and enhanced biocidal efficacy of glutaraldehyde against *D. desulfuricans* biofilms on carbon steel coupons [234]. The biocide ortho-phthalaldehyde has been shown to exert antimicrobial effect at high concentrations for long duration hours at pH 9 [235]. However, this biocide is ineffective for the removal of biofilm from metal surfaces. Moreover, its interaction with the matrix protein has been reported to reduce its antimicrobial effect on biofilm cells [236].

Quaternary ammonium compounds (QAC) such as didecyl dimethyl ammonium chloride, dioctyl dimethyl ammonium chloride and benzalkonium chloride (BC) are most commonly used in antibiofilm formulations [164]. On food contact surfaces, application of BC at a concentration of 10 mg L^{-1} for 30 min has been proven an effective biofilm disinfectant [237]. QAC functionalized poly (propylene imine) dendrimers resulted in quicker lysis of *E. coli* in comparison to *S. aureus* cells by disruption of the cell membrane and precipitation of the phopholipid [238]. A quaternary phosphonium biocide, tributyl tetradecyl phosphonium chloride, effectively decreased *P. florescences* biofilm count in a model oilfield water system. With weekly application at a concentration of 12.5 ppm, the biofilm count was reduced to almost 50% in comparison to the untreated control [239]. Efficacy of biocides is reported to be further enhanced when used in combination with substances which can make the biofilm more pervious to biocides. For example, sodium dodecyl sulfate acts as detergent and denaturing agent and enhances the bactericidal activity of hydrogen peroxide [240]. Similarly calcium chelators have been reported to synergize the biocidal action of hydrogen peroxide, phenol derivatives or sodium hypochlorite [241].

Recent concerns regarding environmental load of biocides and emergence of broad spectrum microbial resistances have pressed the need to devise new environment friendly strategies, such as biocide dosing that takes into account the effect of dose concentration, duration, and frequency of dosing and fluid mechanics on biofilm control. Using the same strategy against *P. florescences* biofilm, lower concentration of 33.6 mg L^{-1} of a proprietary biocide containing 2,2-Di-bromo-3-nitrilopropionamide with short dosing period of less than 15 minutes was found to be effective for biofilm removal [242]. Caution needs to be taken when reducing contact time to decrease the in-use concentration, as it can bring the in-use concentration of biocide to a sub-lethal concentration within the biofilm matrix favoring the development of adaptive and cross resistance [191]. A recent study in this regard has shown that chlorine dioxide at sub-lethal concentration enhanced biofilm formation by *B. cereus* due to overexpression of biofilm matrix related operons [243]. Other environment friendly green antibiofilm agents are enzyme formulations, which have already been discussed in the biological molecules section.

Chitosan is a hydrophilic biopolymer (β-(1-4) linked 2-amino-2-deoxy-D-glucose), obtained by N-deacetylation of crustacean chitin. It has been shown to exhibit antimicrobial activity against various microorganisms and its coating on surfaces has shown antibiofilm activity against bacteria and fungi [244]. The efficiency of its derivatives seems to be greatly influenced by the degree of acetylation and group substitution [245]. Its water soluble derivative, dimethylaminoethyl-chitosan (DMAEC), increases permeability of both outer and inner membrane leading to release of β-galactosidase from the cell. Chitosan binds to anionic antibiotics such as rifampin [246] and ciprofloxacin [247] by formation of anionic complexes. Coating of such complexes has been shown to release antibiotic in sustained manner for longer duration (30 days) and inhibit staphylococcal biofilms. Chitosan, either as coating on packaging materials or preservative, has been used to contain microbial growth in food industry, especially in low fat or no salt protein products [248].

Biocides are the obvious choice as antibiofilm agents due to the versatility of active ingredients and industry specific formulations. However, associated environmental hazards and development of adaptive and cross resistances emphasize the need for cautious usage. Further studies to evolve cheap and effective formulations in synergism with other antibiofilm strategies will be useful in minimizing the associated risks.

Conclusion

Biofilm formation is a dynamic multifaceted process and antibiofilm strategies focus on prevention, reduction and disinfection of preformed biofilms. However, the studies have revealed that killing might exert an evolutionary stress on microorganisms leading to development of more resistant phenotypes. Hence, the conventional antibiofilm strategies based on application of biocidal active ingredients might be aggravating/complicating the problem inadvertently. In the development of new antibiofilm strategies the paradigm is now shifting from killing the microorganism to curtailing the biofilm associated processes such as growth within biofilms, virulence factor expression, matrix production and quorum sensing. In addition, the application of virtual methods such as chemoinformatics and mathematical modeling for high throughput screening of active antibiofilm compounds can be helpful in developing universal and/or industry specific interventions. Conclusively to win the battle against biofilm, a better understanding of the concept coupled with industry specific research initiative to come-up with active antibiofilm compounds/strategies or combination of thereof, is required.

References

[1] Zobell, C. E. & Allen, E. C. (1935). The significance of marine bacteria in the fouling of submerged surfaces. *Journal of Bacteriology, 29,* 239-251.

[2] Costerton, J. W. Geesey, G. G. & Cheng, K. J. (1978). How bacteria stick. *Scientific American, 238*, 86-95.

[3] Costerton, J. W. Stewart, P. S. & Greenberg, E. P. (1999). Bacterial biofilms: A common cause of persistent infections. *Science*, *284*, 1318-1322.

[4] Harro, J. M. Peters, B. M. O'May, G. A. Archer, N. Kerns, P. Prabhakara, & R. Shirtliff, M. E. (2010). Vaccine development in *Staphylococcus aureus*: taking the biofilm phenotype into consideration. *FEMS Immunology and Medical Microbiology*, *59(3)*, 306-23.

[5] Wainwright, M. (1998). Photodynamic antimicrobial chemotherapy (PACT). *Journal of Antimicrobial Chemotherapy, 42,* 13–28.

[6] Friedberg, J. S. Skema, C. Baum, E. D. Burdick, J. Vinogradov, S. A. Wilson, D. F. Horan, A. D. & Nachamkin, I. (2001). In vitro effects of photodynamic therapy on *Aspergillus fumigatus*, *Journal of Antimicrobial Chemotherapy, 48,* 105–107.

[7] Menezes, S. Capella, M. A. M. & Caldas, L. R. (1990). Photodynamic action of methylene blue, repair and mutation in *Escherichia coli. Journal of Photochemistry and Photobiology B, Biology 5,* 505–17.

[8] Bertoloni, G. Lauro, F. M. Cortella, G. & Merchat, M. (2000). Photosensitizing activity of hematoporphyrin on *Staphylococcus aureus* cells. *Biochimica et Biophysica Acta, 1475,* 169–174.

[9] Wilson, M. Burns, T. & Pratten, J. (1996). Killing of *Streptococcus sanguis* in biofilms using a light-activated antimicrobial agent. *Journal of Antimicrobial Chemotherapy, 37,* 377–381.

[10] Lee, C. F. Lee, C. J. Chen, C. T. & Huang, C. T. (2004). δ-Aminolaevulinic acid mediated photodynamic antimicrobial chemotherapy on Pseudomonas aeruginosa planktonic and biofilm cultures. *Journal of Photochemistry and Photobiology B, Biology, 75*, 21–25.

[11] Lin, H. Y. Chen, C. T. & Huang, C.T. (2004). Use of Merocyanine 540 for Photodynamic Inactivation of Staphylococcus aureus Planktonic and Biofilm Cells. *Applied and Environmental Microbiology, 70(11),* 6453-6458.

[12] Sharma, M. Visai, L. Bragheri, F. Cristiani, I. Gupta, P.K. & Speziale, P. (2008). Toluidine blue-mediated photodynamic effects on staphylococcal biofilms. *Antimicrobial Agents and Chemotherapy, 52(1),* 299-305.

[13] Wood, S. D. Metcalf, D. Devine, C. & Robinson. 2006. Erythrosine is a potential photosensitizer for the photodynamic therapy of oral plaque biofilms. *Journal of Antimicrobial Chemotherapy,* 57, 680-684

[14] Zanin, I. C. Gonçalves, R. B. Junior, A. B. Hope, C. K. & Pratten, J. (2005). Susceptibility of *Streptococcus mutans* biofilms to photodynamic therapy, an in vitro study. *Journal of Antimicrobial Chemotherapy, 56,* 324–330.

[15] Demidova, T. N. & Hamblin, M. R. (2005). Effect of cell-photosensitizer binding and cell density on microbial photoinactivation. *Antimicrobial Agents and Chemotherapy, 49,* 2329-2335.

[16] Banin, E. Brady, K. M. & Greenberg, E. P. (2006). Chelator-induced dispersal and killing of *Pseudomonas aeruginosa* cells in a biofilm. *Applied and Environmental Microbiology, 72,* 2064-2069.

[17] Pagonis, T. C. Chen, J. Fontana, C. R. Devalapally, H. Ruggiero, K. Song, X. Foschi, F. Dunham, J. Skobe, Z. Yamazaki, H. Kent, R. Tanner, A. C. R. Amiji, M. M. & Soukos N. S. (2010). Nanoparticle-based endodontic antimicrobial photodynamic therapy. *Journal of Endodontology, 36 (2),* 322–328.

[18] McCarthy, J. R. Perez, J. M. Bruckner, C. & Weissleder, R. (2005). Polymeric nanoparticle preparation that eradicates tumors. *Nano Letters, 5,* 2552–6.

[19] Biel, M. (2006). Advances in photodynamic therapy for the treatment of head and neck cancers. *Lasers in Surgery and Medicine, 38*, 349–355.

[20] Costerton, J. W. Ellis, B. Lam, K. Johnson, F. & Khoury, A. E. (1994). Mechanism of electrical enhancement of efficacy of antibiotics in killing biofilm bacteria. *Antimicrobial Agents and Chemotherapy, 38,* 2803–2809.

[21] Poortinga, A. T. Smit, J. van der Mei, H. C. & Busscher, H. J. (2001). Electric field induced desorption of bacteria from a conditioning film covered substratum. *Biotechnology and Bioengineering, 76,* 395-399.

[22] van der Borden, A. J. van der Werf, H. van der Mei, H. C. & Busscher, H. J. (2004b). Electric current-induced detachment of *Staphylococcus epidermidis* biofilms from surgical stainless steel. *Applied and Environmental Microbiology, 70(11),* 6871-6874.

[23] Khoury, A. E. Lam, K. Ellis, B. & Costerton, J. W. (1992). Prevention and control of bacterial infections associated with medical devices. *American Society for Artificial Internal Organs Journal, 38,* 174–178.

[24] Wellman, N. Fortun, S.M. & Mcleod, B.R. (1996). Bacterial biofilms and the bioelectric effect. *Antimicrobial Agents and Chemotherapy, 40(9),* 2012–2014.

[25] Wattanakaroon, W. & Stewart, P. S. (2000). Electrical enhancement of *Streptococcus gordonii* biofilm killing by gentamicin. *Archives of Oral Biology, 45,* 167– 171.

[26] Giladi, M. Porat, Y. Blatt, A. Wasserman, Y. Kirson, E.D. Dekel, E. & Palti, Y. (2008). Microbial growth inhibition by alternating electric fields. *Antimicrobial Agents and Chemotherapy, 52(10),* 3517–3522.

[27] Hong, S.H. Jeong, J. Shim, S. Kang, H. Kwon, S. Ahn, K.H. & Yoon, J. (2008). Effect of electric currents on bacterial detachment and inactivation. *Biotechnology and Bioengineering, 100 (2),* 379–386.

[28] Stewart, P. S. Wattanakaroon, W. Goodrum, L. Fortun, S. M. & McLeod, B. R. (1999). Electrolytic generation of oxygen partially explains electrical enhancement of tobramycin efficacy against *Pseudomonas aeruginosa* biofilm. *Antimicrobial Agents and Chemotherapy, 43,* 292–296.

[29] Stoodley, P. de Beer, D. & Lappin-Scott, H. (1997). Influence of electrical fields and pH on biofilm structure as related to the bioelectric effect. *Antimicrobial Agents and Chemotherapy, 41,* 1876–1879.

[30] Caubet, R. Pedarros-Caubet, F. Chu, M. Freye, E. de Bele´m Rodrigues, M. Moreau, J. M. & Ellison, W. J. (2004). A radio frequency electric current enhances antibiotic efficacy against bacterial biofilms. *Antimicrobial Agents and Chemotherapy, 48(12),* 4662–4664.

[31] Ehrlich, G. D. Hu, F. Z. & Post, J. C. (2004). Role for biofilms in infectious disease. In: Ghannoum, M. A. and O'Toole, G. (eds). *Microbial Biofilms.* (332–358). Washington, ASM Press.

[32] Yeh, J. I. Chinte, U. & Du, S. (2008). Structure of glycerol-3-phosphate dehydrogenase, an essential monotopic membrane enzyme involved in respiration and metabolism. *Proceedings of the National Academy of Sciences, USA, 105,* 3280–3285.

[33] Massey, T. H. Mercogliano, C. P. Yates, J. Sherratt, D. J. & Lowe, J. (2006). Double-stranded DNA translocation, structure and mechanism of hexameric FtsK. *Molecular Cell, 23,* 457–469.

[34] Del re, B. Bersani, F. Mesirca, P. & Giorgi, G. (2006). Synthesis of DnaK and GroEL in *Escherichia coli* cells exposed to different magnetic field signals. *Bioelectrochemistry, 69(1),* 99–103.

[35] Cellini, L. Grande, R. Di Campli, E. Di Bartolomeo, S. Di Giulio, M. Robuffo, I. Trubiani, O. & Mariggio, M. A. (2008). Bacterial response to the exposure of 50 Hz electromagnetic fields. *Bioelectromagnetics, 29,* 302–311.

[36] Chua, L. Y. & Yeo, S. H. (2005). Surface bio-magnetism on bacterial cells adhesion and surface proteins secretion. *Colloids Surfaces B Biointerfaces,* 40, 45–49.

[37] Stansell, M. J. Winters, W. D. Doe, R. H. & Dart, B. K. (2001). Increased antibiotic resistance of *E. coli* exposed to static magnetic fields. *Bioelectromagnetics, 22(2),* 129–137.

[38] Obermeier, A. Matl, F. D. Friess, W. & Stemberger, A. (2009). Growth inhibition of *Staphylococcus aureus* induced by low-frequency electric and electromagnetic fields. *Bioelectromagnetics, 30(4),* 270–279.

[39] McLeod, B. R. & Sandvik, E. L. (2010). A biofilm growth protocol and the design of a magnetic field exposure setup to be used in the study of magnetic fields as a means of controlling bacterial biofilms. *Bioelectromagnetics, 31(1),* 56–63.

[40] Pickering, S. A. Bayston, R. & Scammell, B. E. (2003). Electromagnetic augmentation of antibiotic efficacy in infection of orthopaedic implants. *Journal of Bone and Joint Surgery, British, 85,* 588–593.

[41] Jass, J. & Lappin-Scott H. (1996). The efficacy of antibiotics enhanced by electrical currents against *Pseudomonas aeruginosa* biofilms. *Journal of Antimicrobial Chemotherapy, 38,* 987–1000.

[42] del Pozo, J. L. Rouse, M. S. Mandrekar, J. N. Sampedro, M. F. Steckelberg, J. M. & Patel, R. (2009). Effect of electrical current on the activities of antimicrobial agents against *Pseudomonas aeruginosa, Staphylococcus aureus,* and *Staphylococcus epidermidis* biofilms. *Antimicrobial Agents and Chemotherapy, 53(1),* 35–40.

[43] Jayaraman, A. & Wood, T. K. (2008). Bacterial quorum sensing, signals, circuits, and implications for biofilms and disease. *Annual Reviews in Biomedical Engineering, 10,* 145-167.

[44] Kobayashi, N. Bauer, T. W. Tuohy, M. J. Fujishiro, T. & Procop, G. W. (2007). Brief ultrasonication improves detection of biofilm-formative bacteria around a metal implant. *Clinical Orthopaedics and Related Research, 457,* 210–213.

[45] Sampedro, M. F. Huddleston, P. M. Piper, K. E. Karau, M. J. Dekutoski, M.B. Yaszemski, M. J. Currier, B. L. Mandrekar, J. N. Osmon, D. R. McDowell, A. Patrick, S. Steckelberg, J. M. & Patel, R. (2010). A biofilm approach to detect bacteria on removed spinal implants. *Spine* (Phila Pa 1976), 35(12), 1218-24.

[46] Padberg, F. T. Jr. Smith, S. M. & Eng, R. H. (1992). Optimal method for culturing vascular prosthetic grafts. *Journal of Surgical Research, 53,* 384–390.

[47] Lim, A. L. & Bai, R. (2003). Membrane fouling and cleaning in microfiltration of activated sludge wastewater. *Journal of Membrane Science, 216,* 279–290.

[48] Kobayashi, H. Oethinger, M. Tuohy, M.J. Procop, G. W. & Bauer, T. W. (2009). Improved detection of biofilm-formative bacteria by vortexing and sonication, a pilot study. *Clinical Orthopaedics and Related Research, 467(5),* 1360-1364.

[49] Stickler, D. J. & Hewitt, P. (1991). The activity of antiseptics against biofilms of mixed bacterial species growing on silicone surfaces. *European Journal of Clinical Microbiology and Infectious Diseases*, 10, 416–421.

[50] Bott, T. R. (2009). Biofouling control in cooling water. *International Journal of Chemical Engineering*, Article ID 619873, 4 pages. doi, 10.1155/2009/619873.

[51] Bott, T. R. (2000). Biofouling control with ultrasound. *Heat Transfer Engineering, 21(3),* 43–49.

[52] Oulahal-Lagsir, N. Martial-Gros, A. Boistier, E. Blum, L. J. & Bonneau, M. (2000). The development of an ultrasonic apparatus for the noninvasive and repeatable removal of fouling in food processing equipment. *Letters in Applied Microbiology*, 30(1), 47-52.

[53] Oulahal-Lagsir, N. Martial-Gros, A. Boistier, E. & Blum, L. J. (2003). "*Escherichia coli*-milk" biofilm removal from stainless steel surfaces, synergism between ultrasonic waves and enzymes. *Biofouling, 19,* 159-168.

[54] Hazan, Z. Zumeris, J. Jacob, H. Raskin, H. Kratysh, G. Vishnia, M. Dror, N. Barliya, T. Mandel, M. & Lavie G. (2006). Effective prevention of microbial biofilm formation on medical devices by low-energy surface acoustic waves. *Antimicrobial Agents and Chemotherapy, 50,* 4144-4152.

[55] Saunders, S. Bocking, A. Challis, J. & Reid, G. (2007). Effect of *Lactobacillus* challenge on *Gardnerella vaginalis* biofilms. *Colloids Surfaces B Biointerfaces, 55(2),* 138-42.

[56] Rodrigues, L. van der Mei, H. C. Teixeira, J. & Oliveira, R. (2004). Influence of biosurfactants from probiotic bacteria on formation of biofilms on voice prostheses. *Applied and Environmental Microbiology, 70(7),* 4408-10.

[57] Walencka, E. Rózalska, S. Sadowska, B. & Rózalska, B. (2008).The influence of *Lactobacillus acidophilus*-derived surfactants on staphylococcal adhesion and biofilm formation. *Folia Microbiologica (Praha), 53(1),* 61-6.

[58] Näse, L. Hatakka, K. Savilahti, E. Saxelin, M. Pönkä, A. Poussa, T. Korpela, R. & Meurman, J. H. (2001). Effect of long-term consumption of a probiotic bacterium, *Lactobacillus rhamnosus* GG, in milk on dental caries and caries risk in children. *Caries Research, 35(6),* 412-20.

[59] Ahola, A. J. Yli-Knuuttila, H. Suomalainen, T. Poussa, T. Ahlström, A. Meurman, J. H. & Korpela, R. (2002). Short-term consumption of probiotic-containing cheese and its effect on dental caries risk factors. *Archives of Oral Biology, 47(11),* 799-804.

[60] Caglar, E. Kavaloglu, S. C. Kuscu, O. O. Sandalli, N. Holgerson, P. L. & Twetman, S. (2007). Effect of chewing gums containing xylitol or probiotic bacteria on salivary mutans streptococci and lactobacilli. *Clinical Oral Investigations, 11(4),* 425-9.

[61] Söderling, E. M. Marttinen, A. M. & Haukioja, A. L. (2011). Probiotic lactobacilli interfere with *Streptococcus mutans* biofilm formation in vitro. *Current Microbiology, 62(2),* 618-22.

[62] Zhao, T. Doyle, M. P. & Zhao, P. (2004). Control of *Listeria monocytogenes* in a biofilm by competitive-exclusion microorganisms. *Applied and Environmental Microbiology,70(7),* 3996-4003.

[63] Kim, Y. Oh, S. & Kim, S. H. (2009). Released exopolysaccharide (r-EPS) produced from probiotic bacteria reduce biofilm formation of enterohemorrhagic *Escherichia coli* O157: H7. *Biochemical and Biophysical Research Communications, 379(2),* 324-9.

[64] Gadekar, S. Nemati, M. & Hill, G. A. (2006). Batch and continuous biooxidation of sulphide by *Thiomicrospira* sp. CVO, reaction kinetics and stoichiometry. *Water Research, 40,* 2436– 2446.

[65] Redner, J. A. His, R. P. & Esfandi, E. J. (1994). Evaluating coatings for concrete in wastewater facilities, update. *The Journal of Protective Coatings and Linings, 11,* 50–61.

[66] Zhang, L. De Schryver, P. De Gusseme, B. De Muynck, W. Boon, N. & Verstraete, W. (2008). Chemical and biological technologies for hydrogen sulphide emission control in sewer systems, a review. *Water Research, 42,* 1–12.

[67] Jayaraman, A. Mansfeld, F. B. & Wood, T. K. (1999). Inhibiting sulphatereducing bacteria in biofilms by expressing the antimicrobial peptides indolicidin and bactenecin. *Journal of Industrial Microbiology and Biotechnology, 22,* 167-75.

[68] Orgaz, B. Kives, J. Pedregosa, A. M. Monistrol, I. F. Laborda, F. & SanJosé, C. (2006). Bacterial biofilm removal using fungal enzymes. *Enzyme and Microbial Technology, 40,* 51-56.

[69] Brouwers, G. J. Vijgenboom, E. Corstjens, P. L. A. M. deVrind, J. P. M. & de Vrine-de Jone E. W. (2000). Bacterial Mn2+ oxidizing systems and multicopper oxidases, an overview of mechanisms and functions. *Geomicrobiology Journal, 17,* 1–24.

[70] Zuo, R. & Wood, T. K. (2004). Inhibiting mild steel corrosion from sulfate-reducing and iron-oxidizing bacteria using gramicidin-S-producing biofilms. *Applied Microbiology and Biotechnology, 65,* 747–753.

[71] Simoes, M. Simoes, L. C. & Vieira, M. J. (2010). A review of current and emergent biofilm control strategies. *LWT - Food Science and Technology, 43,* 573–583.

[72] Deveau, H. Van Calsteren, M. R. & Moineau, S. (2002). Effect of exopolysaccharides on phage-host interactions in *Lactococcus lactis*. *Applied and Environmental Microbiology, 68(9),* 4364-9.

[73] Lu, T. K. & Collins, J. J. (2007). Dispersing biofilms with engineered enzymatic bacteriophage. *Proceedings of the National Academy of Sciences, USA, 104(27),* 11197-202.

[74] Verma, V. Harjai, K. & Chhibber, S. (2010). Structural changes induced by a lytic bacteriophage make ciprofloxacin effective against older biofilm of *Klebsiella pneumoniae*. *Biofouling, 26(6),* 729-737.

[75] Verma, V. Harjai, K. & Chhibber, S. (2009). Restricting ciprofloxacin-induced resistant variant formation in biofilm of *Klebsiella pneumoniae* B5055 by complementary bacteriophage treatment. *Journal of Antimicrobial Chemotherapy, 64(6),* 1212-1218.

[76] Carson, L. Gorman, S. P. & Gilmore, B. F. (2010). The use of lytic bacteriophages in the prevention and eradication of biofilms of *Proteus mirabilis* and *Escherichia coli. FEMS Immunology and Medical Microbiology, 59(3),* 447-455.

[77] May, T. Tsuruta, K. & Okabe, S. (2011). Exposure of conjugative plasmid carrying *Escherichia coli* biofilms to male-specific bacteriophages. *International Society for Microbial Ecology Journal, 5(4),* 771-775.

[78] Fu, W. Forster, T. Mayer, O. Curtin, J. J. Lehman, S. M. & Donlan, R. M. (2010). Bacteriophage cocktail for the prevention of biofilm formation by *Pseudomonas aeruginosa* on catheters in an in vitro model system. *Antimicrobial Agents and Chemotherapy, 54(1),* 397-404.

[79] Sillankorva, S. Neubauer, P. & Azeredo, J. (2010). Phage control of dual species biofilms of *Pseudomonas fluorescens* and *Staphylococcus lentus. Biofouling, 26(5),* 567-75.

[80] Merabishvili, M. Pirnay, J. P. Verbeken, G. Chanishvili, N. Tediashvili, M. Lashkhi, N. Glonti, T. Krylov, V. Mast, J. Van Parys, L. Lavigne, R. Volckaert, G. Mattheus, W. Verween, G. De Corte, P. Rose, T. Jennes, S. Zizi, M. De Vos, D. & Vaneechoutte, M.

(2009). Quality-controlled small-scale production of a well-defined bacteriophage cocktail for use in human clinical trials. *PLoS One, 4(3),* e4944.

[81] Richards, J. J. & Melander, C. (2009). Controlling bacterial biofilms. *Chembiochem. 10(14),* 2287-94.

[82] Sharma, M. Ryu, J. H. & Beuchat, L. R. (2005). Inactivation of *Escherichia coli* O157:H7 in biofilm on stainless steel by treatment with an alkaline cleaner and a bacteriophage. *Journal of Applied Microbiology, 99(3),* 449-59.

[83] Soni, K. A. & Nannapaneni, R. (2010). Removal of *Listeria monocytogenes* biofilms with bacteriophage P100. *Journal of Food Protection, 73(8),* 1519-24.

[84] Goldman, G. Starosvetsky, J. & Armon, R. (2009). Inhibition of biofilm formation on UF membrane by use of specific bacteriophages. *Journal of Membrane Sciences, 342,* 145–152.

[85] Gino, E. Starosvetsky, J. Kurzbaum, E. & Armon, R. (2010). Combined chemical-biological treatment for prevention/rehabilitation of clogged wells by an iron-oxidizing bacterium. *Environmental Science and Technology, 44(8),* 3123-9.

[86] Azeredo, J. & Sutherland, I. W. (2008). The use of phages for the removal of infectious biofilms. *Current Pharmaceutical Biotechnology, 9(4),* 261-6.

[87] Resch, A. Fehrenbacher, B. Eisele, K. Schaller, M. & Götz, F. (2005). Phage release from biofilm and planktonic *Staphylococcus aureus* cells. *FEMS Microbiology Letters, 252(1),* 89-96.

[88] Flemming, H. C. & Wingender, J. (2010). The biofilm matrix. *Nature Reviews Microbiology, 8,* 623–633.

[89] Johansen, C. Falholt, P. & Gram, L. (1997). Enzymatic removal and disinfection of bacterial biofilms. *Applied and Environmental Microbiology, 63(9),* 3724-8.

[90] Loiselle, M. & Anderson, K. W. (2003). The use of cellulase in inhibiting biofilm formation from organisms commonly found on medical implants. *Biofouling, 19(2),* 77-85.

[91] Sass, P. & Bierbaum, G. (2007). Lytic activity of recombinant bacteriophage phi11 and phi12 endolysins on whole cells and biofilms of *Staphylococcus aureus. Applied and Environmental Microbiology,* 73(1), 347-52.

[92] Son, J. S. Lee, S. J. Jun, S. Y. Yoon, S. J. Kang, S. H. Paik, H. R. Kang, J. O. & Choi, Y. J. (2010). Antibacterial and biofilm removal activity of a podoviridae *Staphylococcus aureus* bacteriophage SAP-2 and a derived recombinant cell-wall-degrading enzyme. *Applied Microbiology and Biotechnology, (5),* 1439-49.

[93] Kaplan, J. B. Ragunath, C. Velliyagounder, K. Fine, D. H. & Ramasubbu, N. (2004). Enzymatic detachment of *Staphylococcus epidermidis* biofilms. *Antimicrobial Agents and Chemotherapy, 48(7),* 2633-6.

[94] Izano, E. A. Shah, S. M. & Kaplan, J. B. (2009). Intercellular adhesion and biocide resistance in nontypeable *Haemophilus influenza* biofilms. *Microbial Pathogenesis, 46,* 207–213.

[95] Chaignon, P. Sadovskaya, I. Ragunah, Ch. Ramasubbu, N. Kaplan, J. B. & Jabbouri, S. (2007). Susceptibility of staphylococcal biofilms to enzymatic treatments depends on their chemical composition. *Applied Microbiology and Biotechnology, 75(1),* 125-32.

[96] Leroy, C. Delbarre-Ladrat, C. Ghillebaert, F. Compere, C. & Combes, D. (2008). Effects of commercial enzymes on the adhesion of a marine biofilm-forming bacterium. *Biofouling, 24,* 11–22.

[97] Poele, S. T. & van der Graaf, J. (2005). Enzymatic cleaning in ultrafiltration of wastewater treatment plant effluent. *Desalination, 179,* 73–81.

[98] Whittaker, C. Ridgway, H. & Olson, B. H. (1984). Evaluation of cleaning strategies for removal of biofilms from reverse-osmosis membranes. *Applied and Environmental Microbiology, 48(2),* 395-403.

[99] Molobela, P. I. Cloete, T. E. & Beukes, M. (2010). Protease and amylase enzymes for biofilm removal and degradation of extracellular polymeric substances (EPS) produced by *Pseudomonas fluorescens* bacteria. *African Journal of Microbiology Research, 4(14),* 1515-1524.

[100] Xiong, Y, & Liu, Y. (2010). Biological control of microbial attachment, a promising alternative for mitigating membrane biofouling. *Applied Microbiology and Biotechnology, 86(3),* 825-37.

[101] Moscoso, M. García, E. & López, R. (2006). Biofilm formation by *Streptococcus pneumoniae*, role of choline, extracellular DNA, and capsular polysaccharide in microbial accretion. *Journal of Bacteriology, 188(22),* 7785-7795.

[102] Whitchurch, C. B. Tolker-Nielsen, T. Ragas, P. C. & Mattick, J. S. (2002). Extracellular DNA required for bacterial biofilm formation. *Science, 295(5559),* 1487.

[103] Thomas, V. C. Thurlow, L. R. Boyle, D. & Hancock, L. E. (2008). Regulation of autolysis-dependent extracellular DNA release by *Enterococcus faecalis* extracellular proteases influences biofilm development. *Journal of Bacteriology, 190(16),* 5690-5698.

[104] Samaranayake, Y. H. Cheung, B. P. Parahitiyawa, N. Seneviratne, C. J. Yau, J. Y. Yeung, K. W. & Samaranayake, L. P. (2009). Synergistic activity of lysozyme and antifungal agents against *Candida albicans* biofilms on denture acrylic surfaces. *Archives of Oral Biology, 54(2),* 115-126.

[105] Caro, A. Humblot, V. Mthivier, C. Minier, M. Salmain, M. & Pradier, C. M. (2009). Grafting of lysozyme and/or poly(ethylene glycol) to prevent biofilm growth on stainless steel surfaces. *The Journal of Physical Chemistry B, 113(7),* 2101-2109.

[106] Richards, M. & Cloete, T. E. (2010). Nanozymes in Biofilm Removal. In: Cloete, T. E. de Kwaadsteniet, M. Botes, M. and López-Romero, J. M. (eds). *Nanotechnology in Water Treatment Application.* (89-102). Norfolk, U. K., Caister Academic Press.

[107] Abiko, Y. (2000). Passive immunization against dental caries and periodontal disease, development of recombinant and human monoclonal antibodies. *Critical Reviews in Oral Biology and Medicine, 11(2),* 140-158.

[108] Dallo, S. F. & Weitao, T. (2010). Insights into acinetobacter war-wound infections, biofilms, and control. *Advances in Skin and Wound Care, 23(4),* 169-174.

[109] Scheie, A. A. & Petersen, F. C. (2004). The biofilm concept, consequences for future prophylaxis of oral diseases? *Critical Reviews in Oral Biology and Medicine, 15(1),* 4-12.

[110] Shimazaki, Y. Mitoma, M. Oho, T. Nakano, Y. Yamashita, Y. Okano, K. Nakano, Y. Fukuyama, M. Fujihara, N. Nada, Y. & Koga, T. (2001). Passive immunization with milk produced from an immunized cow prevents oral recolonization by *Streptococcus mutans*. *Clinical and Diagnostic Laboratory Immunology*, *8(6)*, 1136-1139.

[111] Ma, J. K. Hikmat, B. Y. Wycoff, K. Vine, N. D. Chargelegue, D. Yu, L. Hein, M. B. & Lehner, T.(1998). Characterization of a recombinant plant monoclonal secretory antibody and preventive immunotherapy in humans. *Nature Medicine, 4(5)*, 601-606.

[112] Sharma, A. Honma, K. Evans, R. T. Hruby, D. E. & Genco, R. J. (2001). Oral immunization with recombinant *Streptococcus gordonii* expressing *Porphyromonas gingivalis FimA* domains. *Infection and Immunity, 69*, 2928-2934.

[113] Rediske, A. M. Koenig, A. L. Barekzi, N. Ameen, L. C. Slunt, J. B. & Grainger, D. W. (2002). Polyclonal human antibodies reduce bacterial attachment to soft contact lens and corneal cell surfaces. *Biomaterials, 23,* 4565-4572.

[114] Poelstra, K. A. van der Mei, H. C. Gottenbos, B. Grainger, D. W. van Horn, J. R. & Busscher, H. J. (2000). Pooled human immunoglobulins reduce adhesion of *Pseudomonas aeruginosa* in a parallel plate flow chamber. *Journal of Biomedical Materials Research, 51,* 224-232.

[115] Kwok, C. S. Horbett, T. A. & Ratner, B. D. (1999). Design of infection-resistant antibiotic-releasing polymers. II. Controlled release of antibiotics through a plasma-deposited thin film barrier. *Journal of Controlled Release, 62,* 301-311.

[116] Kwok, C. S. Mourad, P. D. Crum, L. A. & Ratner, B. D. (2001). Self-assembled molecular structures as ultrasonically-responsive barrier membranes for pulsatile drug delivery. *Journal of Biomedical Materials Research, 57,* 151-164.

[117] Shahrooei, M. Hira, V. Stijlemans, B. Merckx, R. Hermans, P. W. & Van Eldere, J. (2009). Inhibition of *Staphylococcus epidermidis* biofilm formation by rabbit polyclonal antibodies against the SesC protein. *Infection and Immunity, 77(9),* 3670-8.

[118] Sun, D. Accavitti, M. A. & Bryers, J. D. (2005). Inhibition of biofilm formation by monoclonal antibodies against *Staphylococcus epidermidis* RP62A accumulation-associated protein. *Clinical and Diagnostic Laboratory Immunology, 12(1),* 93-100.

[119] Tashiro, Y. Nomura, N. Nakao, R. Senpuku, H. Kariyama, R. Kumon, H. Kosono, S. Watanabe, H. Nakajima, T. & Uchiyama, H. (2008). Opr86 is essential for viability and is a potential candidate for a protective antigen against biofilm formation by *Pseudomonas aeruginosa. Journal of Bacteriology, 190(11),* 3969-3678.

[120] Sanchez, C. J. Shivshankar, P. Stol, K. Trakhtenbroit, S. Sullam, P. M. Sauer, K. Hermans, P. W. & Orihuela, C. J. (2010).The pneumococcal serine-rich repeat protein is an intra-species bacterial adhesin that promotes bacterial aggregation in vivo and in biofilms. *PLoS Pathogen. 12*, 6-8.

[121] Fujibayashi, T. Nakamura, M. Tominaga, A. Satoh, N. Kawarai, T. Narisawa, N. Shinozuka, O. Watanabe, H. Yamazaki, T. & Senpuku, H. (2009). Effects of IgY against *Candida albicans* and *Candida* spp. adherence and biofilm formation. *Japanese Journal of Infectious Diseases, 62(5),* 337-342.

[122] Robertson, E. J. & Casadevall A. 2009. Antibody-Mediated Immobilization of Cryptococcus neoformans Promotes Biofilm Formation. *Applied and Environmental Microbiology,* 75(8), 2528-2533.

[123] Martinez, L. R. Christaki, E. & Casadevall A. (2006). Specific antibody to *Cryptococcus neoformans* glucurunoxylomannan antagonizes antifungal drug action against cryptococcal biofilms in vitro. *Journal of Infectious Diseases, 194(2),* 261-266.

[124] Desbois, A. P. Mearns-Spragg, A. & Smith, V. J. (2009). A fatty acid from the diatom *Phaeodactylum tricornutum* is antibacterial against diverse bacteria including multiresistant *Staphylococcus aureus* (MRSA). *Marine Biotechnology, 11,* 45-52.

[125] Wille, J. J. & Kydonieus, A. (2003). Palmitoleic acid isomer (C16, 1Δ6) in human skin sebum is effective against Gram-positive bacteria. *Skin Pharmacology and Applied Skin Physiology, 16,* 176-187.

[126] Davies, D. G. & Marques, C. N. H. (2009). A fatty acid messenger is responsible for inducing dispersion in microbial biofilms. *Journal of Bacteriology, 191,* 1393-1403.

[127] Kenny, J. G. Ward, D. Josefsson, E. Jonsson, I. M. Hinds, J. Rees, H. H. Lindsay, J. A. Tarkowski, A. & Horsburgh, M. J. (2009). The *Staphylococcus aureus* response to unsaturated long chain free fatty acids, survival mechanisms and virulence implications. *PLoS One, 4,* e4344.

[128] Beck, V. Jabůrek, M. Demina, T. Rupprecht, A. Porter, R. K. Ježek, P. & Pohl, E. E. (2007). Polyunsaturated fatty acids activate human uncoupling proteins 1 and 2 in planar lipid bilayers. *Federation of American Societies for Experimental Biology, Journal, 21,* 1137-1144.

[129] Sado-Kamdem, S. L. Vannini, L. & Guerzoni, M. E. (2009). Effect 1 of α-linolenic, capric and lauric acid on the fatty acid biosynthesis in *Staphylococcus aureus. International Journal of Food Microbiology, 129,* 288-294.

[130] Schonfeld, P. & Wojtczak, L. (2008). Fatty acids as modulators of the cellular production of reactive oxygen species. *Free Radical Biology and Medicine, 45,* 231-241.

[131] Zheng, C. J. Yoo, J. S. Lee, T. G. Cho, H. Y. Kim, Y. H. & Kim, W. G. (2005). Fatty acid synthesis is a target for antibacterial activity of unsaturated fatty acids. *Federation of European Biochemical Societies Letters, 579,* 5157-5162.

[132] Guil-Guerrero, J. L. Giménez- Giménez, A. Robles-Medina, A. del Mar Rebolloso-Fuentes, M. Belarbi, E. H. Esteban-Cerdán, L. & Molina-Grima, E. (2001). Hexane reduces peroxidation of fatty acids during storage. *European Journal of Lipid Sciences and Technology, 103,* 271-278.

[133] US FDA (1997). Substances affirmed as generally recognized as safe, menhaden oil. Federal Register 62, 30751-30757. Available from URL: http://www.federalregister.gov/articles/1997/06/05/97-14683/substances-affirmed-as-generally-recognized-as-safe-menhaden-oil

[134] Sun, C. Q. O'Connor, C. J. & Roberton, A. M. (2003). Antibacterial actions of fatty acids and monoglycerides against *Helicobacter pylori. FEMS Immunology and Medical Microbiology, 36,* 9-17.

[135] Lukowski, G. Lindequist, U. Mundt, S. Kramer, A. & Jülich, W. D. (2008). Inhibition of dermal MRSA colonisation by microalgal micro- and nanoparticles. *Skin Pharmacology Physiology, 21,* 98-105.

[136] Petschow, B. W. Batema, R. P. & Ford, L. L. (1996). Susceptibility of *Helicobacter pylori* to bactericidal properties of medium-chain monoglycerides and free fatty acids. *Antimicrobial Agents and Chemotherapy, 40,* 302-306.

[137] Ren, D. Zuo, R. Barrios, A. F. G. Bedzyk, L. A. Eldridge, G. R. Pasmore, M. E. & Wood, T. K. (2005). Differential gene expression for investigation of *Escherichia coli* biofilm inhibition by plant extract ursolic acid. *Applied Environmental Microbiology, 71,* 4022–4034.

[138] Garo, E. Eldridge, G. R. Goering, M. G. Pulcini, E. D. Hamilton, M. A. Costerton, J. W. & James, G. A. (2007). Asiatic acid and corosolic acid enhance the susceptibility of *Pseudomonas aeruginosa* biofilms to tobramycin. *Antimicrobial Agents and Chemotherapy, 51,* 1813–1817.

[139] Zeng, Z. Qian, L. Cao, L. Tan, H. Huang, Y. Xue, X. Shen, Y. & Zhou, S. (2008). Virtual screening for novel quorum sensing inhibitors to eradicate biofilm formation of *Pseudomonas aeruginosa*. *Applied Microbiology and Biotechnology, 79,* 119–126.

[140] Calderone, R. A. & Fonzi, W. A. (2001). Virulence factors of *Candida albicans*. *Trends in Microbiology, 9,* 327–335.

[141] Jabra-Rizk, M. A. Falkler, W. A. Jr. Merz, W. G. Baqui, A. A. Kelley, J. I. & Meiller, T. F. (2000). Retrospective identification and characterization of *Candida dubliniensis* isolates among *Candida albicans* clinical laboratory isolates from human immunodeficiency virus (HIV)-infected and non-HIV-infected individuals. *Journal of Clinical Microbiology, 38,* 2423–2426.

[142] McLain, N. Ascanio, R. Baker, C. Strohaver, R. A. & Dolan, J. W. (2000). Undecylenic acid inhibits morphogenesis of *Candida albicans*. *Antimicrobial Agents and Chemotherapy, 44,* 2873–2875.

[143] Clement, M. Tremblay, J., Lange, M., Thibodeau, J. & Belhumeur, P. (2007). Wheyderived free fatty acids suppress the germination of *Candida albicans* in vitro. *FEMS Yeast Research, 7,* 276–285.

[144] Thibane, V. S. Kock, J. L. F. Ells, R. van Wyk, P. W. J. & Pohl, C. H. (2010). Effect of marine polyunsaturated fatty acids on biofilm formation of *Candida albicans* and *Candida dubliniensis*. *Marine Drugs, 8,* 2597-2604.

[145] Huang, T. P. & Wong, A. C. (2007). Extracellular fatty acids facilitate flagella-independent translocation by *Stenotrophomonas maltophilia*. *Research in Microbiology, 158,* 702–711.

[146] Ryan, R. P. Fouhy, Y. Garcia, B. F. Watt, S. A. Niehaus, K. Yang, L. Tolker-Nielsen, T. & Dow, J. M. (2008). Interspecies signalling via the *Stenotrophomonas maltophilia* diffusible signal factor influences biofilm formation and polymyxin tolerance in *Pseudomonas aeruginosa*. *Molecular Microbiology, 68,* 75–86.

[147] Inoue, T. Shingaki, R. & Fukui, K. (2008). Inhibition of swarming motility of *Pseudomonas aeruginosa* by branched-chain fatty acids. *FEMS Microbiology Letters, 281,* 81–86.

[148] Refsgaard, H. H. F. Brockhoff, P. M. B. & Jensen, B. (2000). Free 1 polyunsaturated fatty acids cause taste deterioration of salmon during frozen storage. *Journal of Agriculrural Food Chemistry, 48,* 3280-3285.

[149] Baer, W. S. (1931). The treatment of chronic osteomyelitis with the maggot (larvae of the blowfly). *Journal of Bone and Joint Surgery, 13,* 438–475.

[150] Sherman, R. A. & Pechter, E. A. (1988). Maggot therapy, a review of the therapeutic applications of fly larvae in human medicine, especially for treating osteomyelitis. *Medical and Veterinary Entomology, 2,* 225– 230.

[151] Tantawi, T. I. Gohar, Y. M. Kotb, M. M. Beshara, F. M. & El-Naggar, M. M. (2007). Clinical and microbiological efficacy of MDT in the treatment of diabetic foot ulcers. *Journal of Wound Care, 16,* 379–383.

[152] US FDA, 2004. General Plastic Surgery Devices Panel July 27, 2005. Briefing Information. Panel Classification Memo. Available at, www.fda.gov/ohrms/dockets/ac/05/briefing/2005-4168b1.htm Accessed August 25, 2005.

[153] Huberman, L. Gollop, N. Mumcuoglu, K. Y. Breuer, E. Bhusare, S. R. Shai, Y. & Galun, R. (2007). Antibacterial substances of low molecular weight isolated from the blowfly, *Lucilia sericata*. *Medical and Veterinary Entomology, 21,* 127–31.

[154] Chambers, L. Woodrow, S. Brown, A. P. Harris, P. D. Philips, D. Hall M. Church J. C. & Pritchard, D. I. (2003). Degradation of extracellular matrix components by defined proteinases from the greenbottle larva *Lucilia sericata* used for the clinical debridement of non-healing wounds. *British Journal of Dermatology, 148,* 14–23.

[155] Altincicek, B. & Vilcinskas, A. (2009). Septic injury-inducible genes in medicinal maggots of the green blow fly *Lucilia sericata. Insect Molecular Biology, 18,* 119–125.

[156] Harris, L. G. Bexfield, A. Nigam, Y. Rohde, H. Ratcliffe, N. A. & Mack, D. (2009). Disruption of *Staphylococcus epidermidis* biofilms by medicinal maggot *Lucilia sericata* excretions/secretions. *International Journal of Artificial Organs, 32(9),* 555-564.

[157] van der Plas, M. J. A. Jukema, G. N. Wai, S. W. Dogterom-Ballering, H. C. M. Lagendijk, E. L. van Gulpen, C. van Dissel, J. T. Bloemberg, G. V. & Nibbering, P. H. (2008). Maggot excretions/secretions are differentially effective against biofilms of *Staphylococcus aureus* and *Pseudomonas aeruginosa. Journal of Antimicrobial Chemotherapy,* 61(1), 117-122.

[158] Cazander, G. van Veen, K. E. Bouwman, L. H. Bernards, A. T. & Jukema, G. N. (2009). The influence of maggot excretions on PAO1 biofilm formation on different biomaterials. *Clinical Orthopaedics and Related Research, 467(2),* 536-545.

[159] Cazander, G. van de Veerdonk, M. C. Vandenbroucke-Grauls, C. M. Schreurs, M. W. & Jukema, G. N. (2010). Maggot excretions inhibit biofilm formation on biomaterials. *Clinical Orthopaedics and Related Research, 468(10),* 2789-2796.

[160] Cazander, G. van Veen, K. E. Bernards, A. T. & Jukema, G. N. (2009). Do maggots have an influence on bacterial growth? A study on the susceptibility of strains of six different bacterial species to maggots of *Lucilia sericata* and their excretions/secretions. *Journal of Tissue Viability, 18(3),* 80-7.

[161] Andersen, A. S. Sandvang, D. Schnorr, K. M. Kruse, T. Neve, S. Joergensen, B. Karlsmark, T. & Krogfelt, K. A. (2010). A novel approach to the antimicrobial activity of maggot debridement therapy. *Journal of Antimicrobial Chemotherapy, 65(8),* 1646-1654.

[162] Rogers, J. Dowsett, A. B. Dennis, P. J. Lee, J. V. & Keevil, C. W. (1994). Influence of plumbing materials on biofilm formation and growth of *Legionella pneumophila* in potable water systems. *Applied and Environmental Microbiology, 60,* 1842–1851.

[163] Pagedar, A. Singh, J. & Batish, V. K. (2010). Surface hydrophobicity, nutritional contents affect *Staphylococcus aureus* biofilms and temperature influences its survival in preformed biofilms. *Journal of Basic Microbiology, 50(S1),* S98-106.

[164] de Carvalho, C. C. C. R. (2007). Biofilms, recent developments on an old battle. *Recent Patents on Biotechnology, 1,* 49-57.

[165] Rosenberg, L. E. Cabone, A. L. Römling, U. Uhrich, K. E. & Chikindas, M. L. (2008). Salicylic acid-based poly(anhydride-esters) for control of biofilm formation in *Salmonella enterica* serover Typhimurium. *Letters in Applied Microbiology, 46,* 593-599.

[166] Schmeltzer, R. C. Schmalenberg, K. E. & Uhrich, K. E. (2005). Synthesis and cytotoxicity of salicylate-based poly (anhydride esters). *Biomacromolecules, 6,* 359-367.

[167] Splendiani, A. Livingston, A. G. & Nicolella, C. (2006). Control membrane-attached biofilms using surfactants. *Biotechnology and Bioengineering, 94,* 15–23.

[168] Cloete, T. E. & Jacobs, L. (2001). Surfactants and the attachment of *Pseudomonas aeruginosa* to 3CR12 stainless steel and glass. *Water SA, 27,* 21–26.

[169] Worley, S. D. Eknoian, M. W. and Li, Y. (2002). US20026469177.

[170] Francolini, I. Norris, P. Piozzi, A. Donelli, G. & Stoodley, P. (2004). Usnic acid, a natural antimicrobial agent able to inhibit bacterial biofilm formation on polymer surfaces. *Antimicrobial Agents and Chemotherapy, 48,* 4360-4365.

[171] Jansson, R. E. O'Brien, R. N. & Visaisouk, S. (1992). US5104649.

[172] Fitzpatrick, R. J. Shackett, K. K. & Klinger, J. D. (2005). US20056955806.

[173] Gottenbos, B. Busscher, H. J. Van Der Mei, H. C. & Nieuwenhuis, P. (2002). Pathogenesis and prevention of biomaterial centered infections. *Journal of Materials Science: Materials in Medicine, 13,* 717-722.

[174] Seder, E. V. & Nelson, J. N. (2002). WO02083031A3.

[175] Jamiolkowski, D. D. Rothenburger, S. Spangler, J. & Daniel J. (2003). US20036514517.

[176] Sudarsan, N. Lee, E. R. Weinberg, Z. Moy, R. H. Kim, J. N. Link, K. H. & Breaker, R. R. (2008). Riboswitches in eubacteria sense the second messenger cyclic di-GMP. *Science, 321,* 411–413.

[177] Spangler, C. Kaever, V. & Seifert, R. (2011). Interaction of the di-guanylate cyclase YdeH of *Escherichia coli* with 2',(3')-substituted purine and pyrimidine nucleotides. *Journal of Pharmacology and Experimental Therapeutics. 336(1),* 234-241.

[178] Landini, P. Antoniani, D. Burgess, J. G. & Nijland, R. (2010). Molecular mechanisms of compounds affecting bacterial biofilm formation and dispersal. *Applied Microbiology and Biotechnology, 86,* 813–823.

[179] Ueda, A. Attila, C. Whiteley, M. & Wood, T. K. (2009). Uracil influences quorum sensing and biofilm formation in *Pseudomonas aeruginosa* and fluorouracilnis an antagonist. *Microbial Biotechnology, 2,* 62– 74.

[180] Antoniani, D. Bocci, P. Maciag, A. Raffaelli, N. & Landini, P. (2010). Monitoring of di-guanylate cyclase activity and of cyclic-di-GMP biosynthesis by whole-cell assays suitable for high-throughput screening of biofilm inhibitors. *Applied Microbiology and Biotechnology, 85,* 1095–1104.

[181] Guiton, P. S. Hung, C. S. Kline, K. A. Roth, R. Kau, A. L. Hayes, E. Heuser, J. Dodson, K. W. Caparon, M. G. & Hultgren, S. J. (2009). Contribution of autolysin and Sortase a during *Enterococcus faecalis* DNA dependent biofilm development. *Infection and Immunity, 77,* 3626–3638.

[182] Morgan, R. Kohn, D. Hwang, S. H. Hassett, D. J. & Sauer, K. (2006). BdlA, a chemotaxis regulator essential for biofilm dispersion in Pseudomonas aeruginosa. *Journal of Bacteriology, 188,* 7335–7343.

[183] Barraud, N. Schleheck, D. Klebensberger, J. Webb, J. S. Hassett, D. J. Rice, S. A. & Kjelleberg, S. (2009). Nitric oxide signaling in *Pseudomonas aeruginosa* biofilms mediates phosphodiesterase activity, decreased cyclic diguanosine-5′-monophosphate levels and enhanced dispersal. *Journal of Bacteriology, 191,* 7333–7342.

[184] Thormann, K. Duttler, M. S. Saville, R. M. Hyodo, M. Shukla, S. Hayakawa, Y. & Spormann, A. M. (2006). Control of formation and cellular detachment from *Shewanella oneidensis* MR-1 biofilms by cyclic di-GMP. *Journal of Bacteriology, 188,* 2681–2691.

[185] Holland, L. M. O'Donnell, S. T. Ryjenkov, D. A. Gomelsky, L. Slater, S. R. Fey, P. D. Gomelsky, M. & O'Gara, J. P. (2008). A staphylococcal GGDEF domain protein regulates biofilm formation independently of cyclic dimeric GMP. *Journal of Bacteriology, 190,* 5178–5189.

[186] Galperin, M. Y. (2004). Bacterial signal transduction network in a genomic perspective. *Environmental Microbiology, 6,* 552–567.

[187] Tamayo, R. Pratt, J. T. & Camilli, A. (2007). Roles of cyclic diguanylate in the regulation of bacterial pathogenesis. *Annual Reviews in Microbiology, 61,* 131–148.

[188] Chan, C. Paul, R. Samoray, D. Amiot, N. C. Giese, B. Jenal, U. & Schirmer, T. (2004). Structural basis of activity and allosteric control of diguanylate cyclase. *Proceedings of the National Academy of Sciences, USA, 101,* 17084–17089.

[189] De, N. Pirruccello, M. Krasteva, P. V. Bae, N. Raghavan, R. V. & Sondermann, H. (2008). Phosphorylation-independent regulation of the diguanylate cyclase WspR. *PLoS Biology, 6,* e67.

[190] Kociolek, M. G. (2009). Quorum-sensing inhibitors and biofilms. *Anti-Infective Agents in Medicinal Chemistry, 8,* 315-326.

[191] Pagedar, A. Singh, J. & Batish, V. K. (2011). Efflux mediated adaptive and cross resistance to ciprofloxacin and benzalkonium chloride in *Pseudomonas aeruginosa* of dairy origin. *Journal of Basic Microbiology, 51(3),* 289-295.

[192] Antunes, L. C. Ferreira, R. B. Buckner, M. M. & Finlay, B. B. (2010). Quorum sensing in bacterial virulence. *Microbiology, 156(8),* 2271-82.

[193] Hentzer, M. Riedel, K. Rasmussen, T. B. Heydorn, A. Andersen, J. B. Parsek, M. R. Rice, S. A. Eberl, L. Molin, S. Høiby, N. Kjelleberg, S. & Givskov, M. (2002). Inhibition of quorum sensing in *Pseudomonas aeruginosa* biofilm bacteria by a halogenated furanone compound. *Microbiology, 148,* 87-102.

[194] Kuehl, R. Al-Bataineh, S. Gordon, O. Luginbuehl, R. Otto, M. Textor, M. & Landmann, R. (2009). Furanone at subinhibitory concentrations enhances staphylococcal biofilm formation by luxS repression. *Antimicrobial Agents and Chemotherapy, 53(10),* 4159-66.

[195] Yang, L. Rybtke, M. T. Jakobsen, T. H. Hentzer, M. Bjarnsholt, T. Givskov, M. & Tolker-Nielsen T. (2009). Computer-aided identification of recognized drugs as *Pseudomonas aeruginosa* quorum-sensing inhibitors. *Antimicrobial Agents and Chemotherapy, 53,* 2432-2443.

[196] Rasmussen, T. B. Bjarnsholt, T. Skindersoe, M. E. Hentzer, M. Kristoffersen, P. Kote, M. Nielsen, J. Eberl, L. & Givskov, M. (2005). Screening for quorum-sensing inhibitors (QSI) by use of a novel genetic system, the QSI selector. *Journal of Bacteriology, 187,* 1799–1814.

[197] Vanoyan, N. Walker, S. L. Gillor, O. & Herzberg M. (2010). Reduced bacterial deposition and attachment by quorum-sensing inhibitor 4-nitro-pyridine-N-oxide, the role of physicochemical effects. *Langmuir, 26(14),* 12089-94.

[198] Sintim, H.O. Smith, J. A. I. Wang, J. Nakayama, S. & Yan, L. (2010). Paradigm shift in discovering next generation anti-infective agents, targeting quorum sensing, c-di-GMP signaling and biofilm formation in bacteria with small molecules. *Future Medicinal Chemistry. 2(6),* 1005-1035.

[199] Balaban, N. Cirioni, O. Giacometti, A. Ghiselli, R. Braunstein, J. B. Silvestri, C. Mocchegiani, F. Saba, V. & Scalise, G. (2007). Treatment of *Staphylococcus aureus*

biofilm infection by the quorum-sensing inhibitor RIP. *Antimicrobial Agents and Chemotherapy, 51(6),* 2226-2229.

[200] Gov, Y. Bitler, A. Dell'Acqua, G. Torres, J. V. & Balaban, N. (2001). RNAIII inhibiting peptide (RIP), a global inhibitor of Staphylococcus aureus pathogenesis, structure and function analysis. *Peptides, 22(10),* 1609-1620.

[201] Kiran, M. D. Adikesavan, N. V. Cirioni, O. Giacometti, A. Silvestri, C. Scalise, G. Ghiselli, R. Saba, V. Orlando, F. Shoham, M. & Balaban, N. (2008). Discovery of a quorum-sensing inhibitor of drug-resistant staphylococcal infections by structure-based virtual screening. *Molecular Pharmacology, 73(5),* 1578-1586.

[202] Nakayama, J. Uemura, Y. Nishiguchi, K. Yoshimura, N. Igarashi, Y. & Sonomoto, K. (2009). Ambuic acid inhibits the biosynthesis of cyclic peptide quormones in gram-positive bacteria. *Antimicrobial Agents and Chemotherapy, 53(2),* 580-586.

[203] Qazi, S. Middleton, B. Muharram, S. H. Cockayne, A. Hill, P. O'Shea, P. Chhabra, S. R. Cámara, M. & Williams, P. (2006). N-acylhomoserine lactones antagonize virulence gene expression and quorum sensing in *Staphylococcus aureus. Infection and Immunity, 74(2),* 910-919.

[204] Ponnusamy, K. Paul, D. Kim, Y. S. & Kweon, J. H. (2010). 2(5H)-Furanone, a prospective strategy for biofouling-control in membrane biofilm bacteria by quorum sensing inhibition. *Brazilian Journal of Microbiology, 41,* 227-234.

[205] Kappachery, S. Paul, D. Yoon, J. & Kweon, J. H. (2010). Vanillin, a potential agent to prevent biofouling of reverse osmosis membrane. *Biofouling, 26(6),* 667-672.

[206] Yeon, K. Cheong, W. Oh, H. Lee, W. Hwang, B. Lee, C. Beyenal, H. & Lewandowski, Z. (2009). Quorum Sensing: A new biofouling control paradigm in a membrane bioreactor for advanced wastewater treatment. *Environmental Science and Technology, 43 (2),* 380–385.

[207] Liu, M. Gray, J. M. & Griffiths, M. W. (2006). Occurrence of proteolytic activity and N-acyl-homoserine lactone signals in the spoilage of aerobically chill-stored proteinaceous raw foods. *Journal of Food Protection, 69(11),* 2729-37.

[208] Shobharani, P. & Agarwal, R. (2010). Interception of quorum sensing signal molecule by furanone to enhance shelf life of fermented milk. *Food Control, 21(1),* 61-69.

[209] Saeki, Y. Ito, Y. Shibata, M. Sato, Y. Okuda, K. & Takazoe, I. (1989). Antimicrobial action of natural substances on oral bacteria. *The Bulletin of Tokyo Dental College, 30,* 129–135.

[210] Tichy, J. & Novak, J. (1998). Extraction, assay, and analysis of antimicrobials from plants with activity against dental pathogens (*Streptococcus* sp.). *Journal of Alternative and Complementary Medicine, 4,* 39–45.

[211] Bakri, I. M. & Douglas, C. W. (2005). Inhibitory effect of garlic extract on oral bacteria. *Archives of Oral Biology, 50,* 645–651.

[212] Moulari, B. Lboutounne, H. Chaumont, J. P. Guillaume, Y. Millet, J. & Pellequer, Y. (2006). Potentiation of the bactericidal activity of *Harungana madagascariensis Lam. ex Poir.* (Hypericaceae) leaf extract against oral bacteria using poly (D, L-lactide-co-glycolide) nanoparticles, *in vitro* study. *Acta Odontologica Scandinavica, 64,* 153–158.

[213] Polaquini, S. R. B. Svidzinski, T. I. E. Kemmelmeier, C. & Gasparetto, A. (2006). Effect of aqueous extract from Neem (*Azadirachta indica* A. Juss) on hydrophobicity, biofilm formation and adhesion in composite resin by *Candida albicans. Archives of Oral Biology, 51(6),* 482.

[214] Duarte, S. Gregoire, S. Singh, A. P. Vorsa, N. Schaich, K. Bowen, W. H. & Koo, H. (2006). Inhibitory effects of cranberry polyphenols on formation and acidogenicity of *Streptococcus mutans* biofilms. *FEMS Microbiology Letters, 257,* 50–56.

[215] Johnson-White, B. Buquo, L. Zeinali, M. & Ligler, F. S. Prevention of nonspecific bacterial cell adhesion in immunoassays by use of cranberry juice. *Analytical Chemistry, 78,* 853–857.

[216] Steinberg, D. Feldman, M. Ofek, I. & Weiss, E. I. (2005). Cranberry high molecular weight constituents promote *Streptococcus sobrinus* desorption from artificial biofilm. *International Journal of Antimicrobial Agents, 25,* 247–251.

[217] Kim, H. S. (2005). Do not put too much value on conventional medicines. *Journal of Ethnopharmacology, 100,* 37–39.

[218] Limsong, J. Benjavongkulchai, E. & Kuvatanasuchati, J. (2004). Inhibitory effect of some herbal extracts on adherence of *Streptococcus mutans. Journal of Ethnopharmacology, 92,* 281–289.

[219] Nostro, A. Cannatelli, M. A. Crisafi, G. Musolino, A. D. Procopio, F. & Alonzo, V. (2004). Modifications of hydrophobicity, in vitro adherence and cellular aggregation of *Streptococcus mutans* by *Helichrysum italicum* extract. *Letters in Applied Microbiology, 38,* 423–427.

[220] Palombo, E. A. (2009). Traditional medicinal plant extracts and natural products with activity against oral bacteria, potential application in the prevention and treatment of oral diseases. *Evidence-based Complementary and Alternative Medicine. doi, 10.1093/ecam/nep067*

[221] Ozaki, S. Kano, S. & Shirai, O. (2008). Electrochemical elucidation on the mechanism of uncoupling caused by hydrophobic weak acids. *Physical Chemistry Chemical Physics, 10,* 4449–4455.

[222] Stevenson, B. & Babb, K. (2002). LuxS-mediated quorum sensing in *Borrelia burgdorferi*, the Lyme disease spirochete. *Infection and Immunity, 70,* 4099–4105.

[223] Jain, A. Nishad, K. K. & Bhosle, N. B. (2007). Effects of DNP on the cell surface properties of marine bacteria and its implication for adhesion to surfaces. *Biofouling, 23,* 171–177.

[224] Xu, H. & Liu, Y. (2010). Control of microbial attachment by inhibition of ATP and ATP-mediated autoinducer-2. *Biotechnology and Bioengineering, 107(1),* 31–36.

[225] Jiang, B. & Liu, Y. (2010). Energy uncoupling inhibits aerobic granulation. *Applied Microbiology and Biotechnology, 85,* 589–595.

[226] McLamore, E. S. Zhang, W. Porterfield, D. M. & Banks, M. K. (2010). Membrane-aerated biofilm proton and oxygen flux during chemical toxin exposure. *Environmental Science and Technology, 44(18),* 7050–7057.

[227] Spormann, A. M. (2008). Physiology of Microbes in Biofilms. In: Romeo, T. (ed). *Current Topics in Microbiology and Immunology: Bacterial Biofilms*. (322: 17-36). Verlag Berlin Heidelberg: Germany, Springer Press.

[228] Klebensberger, J. Rui, O. Fritz, E. Schink, B. & Philipp, B. (2006). Cell aggregation of *Pseudomonas aeruginosa* strain PAO1 as an energy-dependent stress response during growth with sodium dodecyl sulfate. *Archives of Microbiology, 185,* 417–427.

[229] Schönfeld, P. & Wojtczak, L. (2008). Fatty acids as modulators of the cellular production of reactive oxygen species. *Free Radical Biology and Medicine, 45,* 231-241.

[230] Baker-Austin, C. & Dopson, M. (2007). Life in acid, pH homeostasis in acidophiles. *Trends in Microbiology, 15(4),* 165-171.

[231] Kuda, T. Yano, T. & Kuda, M. T. (2008). Resistances to benzalkonium chloride of bacteria dried with food elements on stainless steel surface. *LWT – Food Science and Technology, 41,* 988–993.

[232] Maillard, J. Y. (2002). Bacterial target sites for biocide action. *Journal of Applied Microbiology Symposium Supplement, 92,* 16S–27S.

[233] Jones, W. L. Sutton, M. P. McKittrick, L. & Stewart, P. S. (2011). Chemical and antimicrobial treatments change the viscoelastic properties of bacterial biofilms. *Biofouling, 27(2),* 207-215.

[234] Wen, J. Zhao, K. Gu, T. & Raad, I. I. (2009). A green biocide enhancer for the treatment of sulfate-reducing bacteria (SRB) biofilms on carbon steel surfaces using glutaraldehyde. *International Biodeterioration & Biodegradation, 63(8),* 1102-1106.

[235] Pereira, M. O. & Vieira, M. J. (2001). Effects of the interactions between glutaraldehyde and the polymeric matrix on the efficacy of the biocide against pseudomonas fluorescens biofilms. *Biofouling, 17(2),* 93-101.

[236] Simoes, M. Carvalho, H. Pereira, M. O. & Vieira, M. J. (2003). Studies on the behaviour of *Pseudomonas fluorescens* biofilms after Ortho-phthalaldehyde treatment. *Biofouling, 19(3),* 151-157.

[237] Romanova, N. A. Gawande, P. V. Brovko, L. Y. & Griffiths M. W. (2007). Rapid methods to assess sanitizing efficacy of benzalkonium chloride to *Listeria monocytogenes* biofilms. *Journal of Microbiological Methods, 71(3),* 231-237.

[238] Chen, C. Z. S. & Cooper, S. L. (2002). Interactions between dendrimer biocides and bacterial membranes. *Biomaterials, 23,* 3359-3368.

[239] Kramer, J. F. O'Brien, F. & Strba, S. F. (2008). A New High Performance Quaternary Phosphonium Biocide for Microbiological Control in Oilfield Water Systems. Available from URL: http://www.onepetro.org/mslib/servlet/onepetropreview?id=NACE-08660&soc=NACE&speAppNameCookie=ONEPETRO

[240] Prevost, A. Barbeau, J. Cote, L. Charland, R. and Faucher, E. (1998). US5759970.

[241] Day, D.F. (2005). US20056866870.

[242] Grant, D. M. & Bott, T. R. (2005). Biocide dosing strategies for biofilm control. *Heat Transfer Engineering, 26(1),* 44–50.

[243] Shemesh, M. Kolter, R. & Losick, R. (2010). The biocide chlorine dioxide stimulates biofilm formation in *Bacillus subtilis* by activation of the histidine kinase KinC. *Journal of Bacteriology, 192(24),* 6352-6356.

[244] Chung, Y. C. Su, Y. P. Chen, C. C. Jia, G. Wang, H. l. Wu, J. C. G. & J. G. Lin (2004). Relationship between antibacterial activity of chitosan and surface characteristics of cell wall. *Acta Pharmacologica Sinica, 25(7),* 932-936.

[245] Je, J. Y. & Kim, S. K. (2006). Chitosan derivatives killed bacteria by disrupting the outer and inner membrane. *Journal of Agricultural and Food Chemistry,* 54*(18),* 6629-6633.

[246] Cao, Z & Sun, Y. (2009). Chitosan-based rechargeable long-term antimicrobial and biofilm-controlling systems. *Journal of Biomedical Materials Research- Part A, 89(4),* 960-967.

[247] Orhan, Z. Cevher, E. Mülazimoglu, L. Gürcan, D. Alper, M. Araman, A. & Ozsoy, Y. (2006). The preparation of ciprofloxacin hydrochloride-loaded chitosan and pectin

microspheres, their evaluation in an animal osteomyelitis model. *Journal of Bone and Joint Surgery, British, 88(2),* 270-275.

[248] Søltoft-Jensen, J. & Hansen, F. (2005). New chemical and biochemical hurdles. In: Sun, D. W. *Emerging technologies for food processing.* (387-416). London, Elsevier Academic Press.

In: Biofilms in Bioengineering
Editors: Manuel Simões and Filipe Mergulhão

ISBN: 978-1-62948-161-6

Chapter XII

Mechanisms of Microbial Resistance in Biofilms

Sonja Smole Možina*, Anja Klančnik and Peter Raspor
Biotechnical Faculty, Department of Food Science and Technology,
University of Ljubljana, Slovenia

Abstract

With microbial biofilms, the major advantage is that the cell is provided with a protective environment for growth and survival. Consequently, the major biofilm-related properties are cell protection and increased cell resistance against detrimental environmental influences. In biofilms, multiple resistance mechanisms function collectively. In addition, this resistance to adverse environmental conditions depends on single-cell and, in particular, multi-cell strategies that result in physiological and genetic heterogeneity of cells in biofilms. In this chapter, these resistance mechanisms are briefly reviewed, thus highlighting different aspects of resistance. The factors involved include biofilm and cell structure (e.g., biofilm matrix, changed morphology, high cell density), physiology (e.g., decreased diffusion of nutrients and oxygen through the biofilm matrix, starvation and stress responses, formation of viable but nonculturable cells and persister cells), biology (e.g., higher resistance of multispecies biofilms), and genetics (e.g., high frequency of vertical and horizontal gene transfer in biofilm cells). The need for a better understanding of the resistance mechanisms that have evolved in microbial cells in biofilms remains a crucial issue towards new developments in different fields of life and research.

Keywords: Microbial resistance; Biofilm-specific resistance; Persisters; Efflux; Quorum sensing; Multispecies biofilms

* E-mail: sonja.smole@bf.uni-lj.si.

Introduction - Definition of Biofilms and Diversity of Resistance Mechanisms

The vast majority of bacteria in natural environments live in highly organized communities, where they are irreversibly attached to a surface or associated with surface-liquid or liquid-air interfaces, and are encapsulated in a self-produced extracellular polymeric matrix known as the biofilm. Today bacterial biofilms are generally acknowledged as the dominant lifestyle in natural and man-made environments. However, knowledge about biofilms evolved for many years before it was finally accepted that surface-associated and attached bacterial communities live very differently from bacteria growing in a free-swimming planktonic state [1-7].

Cells can thus be embedded in a matrix of self-produced extracellular polymeric substances, in the form of a hydrated gel-like slime, which includes mainly polysaccharides, (glyco)proteins, (phospho)lipids, and nucleic acids, along with many other compounds. Altogether, these provide the mechanical stability and three-dimensional structure of the biofilm, and the adhesion to surfaces [7]. The embedded cells show specific phenotypes that can include observable characteristics, such as changed morphology and growth rate. There are also many other physiological and genetic properties of change, like gene transcription, mutation, and acquisition of extracellular genetic elements [8, 9]. The expanded definition of the biofilm is important for the defining of the resistance mechanisms, as some populations meet certain biofilm criteria (e.g., matrix formation and surface growth) but do not show the distinct biofilm phenotype. These non-biofilm populations, which are often seen for bacteria growing on agar surfaces, behave like planktonic cells and do not show the inherent resistance characteristics of true biofilms [4, 10]. In microbial biofilms, the major advantage for the cells is the provision of a specific environment for growth and survival. Consequently, the major biofilm-related properties are cell protection and promotion of high resistance against potentially detrimental external conditions, such as lack of water, UV radiation, and host immune defenses. Another extremely important biofilm-related advantage is the unusually high resistance to antimicrobial agents [4, 11-15].

Recently, many excellent original studies and review articles have been published that have provided explanations of the different resistance mechanisms in biofilm cells. The approaches for the systematic presentation of several mechanisms that can act together have been different, depending on the different studies [2, 4, 10, 11, 16-21]. These resistance mechanisms increase bacterial survival possibilities under harsh conditions through intrinsic features of the cell, or through acquired cell changes. As well as these intrinsic resistance mechanisms to survive under adverse physical and chemical stress conditions, microbes can acquire resistance by *de-novo* mutation or via acquisition of resistance genes from other microorganisms. They can also adapt physiologically without any genetic changes. This is true for planktonic and biofilm cells; however, in the latter form, the acquisition of genes and/or physiological adaptation can be enhanced through the high cell density, and through other unique biofilm conditions.

Another general view of biofilm resistance mechanisms was given by Anderson and O'Toole [19], who separated the genetic mechanisms of the development of biofilm antibiotic resistance into two general classes: innate resistance factors and induced resistance factors. The mechanisms of resistance in biofilms are different from those of the plasmids, transposons, and mutations that can confer innate resistance to individual bacterial cells. In biofilms, multiple resistance mechanisms operate in concert [22] and this resistance thus depends on multicellular strategies [18]. Innate biofilm resistance mechanisms are activated as part of the biofilm developmental pathway, and these factors are integral parts of the biofilm structure and physiology [12]. Such innate pathways include decreased diffusion of antibiotics through the biofilm matrix, decreased oxygen and nutrient availability accompanied by altered metabolic activity, and the formation of persisters, along with the involvement of other specific molecules that do not fit into one of these groups. Induced resistance factors include those that result from induction by the antimicrobial agent itself [19]. However, it can be difficult to separate out the combined effects of these mechanisms, because many of them mutually interfere and influence the resistance of cells. To provide a more clear picture of these resistance mechanisms of cells in a biofilm, which represents a protected mode of growth in a hostile environment, we will briefly review them consecutively in the following sections.

Slow Penetration through the Biofilm Matrix, Cell Starvation, and Stress Responses

The physical and chemical structure of exopolymeric substances and other aspects of biofilm matrix architecture, such as high viscosity and size exclusion, are common causes of delayed diffusion through the biofilm matrix (see Figure1).

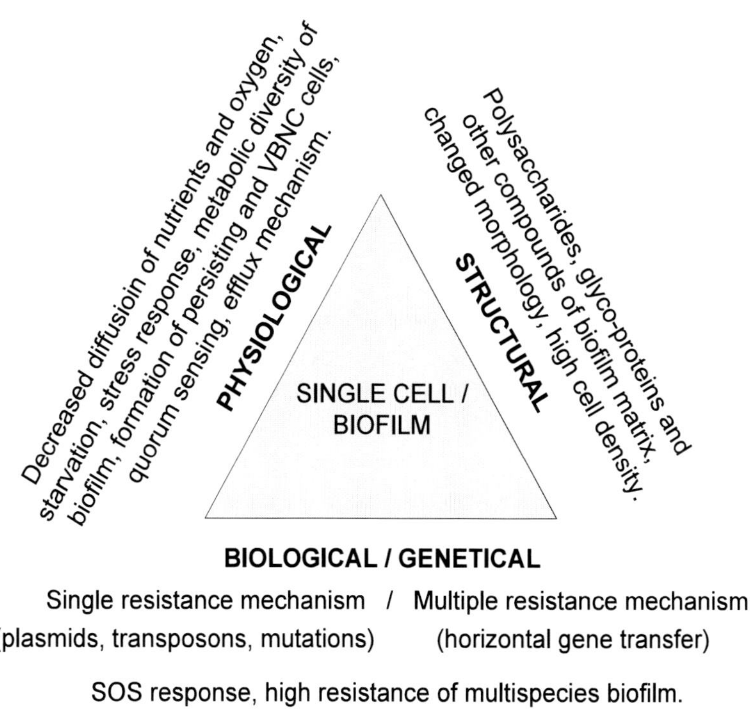

Figure 1. Some aspects of resistance mechanisms of microbial biofilms.

Consequently, localized nutrient limitation directs the biofilm mode of growth, and one of the long-standing hypotheses to explain the resistance of biofilm cells is that biofilms contain slowly growing or nongrowing microorganisms [11, 16, 17]. In the biofilm environment, there are limitations in growth substrates, primarily in electron donors and acceptors, but also in other nutrients, such as phosphorous and ammonia, and there is the accumulation of salts and metabolites, along with a high cell density. The resulting oxidative stress, hyperosmolarity, acidic pH and/or other physical, chemical or biological stress conditions are generally accompanied by an increase in resistance via the induction of the general stress response. The central regulator of this response is the alternate σ factor, RpoS, the expression of which has been confirmed in the stationary phase of growth, and also at high cell densities in biofilms [23-26]. RpoS is considered to function as a master regulator that mediates specific physiological responses to single specific stresses, promoting the general robustness of the stress response and the increased resistance of biofilm cells to antibiotics or other biocides [24, 27-28].

Changes in cell physiology can also include activation of other global regulators. Several strategies are possible, although the common result is a reduction in the growth rate, showing behavior that is similar to that found in sporulating microbes. Prolonged starvation induces loss of culturability under standard conditions, whereas the cells remain metabolically active and structurally intact [29]. This reversible 'viable but nonculturable' state is considered the main reason for the low detection of biofilm infections by routine culture techniques [30]. Transition to the stationary growth phase is accompanied by deep structural and physiological changes that contribute to increased bacterial resistance to lack of nutrients and to different physical and chemical stresses (e.g., UV radiation, high temperature, acidity, salinity, oxidative or osmotic stress, antimicrobial chemicals, and presence of their own secretions or metabolites) [31-34]. Following prolonged treatments or stress exposure, biofilm organisms also increase their capacity to withstand and neutralize monochloramine [35], to stimulate catalase production [36], and to induce expression of chromosomal β-lactamases [37]. As well as pathogens, spoilage bacteria (e.g., lactic acid bacteria biofilms) have also been studied for their resistance to environmental stresses [38]. Bacterial cells in biofilms show greater resistance than planktonic cells under either stationary or logarithmic growth, against various organic acids (acetic, citric, lactic, and malic acids), ethanol, and sodium hypochlorite [39].

Biofilm cells not only grow slowly, but are also subjected to other stresses, such as high osmolarity, high cell density, and accumulation of metabolites. Thus, although other mechanisms cannot be ruled out and might be important, it is believed that most, if not all, of the greatly enhanced phenotypic antibiotic resistance of biofilm cells is simply a consequence of the nonlinear signal integration of multiple stresses by the RpoS circuitry [40].

Similar to the consumption of nutritive substrate compounds, cells at the surface consume oxygen rapidly, contributing to the formation of a decreasing oxygen gradient from the outer to the deepest layers of the biofilm [41]. When *rpoS* expression and RpoS levels were measured as indicators of stationary phase growth within thick, continuously fed *Pseudomonas aeruginosa* biofilms, the levels of *rpoS* expression in a 3-day-old biofilm were shown to be three-fold higher than the average expression in stationary phase planktonic cultures. In planktonic cultures, oxygen limitation does not lead to increased levels of RpoS, which suggests that oxygen limitation is not the environmental signal that causes increased expression of *rpoS*. These results suggest that bacteria within *P. aeruginosa* biofilms can have stationary phase characteristics even when cultured under flow conditions that continually replenish their nutrients [42]. On the other hand, bacteria can alter the expression of respiratory chain components in the biofilm interior, which maximizes its efficiency under microaerophilic conditions, or shifts to oxygen-independent metabolism [10, 32].

Establishment of a biofilm requires regulatory components that are involved in survival under microaerophilic growth conditions, and also arginine metabolism, alkyl-quinolone signaling, pH homeostasis, and the DNA repair system. The urease enzyme has been described as markedly up-regulated in biofilms of *Streptococcus salivarius* [43, 44] and *S. aureus* [32, 45], and it has been speculated that urease is involved in the maintenance of pH homeostasis in biofilm cultures. In oral biofilms, arginine catabolism to polyamines can contribute to caries formation [46]. Maintenance of pH homeostasis is important for *S. mutans* biofilms under acid stress. Biofilm formation can also contribute to the acid resistance of *H. pylori* [47]. Biofilms of nitrifying bacteria appear to balance acid-producing with acid-consuming forms of metabolism [48].

When mathematical modeling has been used to study biofilm dynamics, the results have pointed to a crucial role of nutrient limitation and slow growth in the antibiotic tolerance of microbial cells in biofilms [49]. However, together with delayed diffusion, penetration through the biofilm can also be limited by ionic interactions between the slime layer and antimicrobial substances, by covalent interactions with matrix components, or by enzymatic digestion of trapped molecules by enzymes and other biologically active molecules concentrated within the slime layer; e.g., β-lactamases in *P. aeruginosa* biofilms [10, 50, 51]. However there are also other literature reports that have not confirmed any of these mechanisms. For instance, fluorescently tagged daptomycin accesses the interior of *S. epidermidis* biofilm cell clusters within minutes. The diffusion coefficient of daptomycin in the biofilm was 28% of its value in pure water. So, daptomycin activity against staphylococci embedded in biofilms was unlikely to be limited by penetration of the antibiotic into the biofilm [52]. However, there are also reports in the literature, that reduced antibiotic penetration can vary with the antibiotic used, as was the case in the study of Singh et al., [53], which was carried out to determine the role of reduced antibiotic penetration in the resistance of *S. aureus* and *S. epidermidis* biofilms to different antibiotics. The penetration of oxacillin and cefotaxime (beta-lactams) and vancomycin (a glycopeptide) was significantly reduced through *S. aureus* and *S. epidermidis* biofilms, whereas that of amikacin (an aminoglycoside) and ciprofloxacin (a fluoroquinolone) was unaffected.

Fung et al., [54] determined the degree of tolerance (persistence) to treatment by different antibiotics as a function of growth medium, growth rate, starvation of particular nutrients singly and in combination, and treatment with translation inhibitors and bacteriostatic agents. The effects of a variety of gene deletions that have been reported to reduce persistence were evaluated under a variety of growth conditions. There is no simple course to persistence or tolerance. A variety of combinations of factors can lead to tolerance to one or more antibiotics, with certain pathways predominating under specific conditions [54]. In *Escherichia coli*, mutations of the *hipA* gene (homologs of which are present in a wide variety of Gram-negative organisms [55], increased tolerance, probably by inducing (p)ppGpp synthesis, which potentiates the transition to a dormant state upon application of stress.

Real-time RT-PCR and DNA microarray analyses in studies of *E. coli* biofilms have confirmed that in the early stages of biofilm formation, *rpoS* is expressed in most of the cells, while during the later stages of biofilm formation, its expression occurs only in the outer layers, to shift the physiological state of the cells to stationary growth [56].

Other more recent techniques, such as confocal laser scanning microscopy, have enabled direct and real-time visualization of the penetration of biologically active concentrations of antimicrobial agents into the biofilm interior. These have confirmed the lack of correlation of antimicrobial molecular weight with observed penetration time [57]. In another recent study, cell inactivation within the structure was vizualized with this technique, which was based on monitoring the loss of fluorescence that corresponded to the leakage of a fluorophore out of the cells due to membrane permeabilization by a biocide [58]. Besides classical techniques such as colony plating, regrowth assays, crystal violet staining for total biomass, tetrazolium salt reduction, and ATP content measurements, which provide an indication of overall efficacy in terms of either viability reduction or biofilm removal, such new techniques shed light on the fundamental phenomena that are important in the interactions of the antibiofilm and antimicrobial compound with the biofilm [57]. This is extremely important, as recently some chemicals have been developed that do not kill the cells, but inhibit and disperse bacterial biofilms; e.g., derivatives of 2-aminoimidazoles [59]. Synergistic effects between conventional antibiotics and 2-aminoimidazole-derived antibiofilm agents have been confirmed against multidrug resistant pathogens including methicillin-resistant *S. aureus* (MRSA) and multidrug-resistant *Acinetobacter baumanni* [60]. Also, for the efficacy of the biocides, targeting viable bacteria and the biofilm matrix is of critical importance. A recent study clearly demonstrated that most biocides tested were active only against biofilm bacteria, leaving the matrix undisturbed. Only hydrogen peroxide and sodium hypochlorite were active on both the biofilm matrix and the viable mass of *S. aureus* and *P. aeruginosa* biofilms, making them the better antibiofilm agents [15, 61].

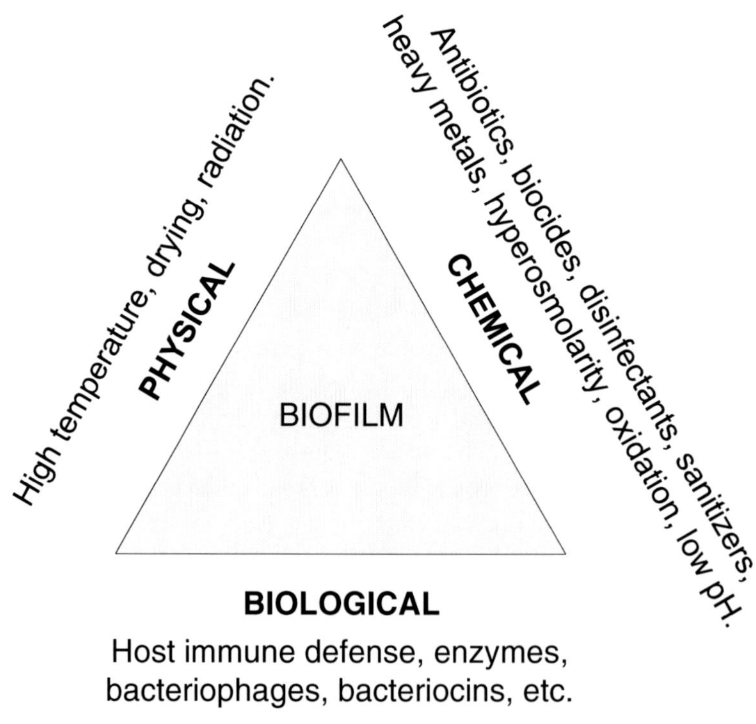

Figure 2. Biofilm potential to sustain environmental impact.

Physiological and Genetic Heterogeneity in Biofilms

The physiological and genetic responses of biofilm microbial cells to external and self-induced stresses, including competition for resources, determine the fate of the biofilm and its diverse subpopulations. Microscale gradients in nutrient concentrations are a well known characteristic of biofilms [18]. Within a biofilm, cells with diverse genotypes and phenotypes that express distinct metabolic pathways, stress responses, and other specific biological activities are joined in the interactive microbial 'society'. The mechanisms responsible for this include microscale chemical gradients, microbial adaptation to local environmental conditions, specific gene expression and genotypic variation that occurs through mutation and selection (see Figure 1). This can enrich for populations of a particular species that are best adapted, through selection for fitter mutant strains that can better survive under the given conditions [20].

In biofilms, the horizontal transfer of genetic material occurs more frequently, and this also increases biofilm stability [62]. However, there is also the possibility of spreading antibiotic resistance, virulence factors, and environmental survival capabilities. Acquisition of new genetic material by bacteria susceptible to antimicrobials from their resistant counterparts can occur through gene transfer, by conjugation (via plasmids and conjugative transposons), transformation, and/or transduction (via bacteriophages and incorporation into the chromosomes of chromosomal DNA or plasmids, see Figure 1). Once acquired, resistance genes are not easily lost. Instead, they can become a relatively stable part of a genome. Additional resistance determinants can join those already prevailing, which will broaden the multidrug resistance phenotype [21].

One of the important and widespread genetic mechanisms that bacteria use to adapt and survive under environmental stress conditions has been shown to also be involved in biofilm formation and the generation of diversity of biofilm cells: the SOS response [63, 64]. This is a conserved inducible pathway involved in DNA repair and the restart of stalled or collapsed replication forks. Activation of the SOS response can result in stress resistance, via induction of stress resistant subpopulations. RecA as the activator, and other SOS response factors, were identified as involved in the formation of biofilm phenotypes of different spoilage and pathogenic bacterial species [65, 66]. As the bacterial SOS response is also one of the most important mechanisms for bacterial adaptive mutagenesis [64], this mechanism presents an important contribution to the presence of diversity in biofilms, and thus to the selection of resistance.

Biofilm-specific Resistant Phenotype and Persister Cells

Although reaction-diffusion limitation and induced nutrition-restricted phenotypes are important resistance mechanisms, they cannot fully explain the observed extreme resistance of biofilms. It is obvious that there are multiple resistance mechanisms at work within a single community (see Figures 1, 2). For example, nutrient depletion can alter the barrier properties of the cell envelope. As binding to the bacterial surface is often the initial step in bacteriocidal action, this can greatly influence bacterial susceptibility to antimicrobials. Furthermore, due to extreme nutrient and oxygen limitation in the deepest layers of the biofilm, bacterial cells become dormant, and so called persister cells show multidrug tolerance (67). Slow-growing biofilms produce substantial numbers of persisters. The fraction of persister cells is much higher in biofilms than during planktonic growth [68]. There is also a higher level of persistence among stationary (by as much as 1%) rather than exponential growth [13, 14].

Stress responses can act as general activators of persister cell formation. However, other stimuli have been identified that give rise to altered gene transcription mediated by quorum sensing mechanisms, such as contact with a solid surface or reaching a certain threshold in cell density [10]. Persister cells have been long believed not to be mutants; rather, it has been hypothesized that they are phenotypic variants of normal cells. However, high persister (*hip*) mutants of *P. aeruginosa* have been identified in patients with cystic fibrosis, and similarly, *hip* mutants of *Candida albicans* were confirmed in patients with an oral thrush biofilm [69]. De Groote et al., [70] presented a small molecule effector as having a role in persister formation in *P. aeruginosa*. The correlation of fosfomycin resistance and lowered persistence led to the proposal of a model in which an organic phosphate or phosphonate that is a substrate of both GlpT transport and FosA glutathione conjugation (as is fosfomycin) can act as a positive regulator of persistence. Thus, those functions that prevent fosfomycin from its antibiotic action would prevent this unknown organic phosph[on]ate from promoting persistence [70].

Identification of persister genes and proteins required for maintaining persisters might also contribute to the identification of realistic targets for the discovery of drugs that can effectively treat chronic infections. Inhibitors of persistence development can be combined with conventional antimicrobials to try to eradicate a biofilm. Recently, the possibility of cross-talk or interplay between resistance and persistence was reported, when fosfomycin resistance led to decreased persistence in *P. aeruginosa* [70], and treatment with fluoroquinolones not only selected for persisters, but specifically induced their formation in *E. coli* [71]. A direct link was provided between the recalcitrance of chronic infections and persistence [72, 73], which represented a major breakthrough in this research area. Indeed, there appears to be a variety of pathways that depending on the growth conditions and the ambient stress, can promote persistence [54, 67, 69].

Up-regulated Efflux and Other Known Mechanisms of Antibiotic Resistance

The higher frequency of acquiring new genetic material in biofilms has already been considered above, which arises as biofilms promote horizontal gene transfer (see Figure 1). The spread of conventional antimicrobial resistance will be increased, as acquired resistance genes can enable bacteria to produce enzymes that inactivate the antibacterial product, modify the target site, or produce an alternative metabolic pathway that bypasses the actions of the antibacterial product; i.e., the classically known mechanisms of antibiotic resistance.

In addition, efflux mechanisms that prevent an antibacterial from reaching its intracellular target can be expressed (see Figure 1). On the other hand, insufficient (i.e., sub-lethal) concentrations of antimicrobial substances might increase selection pressure for antimicrobial resistance and also induce microbial membrane efflux pumps. Efflux pumps are single- or multicomponent systems, the latter of which are found exclusively in Gram-negative bacteria [74]. However, all of these rely on efflux to get rid of toxic substances and waste products. Efflux pumps are widely involved in antibiotic resistance. The pumps can be specific for one substrate or might transport a range of structurally dissimilar compounds, including antibiotics of multiple classes. Such pumps can be associated with multiple drug resistance (MDR) [75].

Efflux pumps are also highly active in bacterial biofilms, and they are important for biofilm-specific resistance to a subset of antibiotics [76, 77]. These efflux pumps are more highly expressed in biofilm cells than in planktonic cells, which provides an explanation for why their genes can be responsible for biofilms, but not also why planktonic cells are resistance to antibiotics. Furthermore, expression of these genes in planktonic cells increases their resistance to antibiotics [77]. It can thus be summarized that efflux pumps contribute to different biofilm-specific resistance mechanisms against antimicrobial substances, including antibiotics, biocides used in industrial hygiene management, food and drug preservatives, and others.

There are several publications about the involvement of bacterial efflux pumps in biofilm resistance and cross resistance [76-79], but there are much fewer studies that have addressed fungal biofilms. For instance, azole resistance of *Aspergillus fumigatus* biofilms was shown to be at least partially associated with efflux pump activity [80]. In general, fungal biofilms have distinct developmental phases, including adhesion, colonization, maturation and dispersal, which are governed by complex molecular events. These are in many aspects not completely understood yet, including the role of efflux pumps and development of antimicrobial resistance in general, like stress response, persisters, overexpression of drug targets and others [81, 82].

As well as this involvement of biofilm cells in antimicrobial resistance, efflux pumps are attractive targets for antibiofilm approaches. A number of efflux pump inhibitors (EPIs) have been identified. EPIs are substances that block the activity of efflux pumps [83], and these have attracted attention over the years due to their ability to block some types of antibiotic resistance in planktonic and biofilm-associated bacteria. EPIs were shown to reduce biofilm formation, and in combination they can abolish biofilm formation completely. Also, EPIs can block the antibiotic tolerance of biofilms. The results of this feasibility study might pave the way for new treatments for biofilm-related problems in clinical, industrial or environmental settings, and might be exploited for prevention of biofilms in general [76].

Quorum Sensing and Resistance Mechanisms

Quorum sensing has been defined as the system of cell-to-cell communication among bacteria. It has been shown to be closely related to some of the important properties of microbes, such as biofilm formation, motility, swarming, antibiotic resistance, sporulation, pathogenicity, and virulence. This issue is part of microbial cooperation and social activities at the microbial population level [84]. The mechanisms of quorum sensing have been shown to be very complex and diverse among microbes, for the coordination of gene expression and group behavior based on population densities. Microbial cells produce and release signaling molecules, e.g., highly species-specific autoinducer-1 (AI-1), the less species-specific autoinducer-2 (AI-2) and others, such as AI-3, which mimic eukaryotic hormones and mediates inter-kingdom signaling in bacteria [85-88], and farnesol, tyrosol and farnesoic acid in fungi [89, 90]. When a critical environmental concentration of the signaling molecule is reached, the cells become induced and undergo changes in gene expression. In principle, a high microbial density can trigger the cells in a cluster to release their communicable signals (e.g., chemicals, radiation) which can activate regulator proteins to switch particular gene expression on or off [91]. The importance of quorum sensing has opened a new toolbox, which can be used for its inhibition or interference. This might destroy or reduce microbial growth efficiency and virulence that is harmful to processes and/or hosts [92, 93]. Consequently, a new field of application appears to have opened, as inhibition of coordinated and collective microbial behavior due to quorum sensing might be an alternative for disinfection in hygiene management, and/or for antibiotic therapy in the future.

The majority of current quorum-sensing information is based on laboratory experiments in controlled environments, and it mirrors the knowledge of molecular signals between neighboring cells scaled to current theoretical models. Due to this, we still lack knowledge on how this and other microbial sensing systems operate in nature. Physical, chemical and biological modifications of signals can be affected in the extracellular space, and these might disrupt intended communication if the signals are no longer recognized as having been transmitted. Information within cell clusters of the biofilm can be disturbed, where signal dispersion might be significantly influenced by the extracellular matrix (i.e., polymers) [94]. These factors span a range of physical, chemical, and biological variables that can impact upon signal production, stability and distribution. Optimizing experimental systems to natural or clinical environments might be crucial for defining when and where quorum sensing occurs. These points have been illustrated in studies of signaling in biofilms of *S. aureus* [95], *Salmonella* spp. [96] and *C. albicans* [97, 98], where signal stability can be affected by the host environment.

Many studies have been performed to define quorum sensing in connection with the synthesis of extracellular materials and motility mechanisms, and for biofilm formation, stability and resistance. Single-gene knockout mutants have been used for the evaluation of the impact of trapping time, surface, and nutritional environment [99]. The basic signaling schemes have been worked out at the molecular level for a few of the major quorum-sensing systems. As these studies continue to refine our understanding of these mechanisms, an emerging challenge is to identify if and when the local environment can affect signaling, particularly if we foresee control of microorganisms, including reduction of biofilm formation and resistance of biofilm cells [91].

While communication via chemical signals has been the focus of most investigations, the use of physical signals for microbial cell–cell communication has received only limited attention so far. This observation is based on experimental evidence of the microbial emission and response to three physical signals: sound waves, electromagnetic radiation, and electric currents. These signals propagate rapidly, even at very low intensities, and they provide useful mechanisms when a rapid response is required [100]. There are still opened questions into the physical nature of microbial signaling networks, and answering these might provide new possibilities of how to control biofilms in desirable processes or in their devastating consequences for the health of the host.

The field of studying microbial cooperation has grown enormously over the last decade, leading to improved experimental techniques and a growing awareness of collective behavior in microbes. Unfortunately, many of the theoretical tools and available concepts for understanding cooperation fail to take into account the peculiarities of the microbial world, namely strong selection strengths, unique population structures, and non-linear dynamics [101]. The relation between a growing biofilm and quorum-sensing behavior requires mathematical modeling of quorum sensing in biofilms that can be applied with nonlinear diffusion reactions. With such approaches, the influence of the hydrodynamic environment and nutrient conditions on biofilm growth and quorum-sensing behavior can be studied. Such models can be extended to include an impact of quorum sensing on biofilm growth, through the explicit description of all of the chemical, physical and biological factors of biofilm formation, and with special attention to the structural variations of biofilms grown on different surfaces and under different natural or man-made conditions [51].

Finally, a number of inhibitors that have been identified through bottom-up and top-down approaches have been proposed, such as surfactin, glucose, mannose halogenated furanones, 4(5)-aryl 2-aminoimidazoles, furocoumarins, and salicylates. Their relationships to cell-to-cell signaling have been explored in an attempt to reduce the ability of cells to form biofilms, attenuate virulence, and modulate other processes influenced by quorum sensing [102]. Inhibition of quorum sensing can be accomplished in several ways; e.g., enzymatic degradation of the signal molecule [103], blocking signal generation [104], and blocking signal reception [105], or combinations of these. By disrupting the biofilm architecture or consistency, cells will lose their natural protection structure, so quorum-sensing inhibitors (QSIs) might be an additional tool to fight biofilms. This can be seen as the newest weapon against infections that involve drug- and biocide-resistant microbes. QSIs can come from a variety of sources and have a wide array of structures, like acylhomoserine lactone (AHL) analogs [106], brominated furanones [107], secondary fungal metabolites [108], plant extracts [109], antibiotics [110, 111], and others. Indeed, this can become a reality when the surfaces are addressed, be they man-made or natural. In both cases, microorganisms can attach and form biofilms and survive. They can contaminate food, water or other organisms, or they can occupy the surface and protect it from contamination by undesired pathogens and spoilage microbes. Regarding this challenge, much is still to be done with microbial quorum sensing in connection with the technology and health issues. However, although the result is not easily predictable, by using biological, chemical and physical tools to disturb quorum sensing in such a way that the biofilm structure is damaged and cannot undergo fast restructuring and recovery, enables the application of real killers, like antibiotics or disinfectants that can easily enter and attack vital points in the structure. Unprotected cells, and those with disturbed communication due to QSIs, will be easy targets for killing agents. This strategy is possible as QSIs will 'disarm' the cells for enough time to allow antimicrobials to penetrate into the biofilms and the cells, and thus to kill them.

Resistance of Multispecies Biofilms

Microbial life in the environment is mostly characterized by multiplicity; e.g., the presence of many species together at the same time in the same ecological niche. This is also true for planktonic and biofilm patterns of growth. Diversity in microbial communities leads to a variety of complex relationships that involve inter- and intraspecies interactions. Cell growth, differentiation, chemotaxis, and cell-to-cell signaling enable biofilm communities to organize structurally in response to external conditions and activities of the different biofilm members. Although the higher resistance of mixed bacterial species biofilms against antiseptics was noted quite early [112, 113], biofilm multispecies interactions have not been thoroughly studied until recently. The regulating mechanisms in many aspects still remain unclear. However, it was shown that mixed populations (e.g., lactic acid bacteria and yeast) can form biofilms, even when neither of these populations can do this in monoculture [114]. In addition, the diversity of biofilm communities can affect the disinfection efficacy, including for chemical and mechanical treatments that have been induced in recent studies of the effects of sequential antimicrobial (with different chemicals) and mechanical (with different shear stress) treatments in single and dual species biofilms. Phenotypic characterization has revealed important differences in the metabolic activities, with about five times higher respiratory activity of *Bacillus cereus* and *P. fluorescens* dual species biofilms in comparison with each single species population [115]. A correlation between metabolic activity and the synthesis of exopolysaccharide substances was indicated [116], and was confirmed again in mixed *B. cereus* and *P. fluorescens* biofilms, with significant proportions of both extracellular proteins and polysaccharides in dual species biofilms [115]. Moreover, dual biofilms have a higher resistance to disinfectant agents and physical removal by shear forces, with higher proportions of viable cells of both bacterial species than their respective single-species biofilms. Also, studies with *Listeria monocytogenes* and *Lactobacillus plantarum* [117] and with multispecies biofilms with six drinking-water-isolated bacterial species [118] have confirmed the impact of microbial diversity and the higher resistance of multispecies biofilms to inactivation and removal.

The resistance mechanisms and survival of complex multispecies biofilms in adverse environments can be understood by well-described biofilm-specific antimicrobial resistance mechanisms [7, 17, 119]. Additionally, the presence of synergistic species association results in higher resistance of multispecies biofilms to inactivation with physico-chemical treatments than for single-species biofilms [120] and also in amplified virulence during co-infection. This has been shown for dual-species biofilms of *C. albicans* and *S. aureus*, which were analyzed using confocal scanning laser microscopy and differential protein expression studies. These findings demonstrated that the hyphae-mediated enhanced pathogenesis of *S. aureus* might not only be due to physical interactions, but can also be attributed to the differential regulation of specific virulence factors induced during polymicrobial growth [121].

Although biofilms are a good example of an evolutionary mechanism by which mixed populations of prokaryotes acquire a collective advantage [122], and it is likely that most natural biofilms exist as multispecies consortia, single-species biofilms have been studied much more extensively than mixed-species biofilms, for their interactions and resistance mechanisms. Early studies indicated that the higher resistance of multispecies biofilms depends on the type of species incorporated and the role of each one [123, 124]. The interactions in multispecies biofilms might influence each other not only with respect to attachment capabilities, but also in terms of the resistance to antimicrobial substances. Multispecies consortia gain advantages compared to single-species biofilms, in terms of the ability of the organisms or biofilm to persist or grow in a given environment or under a particular environmental stress [125, 126]. Interactions beneficial to one or more strains or species include the co-aggregation of cells, and conjugation and protection of one or several species from eradication when the biofilm is exposed to antimicrobial compounds [125, 127, 128]. Such protection can be caused by a variety of factors, including enzyme complementation and the organized spatial distribution of the cells in the biofilm [124, 129]. The co-culturing of two bacteria in a biofilm enhanced the resistance of the individual strains, as has already been shown for different microbes, such as *P. fluorescens, Salmonella enterica* serovar Typhimurium, *S. sciuri*, *Kocuria* spp., *Acinetobacter calcoaceticus* [115, 118, 126, 130, 131]. The increased resistance of multispecies biofilms can be partly explained also by the higher cell densities relative to those of single-species biofilms, although the literature data are different concerning this – e.g., from higher numbers in single-species *L. monocytogenes* biofilms than in mixed ones with *P. fragi* [132], to similar numbers in single and mixed biofilms with *S. aureus* strains [133], and much higher cell densities and survival in mixed species biofilms of *B. cereus* and *P. fluorescens* [115] and of *L. monocytogenes* with *Lactobacillus plantarum* [117]. Moreover, enhanced metabolic activity has been confirmed for *B. cereus* and *P. fluorescens* mixed biofilms, which also results in changed and possibly more resistant matrix composition [115]. Similarly,mixed-culture biofilms of *Campylobacter jejuni* with *P. aeruginosa* were more robust and unaffected by higher liquid flow rates [134].

However, especially in the medicine area and for drinking-water systems, it has been demonstrated that different bacterial species in biofilms affect one another positively as well as negatively. Reports on negative interactions in biofilms include the production of bacteriotoxins and the lowering of the pH by one member of a biofilm consortium [125, 135], as well as real predator interactions, such as *Bdelovibrio* spp. in *E. coli* biofilms [136] and *Micavibrio aeruginosavorus* in *P. aeruginosa*, *Klebsiella pneumonia,* and *Burkholderia cepacia* biofilms [137]. Also, the biocontrol potential of *P. fluorescens* on the planktonic growth of *B. cereus* was indicated for a stratified biofilm structure [138].

As mentioned already, the increased resistance of multispecies biofilms has not been defined completely. However, it is clear that the knowledge of biofilm biodiversities and their multispecies interactions will facilitate the development of improved disinfection and biofilm control processes. Therefore, it remains for future studies to highlight multispecies interactions in mixed-species biofilms and to confirm whether their resistance mechanisms result in activation of similar defense systems, as in planktonic or single-species biofilms, or whether the resistance is conferred via other mechanisms as well.

Conclusion

Recent research has highlighted how and why surface-attached microbial communities can develop resistance to antimicrobial agents and other environmental stresses in clinical, environmental and industrial environments. Many useful methods to study the interactions have been developed and they make it clear that there are multiple resistance mechanisms at work within any single community. Many of these new achievements, especially in the understanding of intercellular signaling and quorum sensing, have begun to reveal how these social behaviors are related in different species. These interactions are even more structured in multispecies communities, which are the most prevalent in natural and man-made environments. As our understanding of biofilm formation, its mature structure, and its resistance mechanisms is largely based on single-species studies in controlled environments, recent scientific contributions still await confirmation from different fields, like for clinically more successful anti-biofilm therapy, or more efficient industrial sanitation procedures against spoilage and pathogenic bacteria. On the other side, more productive biotechnological applications based on improved resistance of surface-attached microbial starters can be developed. A better understanding of the resistance mechanisms that have evolved in microbial cells for biofilms remains the crucial issue for new developments in these different fields of life and research.

References

[1] Costerton, J. W.; Geesey, G. G. & Cheng, K. J. (1978). How bacteria stick. *Scientific American, 238*: 86-95.

[2] Costerton, J.W.; Cheng, K. J.; Geesey, G. G.; Ladd, T. I.; Nickel, J. C.; Dasgupta, M. & Marrie, T. J. (1987). Bacterial biofilms in nature and disease. *Annual Reviews in Microbiology, 41*: 435-464.

[3] Costerton, J. W.; Lewandowski, Z.; Caldwell, D. E.; Korber, D. R. & Lappin-Scott, H. M. (1995). Microbial biofilms. *Annual Reviews in Microbiology, 49*: 711-45.

[4] Donlan, R. M. & Costerton, J. W. (2002). Biofilms: survival mechanisms of clinically relevant microorganisms. *Clinical Microbiology Reviews. 15*: 167-193.

[5] Webb, J. S.; Givskov, M. & Kjelleberg, S. (2003). Bacterial biofilms: prokaryotic adventures in multicellularity. *Current Opinion in Microbiology, 6*:578-585.

[6] Hall-Stoodley, L.; Costerton, J. W. & Stoodley, P. (2004). Bacterial biofilms: from the natural environment to infectious diseases. *Nature Reviews. Microbiology, 2*: 95-108.

[7] Moons, P.; Michiels, C. W. & Aertsen, A. (2009). Bacterial interactions in biofilms. *Critical reviews in Microbiology, 35*: 157-168.

[8] Ghigo, J.-M. (2001). Natural conjugative plasmids induce bacterial biofilm development. *Nature,* 412:442-445.

[9] Donlan R. M. (2002). Biofilms: microbial life on surfaces. *Emerging Infectious Diseases, 8*: 881-890.

[10] Cos P.; Tote, K.; Horemans, T. & Maes, L. (2010). Biofilms: An extra hurdle for effective antimicrobial therapy. *Current Pharmaceutical Design 16*: 2279-2295.

[11] Brown, M. R. W.; Allison, D. G. & Gilbert, P. (1988). Resistance of bacterial biofilms to antibiotics: a growth-rate related effect? *Journal of Antimicrobial Chemotherapy 22*: 777–783.

[12] Stewart, P. S. (2002). Mechanisms of antibiotic resistance in bacterial biofilms. *International Journal of Medical Microbiology, 292*: 107-113.

[13] Keren, I.; Kaldalu, N.; Spoering, A.; Wang, Y. & Lewis, K. (2004a). Persister cells and tolerance to antimicrobials. *FEMS Microbiology Letters, 230*, 13-18.

[14] Keren, I.; Shah, D.; Spoering, A.; Kaldalu, N. & Lewis, K. (2004b). Specialized persister cells and the mechanism of multidrug tolerance in *Escherichia coli*. *Journal of Bacteriology, 186*:8172-8180.

[15] Toté, K.; Horemans, T.; Vanden Berghe, D.; Maes, L. & Cos, P. (2010). Inhibitory effect of biocides on the viable masses and matrices of *Staphylococcus aureus* and *Pseudomonas aeruginosa* biofilms. *Applied and Environmental Microbiology, 76*: 3135-3142.

[16] Brown, M. R. W. & Gilbert, P. (1993). Sensitivity of biofilms to antimicrobial agents. *Journal of Applied Bacteriology 74*: 87S-97S.

[17] Mah, T. F. & O'Toole, G. A. (2001). Mechanisms of biofilm resistance to antimicrobial agents. *Trends in Microbiology, 9*: 34–39.

[18] Stewart, P. S, & Costerton, J. W. (2001). Antibiotic resistance of bacteria in biofilms. *Lancet, 358*: 135-138.

[19] Anderson, G. G. & O'Toole, G. A. (2008). Innate and induced resistance mechanisms of bacterial biofilms. *Current Topics in Microbiology and Immunology, 322*: 85-105.

[20] Stewart, P. S. & Franklin, M. J. (2008). Physiological heterogeneity in biofilms. *Nature Reviews. Microbiology 6*: 199-210.

[21] Simoes, M.; Bennet, R. N. & Rosa, E. A. S. (2009a). Understanding antimicrobial activities of phytochemicals against multidrug resistant bacteria in biofilms. *Natural Products Reports, 26*: 746-757.

[22] Stewart, P. S. (2001). Multicellular resistance: biofilms: *Trends in Microbiology, 9*: p. 204.

[23] Hengge-Aronis, R. (1996). Regulation of gene expression during entry into stationary phase. In: Neidhart, F. C. et al., editors. *Escherichia coli* and *Salmonella*: *Cellular and Molecular Biology*, Washington, D.C, ASM Press, pp. 1497-1512.

[24] Hengge-Aronis, R. (2002). Signal transduction and regulatory mechanisms involved in control of the sigma(S) (RpoS) subunit of RNA polymerase. *Microbiology and Molecular Biology Review, 66*: 373-395.

[25] Liu, X.; Ng, C. & Ferenci, T. (2000) Global adaptations resulting from high population densities in *Escherichia coli* cultures. *Journal of Bacteriology 182:* 4158-4164.

[26] Novick, R. P. (2003). Autoinduction and signal transduction in the regulation of staphylococcal virulence. *Molecular Microbiology, 48*:1429-1449.

[27] O'Toole, G.; Kaplan, H. B. & Kolter, R. (2000). Biofilm formation as microbial development. *Annual Reviews in Microbiology, 54*: 49-79.

[28] Nystrom, T. (2002). Aging in bacteria. *Current Opinion in Microbiology, 5*:596-601.

[29] Rayner, M. G.; Zhang, Y.; Gorry, M. C.; Chen, Y.; Post, J. C. & Ehrlich, G. D. (1998). Evidence of bacterial metabolic activity in culture-negative otitis media with effusion. *JAMA, 279*: 296-299.

[30] Boaretti, M.; Lleò, M. M.; Bonato, B.; Signoretto, C. & Canepari, P. (2003). Involvement of rpoS in the survival of *Escherichia coli* in the viable but non-culturable state. *Environmental Microbiology, 5*: 986-996.

[31] Abee, T, & Wouters, J. A. (1999). Microbial stress response in minimal processing. *International Journal of Food Microbiology, 50*: 65-91.

[32] Beenken, K. E.; Dunman, P. M.; McAleese, F.; Macapagal, D.; Murphy, E., Projan, S. J.; Blevins, J. S.; & Smeltzer, M. S. (2004). Global gene expression in *Staphylococcus aureus* biofilms. *Journal of Bacteriology 186*: 4665-4684.

[33] Klančnik, A.; Guzej, B.; Jamnik, P.; Vučković, D.; Abram, M. & Smole Možina, S. (2009). Stress response and pathogenic potential of *Campylobacter jejuni* cells exposed to starvation. *Research in Microbiology, 160*: 345-352.

[34] Klančnik, A.; Vučković, D.; Plankl, M.; Abram, M. & Smole Možina, S. (2013). *In vivo* modulation of *Campylobacter jejuni* virulence in response to environmental stress. *Foodborne Pathogens and Disease* 10 (6): 566 – 572. DOI: 10.1089/fpd.2012.1298.

[35] Sanderson, S. S. & Stewart, P. S. (1987). Evidence of bacterial adaptation to monochloramine in *Pseudomonas aeruginosa* biofilms and evaluation of biocide action model. *Biotechnology and Bioengineering, 56*: 201-209.

[36] Stewart, P. S.; Roe, F.; Rayner, J.; Elkins, J. G.; Lewandowski, Z.; Ochsner, U. A. & Hassett, D. J. (2000). Effect of catalase on hydrogen peroxide penetration into *Pseudomonas aeruginosa* biofilms. *Applied and Environmental Microbiology, 66*: 836-838.

[37] Bagge, N.; Schuster, M.; Hentzer, M.; Ciofu, O.; Givskov, M.; Greenberg, E. P. & Høiby, N. (2004). *Pseudomonas aeruginosa* biofilms exposed to imipenem exhibit changes in global gene expression and beta-lactamase and alginate production. *Antimicrobial Agents and Chemotherapy, 48*: 1175-1187.

[38] Kubota, H.; Senda, S.; Nomura, N.; Tokuda, H. & Uchiyama, H. (2008). Biofilm formation by lactic acid bacteria and resistance to environmental stress. *Journal of Bioscience and Bioengineering 106*: 381-386.

[39] Kubota, H.; Senda, S.; Tokuda, H.; Uchiyama, H. & Nomura N. (2009). Stress resistance of biofilm and planktonic *Lactobacillus plantarum* subsp. *plantarum* JCM 1149. *Food Microbiology, 26*: 592-597.

[40] Spormann, A. M. (2009). Physiology of microbes in biofilms. In: Romeo T. editor. *Bacterial Biofilms* Heidelberg: Springer, pp. 17-36.

[41] Xu, K. D.; Stewart, P. S.; Xia, F.; Huang, C. T. & McFeters, G. A. (1998). Spatial physiological heterogeneity in *Pseudomonas aeruginosa* biofilm is determined by oxygen availability. *Applied and Environmental Microbiology, 64*: 4035-4039.

[42] Xu, K. D.; Franklin, M. J.; Park, C. H.; McFeters, G. A. & Stewart, P. S. (2001). Gene expression and protein levels of the stationary phase sigma factor, RpoS, in continuously-fed *Pseudomonas aeruginosa* biofilms. *FEMS Microbiology Letters, 15*: 67-71.

[43] Li, Y. H.; Chen, Y. Y. & Burne, R. A. (2000). Regulation of urease gene expression by *Streptococcus salivarius* growing in biofilms. *Environmental Microbiology, 2*: 169-177.

[44] Whiteley, M.; Bangera, M. G.; Bumgarner, R. E.; Parsek, M. R.; Teitzel, G. M.; Lory, S. & Greenberg, E. P. (2001). Gene expression in *Pseudomonas aeruginosa* biofilms. *Nature 413*: 860-864.

[45] Müsken, M.; Di Fiore, S.; Dötsch, A.; Fischer, R. & Häussler, S. (2010). Genetic determinants of *Pseudomonas aeruginosa* biofilm establishment. *Microbiology, 156*: 431-441.

[46] Burne, R. A. & Marquis, R. E. (2000) Alkali production by oral bacteria and protection against dental caries. *FEMS Microbiology Letters, 193*: 1-6.

[47] Stark, R. M.; Gerwig, G. J.; Pitman, R. S.; Potts, L. F.; Williams, N. A.; Greenman, J.; Weinzweig, I. P.; Hirst, T. R. & Millar, M. R. (1999). Biofilm formation by *Helicobacter pylori*. *Letters in Applied Microbiology, 28*: 121-126.

[48] Gieseke, A.; Tarre, S.; Green, M. & de Beer, D. (2006). Nitrification in a biofilm at low pH values: role of *in situ* microenvironments and acid tolerance. *Applied and Environmental Microbiology, 72*: 4283-4292.

[49] Roberts, M. E. & Stewart, F. S. (2004). Modeling antibiotic tolerance in biofilms by accounting for nutrient limitation. *Antimicrobial Agents and Chemotherapy*, *48*: 48–52.

[50] Dibdin, G. H., Assinder, S. J.; Nichols, W. W. & Lambert, P. A. (1996). Mathematical model of β-lactam penetration into a biofilm of *Pseudomonas aeruginosa* while undergoing simultaneous inactivation by released β-lactamases. *Journal of Antimicrobial Chemotherapy, 38*: 757-769.

[51] Høiby, N.; Bjarnsholt, T.; Givskov, M.; Molin, S. & Ciofu, O. (2010). Antibiotic resistance of bacterial biofilms. *International Journal of Antimicrobial Agents, 35*: 322-332.

[52] Stewart, P. S.; Davison, W. M. & Steenbergen, J. N. (2009). Daptomycin rapidly penetrates a *Staphylococcus epidermidis* biofilm. *Antimicrobial Agents and Chemotherapy 53*: 3505-3507.

[53] Singh, R.; Ray, P.; Das, A. & Sharma, M. (2010). Penetration of antibiotics through *Staphylococcus aureus* and *Staphylococcus epidermidis* biofilms. *Journal of Antimicrobials and Chemotherapy, 65*: 1955-1958.

[54] Fung, D. K. C.; Chan, E. W. C.; Chin, M. L. & Chan, R. C. Y. (2010). Delineation of a bacterial starvation stress response network which can mediate antibiotic tolerance development. *Antimicrobial Agents and Chemotherapy 54*:1082 -1093.

[55] Korch, S. B.; Henderson, T. A.; Hill, T. M. (2003). Characterization of the hipA7 allele of *Escherichia coli* and evidence that high persistence is governed by (p)ppGpp synthesis. *Molecular Microbiology, 50*: 1199-1213.

[56] Ito, A.; May, T.; Taniuchi, A.; Kawata, K. & Okabe, S. (2009). Localized expression profiles of rpoS in *Escherichia coli* biofilms. *Biotechnology and Bioengineering, 103*: 975-983.

[57] Davison,W. M.; Pitts, B. & Stewart, P. S. (2010). Spatial and temporal patterns of biocide action against *Staphylococcus epidermidis* biofilms. *Antimicrobial Agents and Chemotherapy, 54*: 2920-2927.

[58] Bridier, A.; Dubois-Brissonnet, F.; Greub, G.; Thomas, V. & Briandet, R. (2011). Dynamics of the action of biocides in *Pseudomonas aeruginosa* biofilms. *Antimicrobial Agents and Chemotherapy, 55*: 2648-2654.

[59] Richards, J. J. & Melander, C. (2009). Controlling bacterial biofilms. *Chemistry and Biochemistry, 10*: 2287-2294.

[60] Rogers, S. A.; Huigens, R. V.; Cavanagh, J. & Melander, C. (2010). Synergistic effects between conventional antibiotics and 2-aminoimidazole-derived antibiofilm agents. *Antimicrobial Agents and Chemotherapy, 54*: 2112-2118.

[61] Toté, K.; Vanden Berghe, D.; Deschacht, M.; de Wit, K.; Maes, L. & Cos P. (2009). Inhibitory efficacy of various antibiotics on matrix and viable mass of *Staphylococcus aureus* and *Pseudomonas aeruginosa* biofilms. *International Journal of Antimicrobial Agents, 33*: 525-531.

[62] Molin, S. & Tolker-Nielsen, T. (2003). Gene transfer occurs with enhanced efficiency in biofilms and induces enhanced stabilisation of the biofilm structure. *Current Opinion in Biotechnology, 14*: 255–261.

[63] Taddei, F.; Matic, I. & Radman, M. (1995). cAMP-dependent SOS induction and mutagenesis in resting bacterial populations. *Proceedings of the National Academy of Science USA, 92:*11736-11740.

[64] van der Veen, S. & Abee, T. (2011). Bacterial SOS response: a food safety perspective. *Current Opinion in Biotechnology, 22*: 136-142.

[65] Gotoh, H.; Kasaraneni, N.; Devineni, N.; Dallo, S.F. & Weitao, T. (2010). SOS involvement in stress-inducible biofilm formation. *Biofouling. 26*: 603-611.

[66] van der Veen S. & Abee, T. (2010a). Dependence of continuous-flow biofilm formation by *Listeria monocytogenes* EGD-e on SOS response factor YneA. *Applied and Environmental Microbiology, 76*: 1992-1995.

[67] Lewis, K. (2008). Multidrug tolerance of biofilms and persister cells. *Current Topics in Microbiology and Immunology 322*: 107-131.

[68] Spoering, A. L. & Lewis, K. (2001). Biofilms and planktonic cells of *Pseudomonas aeruginosa* have similar resistance to killing by antimicrobials. ***Journal of Bacteriology*** *183:* 6746-6751.

[69] Lewis, K. (2010). Persister cells. *Annual Reviews in Microbiology, 13*: 357-372.

[70] De Groote, V. N.; Fauvart, M.; Kint, C. I.; Verstraeten, N.; Jans, A.; Cornelis, P. & Michiels, J. (2011). *Pseudomonas aeruginosa* fosfomycin resistance mechanisms affect noninherited fluoroquinolone tolerance. *Journal of Medical Microbiology, 60*: 329-336.

[71] Dörr, T., Lewis, K. & Vulić, M. (2009). SOS response induces persistence to fluoroquinolones in *Escherichia coli*. *PLoS Genetics 5*:e1000760.

[72] Lafleur, M. D.; Qi, Q. & Lewis, K. (2010). Patients with long-term oral carriage harbor high-persister mutants of *Candida albicans*. *Antimicrobial Agents and Chemotherapy, 54*: 39-44.

[73] Mulcahy, L. R.; Burns, J. L.; Lory, S. & Lewis, K. (2010). Emergence of *Pseudomonas aeruginosa* strains producing high levels of persister cells in patients with cystic fibrosis. *Journal of Bacteriology 192*: 6191-6199.

[74] Lee, A.; Mao, W.; Warren, M. S.; Mistry, A.; Hoshino, K.; Okumura, R.; Ishida, H. & Lomovskaya, O. (2000). Interplay between efflux pumps may provide either additive or multiplicative effects on drug resistance. *Journal of Bacteriology, 182*: 3142-3150.

[75] Webber M. A. & Piddock, L. J. V. (2003). The importance of efflux pumps in bacterial antibiotic resistance. *Journal of Antimicrobial Chemotherapy 51*: 9-11.

[76] Kvist, M.; Hancock, V. & Klemm, P. (2008). Inactivation of efflux pumps abolishes bacterial biofilm formation. *Applied and Environmental Microbiology, 74:* 7376–7382.

[77] Zhang, L. & Mah, T.-F. (2008). Involvement of a novel efflux system in biofilm-specific resistance to antibiotics. *Journal of Bacteriology, 190*: 4447-4452.

[78] Pagedar, A., Singh, J. & Batish, V.K. (2011). Efflux mediated adaptive and cross resistance to ciprofloxacin and benzalkonium chloride in *Pseudomonas aeruginosa* of dairy origin. *Journal of Basic Microbiology, 51*: 289-295.

[79] Soto, S. M. (2013). Role of efflux pumps in the antibiotic resistance of bacteria embedded in a biofilm. *Virulence 4*: 223-229.

[80] Rajendran, R.; Mowat, E.; McCulloch, C.; Lappin, D. F., Jones, B.; Lang, S.; Majithiya, J. B.; Warn, P.; Williams, C. & Ramage, G. (2011). *Aspergillus fumigatus* biofilms is partially associated with efflux pump activity. *Antimicrobial Agents and Chemotherapy, 55*: 2092-2097.

[81] Ramage, G.; Mowat, E.; Jones, B.; Willims, C. & Lopez-Ribot, J. (2009). Our current understanding of fungal biofilms. *Critical Reviews in Microbiology, 35*: 340-355.

[82] Ramage, G.; Rajendran, R.; Sherry, L. & Williams, C. (2012). Fungal biofilm resistance. *International Journal of Microbiology*, doi:10.1155/2012/528521.

[83] Marquez, B. (2005). Bacterial efflux systems and efflux pump inhibitors. *Biochimie 87*: 1137-1147.

[84] Parsek, M. R. & Greenberg, E. P. (2005). Sociomicrobiology: the connections between quorum sensing and biofilms. *Trends in Microbiology, 13*: 27-33.

[85] Bassler B. L. (1999). How bacteria talk to each other: regulation of gene expression by quorum sensing. *Current Opinion in Microbiology, 2*: 582-587.

[86] Miller, M. B. & Bassler, B. L. (2001). Quorum sensing in bacteria. *Annual Reviews in Microbiology 55*: 165-199.

[87] Waters C. M. & Bassler B. L. (2005). Quorum sensing: cell-to-cell communication in bacteria. *Annual Review of Cell and Developmental Biology, 21*: 319-346.

[88] Sperandio, V., Torres, A. G., Jarvis, B., Nataro, J. P. & Kaper J. B. (2003). Bacteria–host communication: The language of hormones. *Proc. Natl Acad Sci U S A, 100*: 8951-8956.

[89] *Nickerson, K. W.; Atkin, A. L. & Hornby, J. M. (2006).* Quorum sensing in dimorphic fungi: Farnesol and beyond. *Applied and Environmental Microbiology,* 72:6 3805-3813.

[90] Chen, H. & Fink, G.R. (2006). Feedback control of morphogenesis in fungi by aromatic alcohols. *Genes and Development, 20*, 9:1150-1161.

[91] Nadell, C. D.; Xavier, J. B.; Levin, S. A. & Foster, K. R. (2008). The evolution of quorum sensing in bacterial biofilms. *PLoS Biology, 6*: 171-179.

[92] Li Y. H. & Tian X. L. (2012). Quorum sensing and bacterial social interactions in biofilms. *Sensors, 12,* 3: 2519-2538.

[93] Plummer P.J. (2012). *LuxS* and quorum-sensing in *Campylobacter. Frontiers in Cellular Infection Microbiology, 2*: 22 doi: 10.3389/fcimb.2012.00022.

[94] Decho, A.W.; Norman, R. S. & Visscher, P. T. (2010). Quorum sensing in natural environments: emerging views from microbial mats. *Trends in Microbiology, 18*: 73-80.

[95] Horswill, A. R.; Stoodley, P.; Stewart, P. S. & Parsek, M. R. (2007). The effect of the chemical, biological, and physical environment on quorum sensing in structured microbial communities. *Analytical and Bioanalytical Chemistry, 387*: 371-380.

[96] Steenackers, H.; Hermans, K.; Vanderleyden, J. & De Keersmaecker, S. C. J. (2012). *Salmonella* biofilms: An overview on occurrence, structure, regulation and eradication. *Food Research International, 45*: 502-531.

[97] Nigam, S.; Ciccoli, R.; Ivanov, I.; Sczepanski, M. & Deva, D. (2011). On mechanism of quorum sensing in *Candida albicans* by 3(R)-hydroxy-tetradecaenoic acid. *Current Microbiology, 62*: 55–63.

[98] Han, T.-L.; Cannon, R.D. & Villas-Bôas, S. G. (2011). The metabolic basis of *Candida albicans* morphogenesis and quorum sensing. *Fungal Genetics and Biology*, *48*: 747-763.

[99] Gordon, V.; Butler, J.; Smalyukh, I.; Parsek, M.R. & Wong, G. (2008). Quorum sensing and biofilm formation investigated using laser-trapped bacterial arrays. *Bulletin of the American Physical Society, APS Meeting 53,* 2, p.16.

[100] Reguera, G. (2011). When microbial conversations get physical. *Trends in Microbiology, 19*: 105-113.

[101] Damore, J. A. & Gore, J., (2011). Understanding microbial cooperation. *Journal of Theoretical Biology*, *299*: 31-41.

[102] Kociolek, M. G. (2009). Quorum-sensing inhibitors and biofilms. *Anti-Infective Agents in Medicinal Chemistry, 8*: 315-326.

[103] Roche, D. M.; Byers, J. T.; Smith, D. S.; Glansdorp, F. G.; Spring, D. R. & Welch, M. (2004). Communications blackout? Do N-acylhomoserinelactone-degrading enzymes have any role in quorum sensing? *Microbiology, 150*: 2023–2028.

[104] Hentzer, M,& Givskov, M. (2003). Pharmacological inhibition of quorum sensing for the treatment of chronic infections. *Journal of Clinical Investigations, 112*:1300–1307.

[105] Kjelleberg, S.; McDougald, D.; Rasmussen, T.B. & Givskov, M. (2008). Quorum-sensing inhibition. In: Winans, S. C. editor; Bassler, B. L., editor. *Chemical communication among bacteria.* Washington, D.C. : ASM Press., pp 393–416.

[106] Whitehead, N. A.; Barnard, A. M. L.; Slater, H.; Simpson, N. J. L. & Salmond, G. P. C. (2001). Quorum-sensing in Gram-negative bacteria. *FEMS Microbiology Reviews, 25*: 365-404.

[107] Givskov, M.; de Nys, R.; Manefield, M.; Gram, L.; Maximilien, R.; Eberl, L.; Molin, S.; Steinberg, P. D. & Kjelleberg, S. (1996). Eukaryotic interference with homoserine lactone-mediated prokaryotic signaling. *Journal of Bacteriology, 178*: 6618-6622.

[108] Rasmussen, T. B.; Skindersoe, M. E.; Bjarnsholt, T.; Phipps, R. K.; Christensen, K. B.; Jensen, P. O.; Andersen, J. B.; Koch, B.; Larsen, T. O.; Hentzer, M.; Eberl, L.; Hoiby, N.; Givskov, M. (2005). Identity and effects of quorum-sensing inhibitors produced by *Penicillium* species. *Microbiology, 151*: 1325-1340.

[109] Al-Hussaini, R. & Mahasneh, A. M. (2009). Microbial growth and quorum sensing antagonist activities of herbal plants extracts. *Molecules, 14*: 3425-3435.

[110] Tateda, K.; Comte, R.; Pechere, J.-C.; Kohler, T.; Yamaguchi, K.; Van Delden, C. (2001). Azithromycin inhibits quorum sensing in *Pseudomonas aeruginosa. Antimicrobial Agents and Chemotherapy, 45*: 1930-1933.

[111] Heeb, S.; Fletcher, M. P.; Chhabra, S. R.; Diggle, S. P.; Williams, P. & Cámara, M. (2011). Quinolones: from antibiotics to autoinducers. *FEMS Microbiology Reviews, 35*: 247-274.

[112] Stickler, D. & Hewett, P. (1991). Activity of antiseptics against biofilms of mixed bacterial species growing on silicone surfaces. European Journal of Clinical Microbiology and Infectious Diseases, 10: 416-421.

[113] Stickler, D.; Dolman, J.; Rolfe, S. & Chawla, J. (1991). Activity of some antiseptics against urinary tract pathogens growing as biofilms on silicone surfaces. European Journal of Clinical Microbiology and Infectious Diseases, 10: 410-415.

[114] Kawarai, T.; Furukawa, S.; Ogihara, H. & Yamasaki, M. (2007). Mixed-species biofilm formation by lactic acid bacteria and rice wine yeasts. *Applied and Environmental Microbiology, 73*: 4673-4676.

[115] Simoes, M.; Simoes, L. C. & Vieira, M. J. (2009b). Species association increases biofilm resistance to chemical and mechanical treatments. *Water Research, 43*: 229–237.

[116] Simoes, M.; Pereira, M. O.; Sillankorva, S.; Azeredo, S. & Vieira, M. J., (2007b). The effect of hydrodynamic conditions on the phenotype of *Pseudomonas fluorescens* biofilms. *Biofouling 23*: 249-258.

[117] van der Veen, S. & Abee, T. (2010b). Mixed species biofilms of *Listeria monocytogenes* and *Lactobacillus plantarum* show enhanced resistance to

benzalkonium chloride and peracetic acid. *International Journal of Food Microbiology 144*: 421-431.

[118] Simoes L. C.; Simoes, M. & Vieira, M. J. (2010a). Influence of the diversity of bacterial isolates from drinking water on resistance of biofilms to disinfection. *Applied and Environmental Microbiology, 76*: 6673–6679.

[119] Davies, D. (2003). Understanding biofilm resistance to antibacterial agents. *Nature Reviews Drug Discovery 2*: 114–122.

[120] Simoes L. C.; Simoes, M. & Vieira, M. J. (2010b). A review of current and emergent biofilm control strategies. *LWT - Food Science and Technology 43*: 573-583.

[121] Peters, B. M.; Jabra-Rizk, M. A.; Scheper, M. A.; Leid, J. G.; Costerton, J. W. & Shirtliff, M. E. (2010). Microbial interactions and differential protein expression in *Staphylococcus aureus-Candida albicans* dual-species biofilms. *FEMS Immunology and Medical Microbiology, 59*: 493-503.

[122] Parsek, M. R.; & Fuqua, C. (2004). Biofilms 2003: Emerging themes and challenges in studies of surface-associated microbial life. *Journal of Bacteriology 186*: 4427-4440.

[123] Leriche, V. & Carpentier, B. (1995). Viable but non culturable *Salmonella typhimurium* within single and binary species biofilms in response to chlorine treatment. *Journal of Food Protection, 58*: 1186–1191.

[124] Leriche, V.; Briandet, R. & Carpentier, B. (2003). Ecology of mixed biofilms subjected daily to a chlorinated alkaline solution: spatial distribution of bacterial species suggests a protective effect of one species to another. *Environmental Microbiology, 5*: 64–71.

[125] Burmølle, M.; Webb, J. S.; Rao, D.; Hansen, L. H.; Sørensen, S. J. & Kjelleberg. S. (2006). Enhanced biofilm formation and increased resistance to antimicrobial agents and bacterial invasion are caused by synergistic interactions in multispecies biofilms. *Applied and Environmental Microbiology, 72*: 3916–3923.

[126] Shakeri, S.; Kermanshahi, R. K.; Moghaddam, M. M. & Emtiazi, G. (2007). Assessment of biofilm cell removal and killing and biocide efficacy using the microtiter plate test. *Biofouling 23*:79–86.

[127] Rickard, A. H.; Gilbert, P.; High, N. J.; Kolenbrander P. E. & Handley, P. S. (2003). Bacterial coaggregation: an integral process in the development of multi-species biofilms. *Trends in Microbiology, 11*: 94–100.

[128] Sharma, A.; Inagaki, S.; Sigurdson, W. & Kuramitsu, H. K. (2005). Synergy between *Tannerella forsythia* and *Fusobacterium nucleatum* in biofilm formation. *Oral Microbiology and Immunology, 20*: 39–42.

[129] Tielen, P.; Rosenau, F.; Wilhelm, S.; Jaeger, K. E.; Flemming, H. K. & Wingender, J. (2010). Extracellular enzymes affect biofilm formation of mucoid *Pseudomonas aeruginosa*. *Microbiology 156*, 2239–2252.

[130] Simoes, L. C.; Simoes, M. & Vieira, M. J. (2007a). Biofilm interactions between distinct bacterial genera isolated from drinking water. *Applied and Environmental Microbiology, 73*: 6192–6200.

[131] Hibbing, C. F.; Parsek, M. R. & Peterson, S. B. (2010). Bacterial competition: surviving and thriving in the microbial jungle. *Nature Reviews Microbiology, 8*: 15–25.

[132] Norwood, D. E. & Gilmour, A. (2001). The differential adherence capabilities of two *Listeria monocytogenes* strains in monoculture and multispecies biofilms as a function of temperature. *Letters in Applied Microbiology 33*: 320–324.

[133] Rieu, A.; Lemaitre, J. P.; Guzzo, J. & Piveteau, P. (2008). Interactions in dual species biofilms between *Listeria monocytogenes* EGD-e and several strains of *Staphylococcus aureus*. *International Journal of Food Microbiology 126*, 76–82.

[134] Ica, T.; Caner, V.; Istanbullu, O.; Nguyen, H. D.; Ahmed, B.; Call, D. R. & Beyenal, H. (2012). Characterization of mono- and mixed-culture *Campylobacter jejuni* biofilms. *Applied and Environmental Microbiology,* 78: 1033-1038.

[135] Rao, D., Webb, J. S. & Kjelleberg, S. (2005). Competitive interactions in mixed-species biofilms containing the marine bacterium *Pseudoalteromonas tunicata*. *Applied and Environmental Microbiology, 71*: 1729–1736.

[136] Kadouri, D. & O'Toole, G. A. (2005). Susceptibility of biofilms to *Bdellovibrio bacteriovorus* attack. *Applied and Environmental Microbiology, 71*: 4044–4051.

[137] Kadouri, D.; Venzon, N. C. & O'Toole, G. A. (2007). Vulnerability of pathogenic biofilms to *Micavibrio aeruginosavorus*. *Applied and Environmental Microbiology, 73*: 605–614.

[138] Simoes, M.; Simoes, L. C.; Pereira, M. O. & Vieira, M. J. (2008). Antagonism between *Bacillus cereus* and *Pseudomonas fluorescens* in planktonic system and biofilms. *Biofouling 24*: 339–349.

Index

B

C

D

E

F

G

H

I

J

K

L

M

N

O

P

Q

R

S

T

U

V

W

X

Y

Z